Yu-Jin Zhang
Image Engineering 2
De Gruyter Graduate

Also of Interest

Image Engineering Vol. 1: Image Processing
Y-J. Zhang, 2017
ISBN 978-3-11-052032-3, e-ISBN 978-3-11-052422-2,
e-ISBN (EPUB) 978-3-11-052411-6

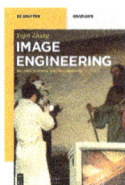

Image Engineering Vol. 3: Image Understanding
Y-J. Zhang, 2017
ISBN 978-3-11-052034-7, e-ISBN 978-3-11-052413-0,
e-ISBN (EPUB) 978-3-11-052423-9

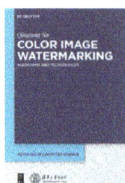

Color Image Watermarking
Q. Su, 2016
ISBN 978-3-11-048757-2, e-ISBN 978-3-11-048773-2,
e-ISBN (EPUB) 978-3-11-048763-3, Set-ISBN 978-3-11-048776-3

Modern Communication Technology
N. Zivic, 2016
ISBN 978-3-11-041337-3, e-ISBN 978-3-11-041338-0,
e-ISBN (EPUB) 978-3-11-042390-7

Yu-Jin Zhang

Image Engineering

Volume II: Image Analysis

DE GRUYTER

清華大學出版社
TSINGHUA UNIVERSITY PRESS

Author

Yu-Jin ZHANG
Department of Electronic Engineering
Tsinghua University, Beijing 100084
The People's Republic of China
E-mail: zhang-yj@tsinghua.edu.cn
Homepage: http://oa.ee.tsinghua.edu.cn/~zhangyujin/

ISBN 978-3-11-052033-0
e-ISBN (PDF) 978-3-11-052412-3
e-ISBN (EPUB) 978-3-11-052428-4

Library of Congress Cataloging-in-Publication Data
A CIP catalog record for this book has been applied for at the Library of Congress.

Bibliographic information published by the Deutsche Nationalbibliothek
The Deutsche Nationalbibliothek lists this publication in the Deutsche Nationalbibliografie; detailed bibliographic data are available on the Internet at http://dnb.dnb.de.

© 2017 Walter de Gruyter GmbH, Berlin/Boston
Typesetting: Integra Software Services Pvt. Ltd.
Printing and binding: CPI books GmbH, Leck
Cover image: Victorburnside/Getty images
♾ Printed on acid-free paper
Printed in Germany

www.degruyter.com

Preface

This book is the Volume II of "Image Engineering," which focuses on "Image Analysis," the middle layer of image engineering.

This book has grown out of the author's research experience and teaching practices for full-time undergraduate and graduate students at various universities, as well as for students and engineers taking summer courses, in more than 20 years. It is prepared keeping in mind the students and instructors with the principal objective of introducing basic concepts, theories, methodologies, and techniques of image engineering in a vivid and pragmatic manner.

Image engineering is a broad subject encompassing other subjects such as computer science, electrical and electronic engineering, mathematics, physics, physiology, and psychology. Readers of this book should have some preliminary background in one of these areas. Knowledge of linear system theory, vector algebra, probability, and random process would be beneficial but may not be necessary.

This book consists of eight chapters covering the main branches of image analysis. It has totally 58 sections, 104 subsections, with 244 figures, 21 tables, and 394 numbered equations, in addition to 48 examples and 96 problems (the solutions for 16 of them are provided in this book). Moreover, over 200 key references are given at the end of book for further study.

This book can be used for the second course "Image Analysis" in the course series of image engineering, for undergraduate and graduate students of various disciplines such as computer science, electrical and electronic engineering, image pattern recognition, information processing, and intelligent information systems. It can also be of great help to scientists and engineers doing research and development in connection within related areas.

Special thanks go to De Gruyter and Tsinghua University Press, and their staff members. Their kindness and professional assistance are truly appreciated.

Last but not least, I am deeply indebted to my wife and my daughter for their encouragement, patience, support, tolerance, and understanding during the writing of this book.

Yu-Jin ZHANG

Contents

1 Introduction to Image Analysis

This book is the second volume of the book set "Image Engineering" and focuses on image analysis.

Image analysis concerns with the extraction of information from an image, that is, it yields data out for an image in. It is based on the results of image processing and provides the foundation for image understanding. It thus establishes the linkage between image processing and image understanding.

The sections of this chapter are arranged as follows:

Section 1.1 reviews first some concepts and definitions of image and then provides an overview of image engineering—the new disciplines for the combination of various image technology for the entire image field of research and application.

Section 1.2 recapitulates the definition and research contents of image analysis, discusses the connection and difference between image analysis and image processing as well as image analysis and pattern recognition, and presents the functions and features of each working module in the framework of an image analysis system.

Section 1.3 introduces a series of digital concepts commonly used in image analysis, including discrete distances, connected components, digitization models, digital arcs, and digital chords.

Section 1.4 focuses on the distance transforms (DTs) used extensively in image analysis. First, the definition and properties of DT are given. Then, the principle of using local distance to calculate the global distance is introduced. Finally, serial implementation and parallel implementation of discrete distance transformation are discussed separately.

Section 1.5 overviews the main contents of each chapter in the book and indicates the characteristics of the preparation and some prerequisite knowledge for this book.

1.1 Image and Image Engineering

First, a brief review on the basic concepts of image is introduced, and then, the main terms, definitions, and development of the three levels of image engineering, as well as the relationship between different disciplines, are summarized.

1.1.1 Image Foundation

Images are obtained by observing the objective world in different forms and by various observing systems; they are entities that can be applied either directly or indirectly to the human eye to provide visual perception (Zhang, 1996a). Here, the concept of the **image** is more general, including photos, pictures, drawings, animation, video, and

DOI 10.1515/9783110524123-001

even documents. The image contains a rich description of the objects it expresses, which is our most important source of information.

The objective world is three dimensional (3-D) in space, and two-dimensional (2-D) images are obtained by projection, which is the main subject of this book. In the common observation scale, the objective world is continuous. In image acquisition, the requirements of computer processing are taken into account, and hence, both in the coordinate space and in the property space, the discretizations are needed. This book mainly discusses discreted digital images. In the case of no misunderstanding, the word "image" is simply used. An image can be represented by a 2-D array $f(x, y)$, where x and y represent the positions of a discrete coordinate point in the 2-D space XY, and f represents the discrete value of the property F at the point position (x, y) of the image. For real images, x and y and f all have finite values.

From a computational point of view, a 2-D $M \times N$ matrix $\mathbf{F}$, where M and N are the total number of rows and the total number of columns, respectively, is used to represent a 2-D image:

$$\mathbf{F} = \begin{bmatrix} f_{11} & f_{12} & \cdots & f_{1N} \\ f_{21} & f_{22} & \cdots & f_{2N} \\ \vdots & \vdots & \ddots & \vdots \\ f_{M1} & f_{M2} & \cdots & f_{MN} \end{bmatrix} \tag{1.1}$$

Equation (1.1) also provides a visual representation of a 2-D image as a spatial distribution pattern of certain properties (e. g., brightness). Each basic unit (corresponding to each entry in the matrix of eq. (1.1)) in 2-D image is called an image element (picture element in early days, so it is referred to as **pixel**). For a 3-D image in space XYZ, the basic unit is called **voxel** (comes from volume element). It can also be seen from eq. (1.1) that a 2-D image is a 2-D spatial distribution of a certain magnitude pattern. The basic idea of 2-D image display is often to regard a grayscale 2-D image as a luminance distribution at the 2-D spatial positions. Generally, a display device is used to display images at different spatial positions with different gray values, as in Example 1.1.

Example 1.1 Image display.
Figure 1.1 shows two typical grayscale images (Lena and Cameraman), which are displayed in two different forms, respectively. The coordinate system shown in Figure 1.1(a) is often used in the screen display. The origin O of the coordinate system is at the upper left corner of the image. The vertical axis marks the image row and the horizontal axis marks the image column. The coordinate system shown in Figure 1.1(b) is often used in image computation. Its origin O is at the lower left corner of the image, the horizontal axis is the X-axis, and the vertical axis is the Y-axis (the same as the Cartesian coordinate system). Note that $f(x, y)$ can be used to represent this image and the pixel value at the pixel coordinates (x, y). ◻

Figure 1.1: Image and pixel.

1.1.2 Image Engineering

Image engineering is a new discipline, which combines the basic theory of mathematics, optics, and other basic principles with the accumulated experience in the development of the image applications. It collects the various image techniques to focus on the research and application of the entire image field.

    Image technique, in a broad sense, is a general term for a variety of image-related technologies. In addition to the image acquisition and a variety of different operations on images, it can also include decision making and behavior planning based on the operating results, as well as the design, implementation, and production of hardware for the completion of these functions, as well as other related aspects.

1.1.2.1 Three Levels of Image Engineering

Image engineering is very rich in content and has a wide coverage. It can be divided into three levels, namely image processing, image analysis, and image understanding, according to the degree of abstraction, data volume, and semantic level and operands, as shown in Figure 1.2. The first book in this set focused on image processing with relatively low level of operations. It treats mainly the image pixel with

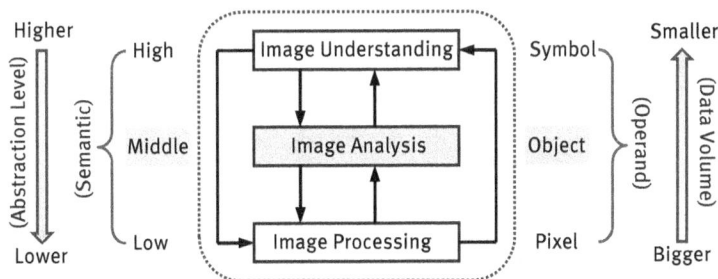

Figure 1.2: Three levels of image engineering.

a very large volume of data. This book focuses on the image analysis that is in the middle level of image engineering, and both segmentation and feature extraction have converted the original pixel representation of the image into a relatively simple non-graphical description. The third book in this set focuses on image understanding, that is, mainly high-level operations. Basically it carries on the symbols representing the description of image. In this level, the process and method used can have many similarities with the human reasoning.

As shown in Figure 1.2, as the level of abstraction increases, the amount of data will be gradually reduced. Specifically, the original image data, through a series of processing, undergo steady transformation and become more representative and better organized useful information. In this process, on the one hand, the semantics get introduced, the operands are changed, and the amount of data is compressed. On the other hand, high-level operations have a directive effect on lower-level operations, thereby improving the performance of lower-level operations.

Example 1.2 From image representation to symbolic expression.
The original collected image is generally stored in the form of a raster image. The image area is divided into small cells, each of which uses a grayscale value between the maximum and minimum values to represent the brightness of the image at the cell. If the grating is small enough, that is, the cell size is small enough, a continuous image can be seen.

The purpose of **image analysis** is to extract data that adequately describe the characteristics of objects and use as little memory as possible. Figure 1.3 shows the change in expression that occurs in the raster image analyzed. Figure 1.3(a) is the original raster form of the image, and each small square corresponds to a pixel. Some pixels in the image have lower gray levels (shadows), indicating that they differ (in property) from other pixels around them. These pixels with lower grayscale form an approximately circular ring. Through image analysis, the circular object (a collection of some pixels in the image) can be segmented to get the contour curve as shown in Figure 1.3(b). Next, a (perfect) circle is used to fit the above curve, and the circle vector is quantized using geometric representation to obtain a closed smooth curve. After extracting the circle, the symbolic expression can be obtained. Figure 1.3(c) quantitatively and concisely gives the equation of the circle. Figure 1.3(d) abstractly

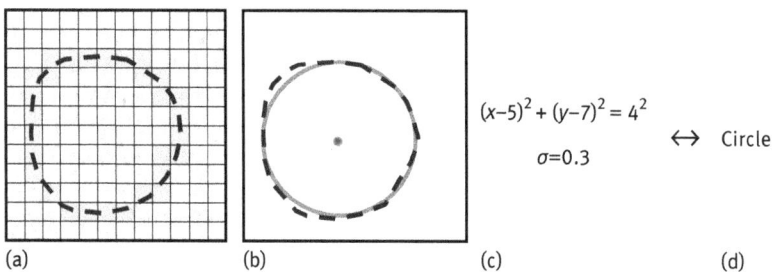

$$(x-5)^2 + (y-7)^2 = 4^2$$
$$\sigma = 0.3$$

$\leftrightarrow$ Circle

(a) (b) (c) (d)

Figure 1.3: From image representation to symbolic expression.

gives the concept of "circle." Figure 1.3(d) can be seen as a result or interpretation of the symbolic expression given by the mathematical formula in Figure 1.3(c). Compared to the quantitative expression (describing the exact size and shape of the object) of Figure 1.3(c), the interpretation in Figure 1.3(d) is more qualitative and more abstract.

Obviously, from image to object, from object to symbolic expression, and from symbolic expression to abstract concept, the amount of data are gradually reduced, and the semantic levels are gradually increased, along with the image processing, analysis, and understanding. ◙

1.1.2.2 Papers Classification for Three Levels

The research and application of the three levels of image engineering are constantly developing. In the survey series of the image engineering that started since 1996, some statistical classifications for papers published are performed (Zhang, 2015e; 2017).

A summary of the number of publications concerning image engineering from 1995 to 2016 is shown in Table 1.1. In Table 1.1, the total number of papers published in these journals (#T), the number of papers selected for survey as they are related to image engineering (#S), and the selection ratio (SR) for each year, are provided.

Table 1.1: Summary of image engineering over 22 years.

Year	#T	#S	SR	#IP	#IA
1995	997	147	14.7%	35 (23.8%)	51 (34.7%)
1996	1,205	212	17.6%	52 (24.5%)	72 (34.0%)
1997	1,438	280	19.5%	104 (37.1%)	76 (27.1%)
1998	1,477	306	20.7%	108 (35.3%)	96 (31.4%)
1999	2,048	388	19.0%	132 (34.0%)	137 (35.3%)
2000	2,117	464	21.9%	165 (35.6%)	122 (26.3%)
2001	2,297	481	20.9%	161 (33.5%)	123 (25.6%)
2002	2,426	545	22.5%	178 (32.7%)	150 (27.5%)
2003	2,341	577	24.7%	194 (33.6%)	153 (26.5%)
2004	2,473	632	25.6%	235 (37.2%)	176 (27.8%)
2005	2,734	656	24.0%	221 (33.7%)	188 (28.7%)
2006	3,013	711	23.60	239 (33.6%)	206 (29.0%)
2007	3,312	895	27.02	315 (35.2%)	237 (26.5%)
2008	3,359	915	27.24	269 (29.4%)	311 (34.0%)
2009	3,604	1,008	27.97	312 (31.0%)	335 (33.2%)
2010	3,251	782	24.05	239 (30.6%)	257 (32.9%)
2011	3,214	797	24.80	245 (30.7%)	270 (33.9%)
2012	3,083	792	25.69	249 (31.4%)	272 (34.3%)
2013	2,986	716	23.98	209 (29.2%)	232 (32.4%)
2014	3,103	822	26.49	260 (31.6%)	261 (31.8%)
2015	2,975	723	24.30	199 (27.5%)	294 (40.7%)
2016	2,938	728	24.78	174 (23.9%)	266 (36.5%)
Total	56,391	13,577		4,296 (31.64)	4,285 (31.56)
Average	2,563	617	24.08	195	195

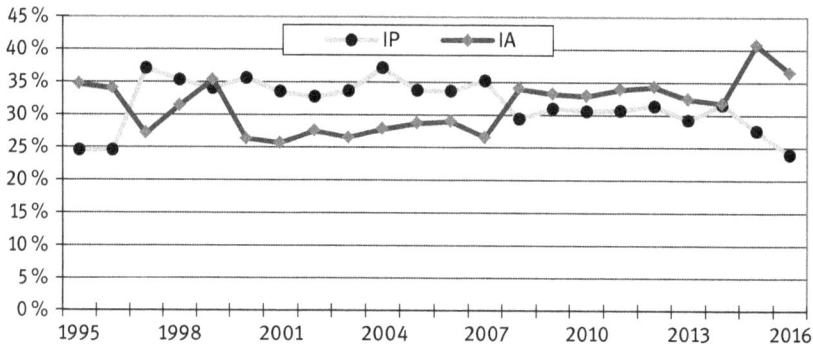

Figure 1.4: The percentage of papers in IP and IA categories for the last 22 years.

It is observed that image engineering is an (more and more) important topic for electronic engineering, computer science, automation, and information technology. The average SR is more than 1/5 (it is more than 1/4 in several recent years), which is remarkable in considering the wide coverage of these journals.

The selected papers on image engineering have been further classified into five categories: image processing (IP), image analysis (IA), image understanding (IU), technique applications (TA), and surveys.

In Table 1.1, the statistics for IP papers and IA papers are also provided for comparison. The listed numbers under #IP and #IA are the numbers of papers published each year belonging to image processing and to image analysis, respectively. The percentages in parenthesis are the ratios of #IP/#S and #IA/#S, respectively. The total number of papers related to image analysis is almost equal to the total number of papers related to image processing in these 21 years.

The proportions and variations of the number of papers in these years are shown in Figure 1.4. The papers related to image processing are generally the most in number, but the papers related to image analysis are on rise for many years steadily and occupies the top spot in recent years. At present, image analysis is considered the prime focus of image engineering.

1.2 The Scope of Image Analysis

As shown in Figure 1.2, image analysis is positioned in the middle level of the image engineering. It plays an important role in connecting the upper level and lower level by using the results of image processing and by foundering image understanding.

1.2.1 Definition and Research Content of Image Analysis

The definition of image analysis and its research contents are enclosed by the upper and lower layers of image engineering.

1.2.1.1 Definition of Image Analysis

The definition and description of image analysis have different versions. Here are a few examples:

1. The purpose of image analysis is to construct a description of the scene based on the information extracted from the image (Rosenfeld, 1984).
2. Image analysis refers to the use of computers for processing images to find out what is in the image (Pavlidis, 1988).
3. Image analysis quantifies and classifies images and objects in images (Mahdavieh, 1992).
4. Image analysis considers how to extract meaningful measurement data from multidimensional signals (Young, 1993).
5. The central problem with image analysis is to reduce the grayscale or color images with a number of megabytes to only a few meaningful and useful digits (Russ, 2002).

In this book, image analysis is seen as the process and technique starting from an image, and detecting, extracting, representing, describing, and measuring the objects of interest inside, to obtain objective information and output data results.

1.2.1.2 Differences and Connections Between Image Processing and Image Analysis

Image processing and image analysis have their own characteristics, which have been described separately in Section 1.1.2. Image processing and image analysis are also closely related. Many techniques for image processing are used in the process from collecting images to analyze the results. In other words, the work of image analysis is often performed on the basis of the results of image (pre)processing.

Image processing and image analysis are different. Some people think that image processing (such as word processing and food processing) is a science of reorganization (Russ, 2016). For a pixel, its attribute value may change according to the values of its neighboring pixels, or it can be itself moved to other places in the image, but the absolute number of pixels in the whole image does not change. For example, in word processing, the paragraphs can be cut or copied to perform phonics checks, or the fonts can be changed without reducing the amount of text. As in food processing, it is necessary to reorganize the ingredients to produce a better combination, rather than to extract the essence of various ingredients. However, the image analysis is different, and its goal is to try to extract the descriptive coefficients from the image (similar as the extraction of the essence of various ingredients in food processing), which can succinctly represent the important information in the image and can express quantitatively and effectively the content of the image.

Example 1.3 Process flow of image analysis.
Figure 1.5 shows a practical process flow and the main steps involved in a 3-D image analysis program:

1. Image acquisition: Obtaining the image from the original scene;

2. Preprocessing: Correcting the image distortion produced during the process of image acquisition;
3. Image restoration: Filtering the image after correction to reduce the influence of noise and other distortion;
4. Image segmentation: Decomposing the image into the target and other background for analyzing the image;
5. Object measurement: Estimating the "analog" property of the original scene from the digital image data;
6. Graphic generation: Providing the results of the measurement in a user-friendly and easy to understand way in a visualized display.

From the above steps the links between image processing and image analysis can be understood.

1.2.1.3 Image Analysis and Pattern Recognition

The main functional modules of image analysis are shown in Figure 1.6.

As shown in Figure 1.6, image analysis mainly includes the segmentation of the image, the representation/description of the object, as well as the measurement and analysis of the characteristic parameters of objects. In order to accomplish these tasks, many corresponding theories, tools, and techniques are needed. The work of image analysis is carried out around its operand—object. Image segmentation is to separate and extract the object from the image. The representation/description is to express and explain the object effectively, and the characteristic measurement and analysis

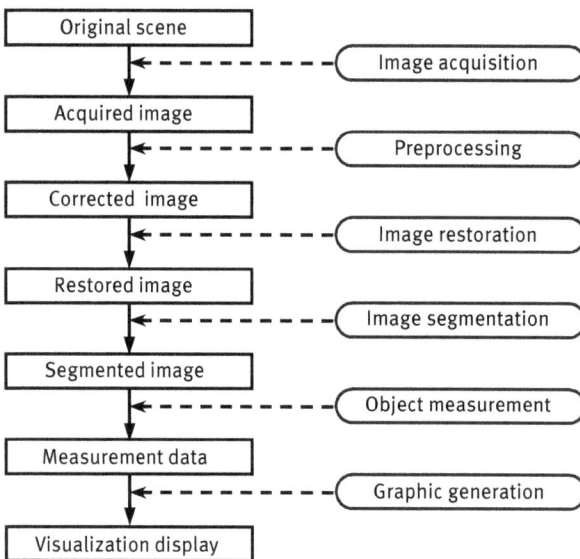

Figure 1.5: An image analysis flow diagram with multiple steps.

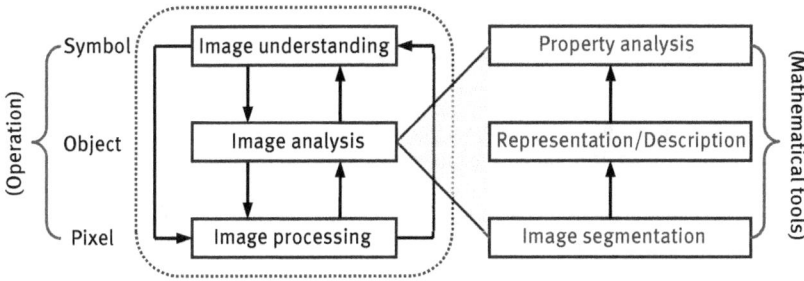

Figure 1.6: The main functional modules of image analysis.

are to obtain the properties of the object. Since the object is a set of pixels, image analysis places more emphasis on the relationships and connections between pixels (*e. g.*, neighborhood and connectivity).

Image analysis and pattern recognition are closely related. The purpose of **pattern recognition** is to classify the different patterns. The separation of the objects from the background can be considered as a region classification. To further classify the regions, it is necessary to determine the parameters that describe the characteristics of the regions. A process flow of (image) pattern recognition is shown in Figure 1.7.

The input image is segmented to obtain the object image, the features of the object are detected, measured, and calculated, and the object can be classified according to the obtained feature measurements. In conclusion, image analysis and (image) pattern recognition have the same input, similar process steps or procedures, and convertible outputs.

1.2.1.4 Subcategories of Image Analysis

From the survey series mentioned in Section 1.1.2, the selected papers on image engineering have been further classified into different categories such as image processing (IP) and image analysis (IA). The papers under the category IP have been

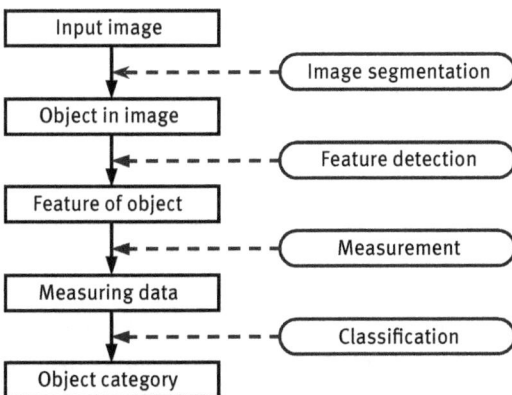

Figure 1.7: The process flow of image pattern recognition.

Table 1.2: The classification of image analysis category.

Subcategory	#/Year
B1: Image segmentation, detection of edge, corner, interest points	71
B2: Representation, description, measurement of objects (bi-level image processing)	12
B3: Feature measurement of color, shape, texture, position, structure, motion, etc.	17
B4: Object (2-D) extraction, tracking, discrimination, classification, and recognition	53
B5: Human organ (biometrics) detection, location, identification, categorization, etc.	53

further classified into five subcategories. Their contents and numbers of papers per year are listed in Table 1.2.

1.2.2 Image Analysis System

Image analysis systems can be constructed using a variety of image analysis devices and techniques and can be used to perform many image analysis tasks.

1.2.2.1 History and Development

Image analysis has a long history. The following are some of the early events of the image analysis system (Joyce, 1985):

1. The first system to analyze images using a television camera was developed by a metal research institute. Its earliest model system was developed in 1963.
2. Electronic era really began in 1969, when a company in the United States (Bausch and Lomb) produced an image analyzer that could store a complete black-and-white image in a small computer for analysis.
3. In 1977, the company Joyce Loebl of the United Kingdom proposed a third-generation image analysis system, which uses software instead of hardware. The analysis function is separated from the image acquisition and becomes more general.

Image analysis systems have already obtained in-depth research, rapid development, and wide application, in the 1980s and 1990s of twentieth century. During this period, many typical system structures were proposed, and many practical application systems were established (Zhang, 1991a; Russ, 2016).

Image analysis system takes the hardware as the physical basis. A variety of image analysis tasks can generally be described in the form of algorithms, and most of the algorithms can be implemented in software, so there are many image analysis systems now only need to use a common general-purpose computer. Special hardware can be used in order to increase the speed of operation or to overcome the limitations of general-purpose computers. In the 1980s and 1990s of twentieth century, people have designed a variety of image cards compatible with industry-standard bus, which can be inserted into the computer or workstation. These image cards include image capture cards for image digitization and temporary storage, arithmetic and logic units for arithmetic and logic operations at video speeds, and memories such as frame buffers. Getting into the twenty-first century, the system's integration is further improved, and system-on-chip (SOC) has also been rapidly developed. The progress of these hardware not only reduces the cost, but also promotes the development of special software for image processing and analysis. There are a number of image analysis systems and image analysis packages commercially available.

1.2.2.2 Diagram of Analysis System

The structure of a basic image analysis system can be represented by Figure 1.8. It is similar to the framework of image-processing system, in which the analysis module is continued after the (optional) processing module. It has also specific modules for collecting, synthesizing, inputting, communicating, storing, and outputting functions. The final output of the analysis is either data (the data measured from the object in image and/or the resulting data of analysis) or the symbolic representation for further understanding, which is different from the image processing system.

Because that the basic information about processing modules has been covered in the first book of this set, the techniques of the analysis module will be discussed in detail in this (second) book. According to the survey on the literature of image engineering, the current image analysis researches are mainly concentrated in the following five aspects:

1. Edge detection, image segmentation;
2. Object representation, description, measurement;
3. Analysis of color, shape, texture, space, movement, and other characteristics of the objects;
4. Object detection, extraction, tracking, identification, and classification;
5. Human face and organ detection, positioning, and identification.

The following two sections give a brief introduction to some fundamental concepts and terminologies of the image analysis.

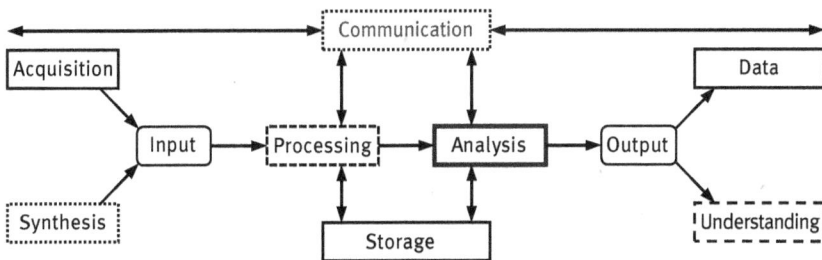

Figure 1.8: Schematic diagram of the image analysis system.

1.3 Digitization in Image Analysis

The goal of image analysis is to obtain the measurement data of the interest objects in the image; these objects of interest are obtained by discretizing successive objects in the scene. **Object** is an important concept and the main operand in image analysis, which is composed of related pixels. Related pixels are closely linked both in the spatial locations and in the attribute characteristics. They generally constitute the connected components in image. To analyze the position link of the pixels, the spatial relationship among the pixels must be considered, where the discrete distance plays an important role. While to analyze the connectivity, not only the distance relationship but also the attribute relationship should be taken into account. In order to obtain accurate measurement data from the objects, an understanding of the digitization process and characteristics in analysis is required.

1.3.1 Discrete Distance

In digital images, the pixels are discrete, the spatial relationship between them can be described by the **discrete distance**.

1.3.1.1 Distance and Neighborhood

If assuming that p, q, and r are three pixels and their coordinates are (x_p, y_p), (x_q, y_q), and (x_r, y_r), respectively, then the **distance function** d should satisfy the following three metric conditions:

1. $d(p, q) \geq 0$ ($d(p, q) = 0$, if and only if $p = q$);
2. $d(p, q) = d(q, p)$;
3. $d(p, r) \leq d(p, q) + d(q, r)$.

The neighborhood of the pixels can be defined by distance. The four-neighborhood $N_4(p)$ of a pixel $p(x_p, y_p)$ is defined as (see the pixel $r(x_r, y_r)$ in Figure 1.9(a)):

$$N_4(p) = \{r \,|\, d_4(p, r) = 1\} \tag{1.2}$$

The **city block distance** $d_4(p, r) = |x_p - x_r| + |y_p - y_r|$.

The eight-neighborhood $N_8(p)$ of a pixel $p(x_p, y_p)$ is defined as the AND set of its four-neighborhood $N_4(p)$ and its diagonal neighborhood $N_D(p)$ (see pixels marked by s in Figure 1.9(b)), that is:

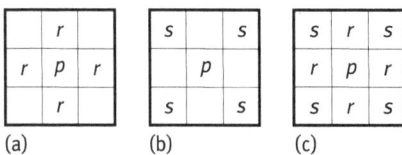

	r	
r	p	r
	r	

(a)

s		s
	p	
s		s

(b)

s	r	s
r	p	r
s	r	s

(c)

Figure 1.9: The different neighborhoods of a pixel p.

$$N_8(p) = N_4(p) \bigcup N_D(p) \tag{1.3}$$

So the eight-neighborhood $N_8(p)$ of a pixel $p(x_p, y_p)$ can also be defined as (see the pixels marked by r and s in Figure 1.9(c)):

$$N_8(p) = \{r \,|\, d_8(p, r) = 1\} \tag{1.4}$$

The **chessboard distance** $d_8(p, r) = \max(|x_p - x_r|, |y_p - y_r|)$.

1.3.1.2 Discrete Distance Disc

Given a discrete distance metric d_D, a disc with radius R ($R \geq 0$) and centered on pixel p is a point set satisfying $\Delta_D(p, R) = \{q | d_D(p, q) \leq R\}$. When the position of the center pixel p can be disregarded, the disc with radius R can also be simply denoted by $\Delta_D(R)$.

Let $\Delta_i(R)$, $i = 4, 8$, represent an equidistant disc with the distance d_i from the center pixel less than or equal to R, and $\#[\Delta_i(R)]$ represent the number of pixels, except center pixel, in $\Delta_i(R)$, then the number of pixels increases in proportion with distance. For the city block disc, there is

$$\#[\Delta_4(R)] = 4 \sum_{j=1}^{R} j = 4(1 + 2 + 3 + \cdots + R) = 2R(R + 1) \tag{1.5}$$

Similarly, for the chessboard disc, there is

$$\#[\Delta_8(R)] = 8 \sum_{j=1}^{R} j = 8(1 + 2 + 3 + \cdots + R) = 4R(R + 1) \tag{1.6}$$

In addition, the chessboard disc is actually a square, and so the following formula can also be used to calculate the number of pixels, except the central pixel, in the chessboard disc

$$\#[\Delta_8(R)] = (2R + 1)^2 - 1 \tag{1.7}$$

Here are a few different values of $\Delta_i(R)$: $\#[\Delta_4(5)] = 60$, $\#[\Delta_4(6)] = 84$, $\#[\Delta_8(3)] = 48$, $\#[\Delta_8(4)] = 80$.

1.3.1.3 The Chamfer Distance

The **chamfer distance** is an integer approximation to the Euclidean distance in the neighborhood. Traveling from pixel p to its 4-neighbors needs simply a horizontal or vertical move (called a-move). Since all displacements are equal according to the symmetry or rotation, the only possible definition of the discrete distance is d_4 distance, where $a = 1$. Traveling from pixel p to its 8-neighborhood cannot be achieved by only a horizontally or vertically move, a diagonal move (called b-move) is also required. Combining these two moves, the chamfer distance is obtained, and can be recorded as $d_{a,b}$.

The most natural value for b is $2^{1/2}a$, but for the simplicity of calculation and for reducing the amount of memory, both a and b should be integers. The most common set of values is $a = 3$ and $b = 4$. This set of values can be obtained as follows. Considering that the number of pixels in the horizontal direction between two pixels p and q is n_x and the number of pixels in vertical direction between two pixels p and q is n_y (without loss of generality, let $n_x > n_y$), then the chamfer distance between pixels p and q is

$$D(p, q) = (n_x - n_y)a + n_y b \tag{1.8}$$

The difference between the chamfer distance and Euclidean distance is

$$\Delta D(p, q) = \sqrt{n_x^2 + n_y^2} - [(n_x - n_y)a + n_y b] \tag{1.9}$$

If $a = 1$, $b = \sqrt{2}$, computing the derivative of $\Delta D(p, q)$ for n_y gives

$$\Delta D'(p, q) = \frac{n_y}{\sqrt{n_x^2 + n_y^2}} - (\sqrt{2} - 1) \tag{1.10}$$

Let the above derivative be zero, computing the extreme of $\Delta D(p, q)$ gives

$$n_y = \sqrt{(\sqrt{2} - 1)/2} n_x \tag{1.11}$$

The difference between the two distances, $\Delta D(p, q)$, will take the maximum value $(2^{3/2} - 2)^{1/2} n_x \approx -0.09 n_x$ when meeting the above equation (the angle between the straight line and the horizontal axis is about 24.5°). It can be further proved that the maximum of $\Delta D(p, q)$ can be minimized when $b = 1/2^{1/2} + (2^{1/2} - 1)^{1/2} = 1.351$. As $4/3 \approx 1.33$, so in the chamfer distance $a = 3$ and $b = 4$ are taken.

Figure 1.10 shows two examples of equidistant discs based on the chamfer distance, where the left disc is for $\Delta_{3,4}(27)$, and the right disc is for $\Delta_{a,b}$.

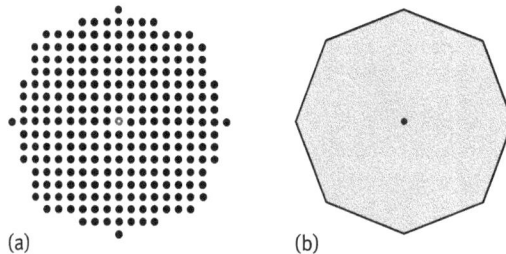

(a)   (b)

Figure 1.10: Two example of equidistant discs.

1.3.2 Connected Component

The object in image is a **connected component** composed of pixels. A connected component is a set of connected pixels. Connection needs to be defined according to the connectivity. A connectivity is a relationship between two pixels, in which both their positional relation and their magnitude relation are considered. Two pixels having connectivity are spatially adjacent (*i. e.,* one pixel is in the neighborhood of another pixel) and are similar in magnitude under certain criteria (for grayscale images, their gray values should be equal, or more generally in a same grayscale value set V).

It can be seen that the adjacency of two pixels is one of the necessary conditions for these two pixels to be connected (the other condition is that they are taken values from a same grayscale value set V). As shown in Figure 1.10, two pixels may be 4-adjacent (one pixel is within a 4-neighborhood of another pixel), or 8-contiguous (one pixel is within an 8-neighborhood of another pixel). Correspondingly, two pixels may be 4-connected (these two pixels are 4-adjacent), or 8-connected (these two pixels are 8-adjacent).

In fact, it is also possible to define another kind of connection, m-connection (mixed connection). If two pixels, p and r, take values in V and satisfy one of the following conditions, then they are m-connected: (1) r is in $N_4(p)$; (2) r is in $N_D(p)$ and there is no pixel with value in V included in $N_4(p) \cap N_4(r)$. For a further explanation of the condition (2) in the mixed connection, see Figure 1.11. In Figure 1.11(a), the pixel r is in $N_D(p)$, $N_4(p)$ consists of four pixels denoted by a, b, c, d, $N_4(r)$ consists of four pixels denoted by c, d, e, f, and $N_4(p) \cap N_4(r)$ includes two pixels labeled c and d. Let $V = \{1\}$, then Figure 1.11(b, c) gives an example of satisfying and no-satisfying condition (2), respectively. In the two figures, the two shaded pixels are each other in the opposite diagonal neighborhoods, respectively. However, the two pixels in Figure 1.11(b) do not have a common neighbor pixel with value 1, while the two pixels in Figure 1.11(c) do have a common neighbor pixel with value 1. Thus, the pixels p and r are m-connected in Figure 1.11(b) and the m-connection between the pixels p and r in Figure 1.11(c) does not hold (they are connected via pixel c).

As discussed earlier, the mixed connection is essentially to take a 4-connection when both 4-connection and 8-connection are possible for two pixels, and to shield the 8-connection between the two pixels having 4-connections with a same pixel.

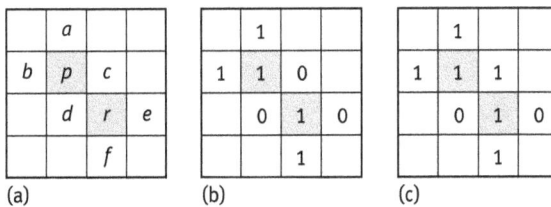

(a)

	a		
b	p	c	
	d	r	e
		f	

(b)

	1		
1	1	0	
	0	1	0
		1	

(c)

	1		
1	1	1	
	0	1	0
		1	

Figure 1.11: Further explanation of the condition (2) in the mixed connection.

```
1····1—1—1···1        1—1—1—1—1
1  0  0  0  1         1  0  0  0  1
1···1—1—1···1         1—1—1—1—0
(a)                    (b)
```

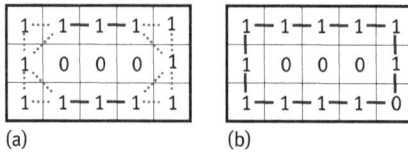

Figure 1.12: Mixed connection eliminates multipath ambiguity problems.

A mixed connection can be thought of as a variant of the 8-connection, whose introduction is to eliminate the multiplexing problem that often occurs with 8-connections. Figure 1.12 shows an example. In Figure 1.12(a), the pixel labeled 1 forms the boundary of a region consisting of three pixels with values 0. When this boundary is considered as a path, there are two paths between the three pixels at each of the four corners, which is the ambiguity caused by the use of an 8-connection. This ambiguity can be eliminated by using m-connection, and the results are shown in Figure 1.12(b). Since the direct m-connection between the two diagonal pixels cannot be established (the two conditions for the mixed connection are not satisfied), there is now only one path, which has no ambiguous problem.

Connection can be seen as an extension of connectivity. It is a relationship created by two pixels with the help of other pixels. Consider a series of pixels $\{p_i(x_i, y_i), i = 0, n\}$. When all $p_i(x_i, y_i)$ are connected with $p_{i-1}(x_{i-1}, y_{i-1})$, then two pixels $p_0(x_0, y_0)$ and $p_n(x_n, y_n)$ are connected through other pixels $p_i(x_i, y_i)$. If $p_i(x_i, y_i)$ has 4-connectivity with $p_{i-1}(x_{i-1}, y_{i-1})$, then $p_0(x_0, y_0)$ and $p_n(x_n, y_n)$ are 4-connected. If $p_i(x_i, y_i)$ has 8-connectivity with $p_{i-1}(x_{i-1}, y_{i-1})$, then $p_0(x_0, y_0)$ and $p_n(x_n, y_n)$ are 8-connected. If $p_i(x_i, y_i)$ has m-connectivity with $p_{i-1}(x_{i-1}, y_{i-1})$, then $p_0(x_0, y_0)$ and $p_n(x_n, y_n)$ are m-connected.

A series of pixels connected as above constitutes a connected component in the image, or an image subset, or a region. In a connected component, any two pixels are connected by the connectivity of other pixels in the component. For two connected components, if one or several pixels in a connected component are adjacent to one or several pixels in another connected component, these two connected components are adjacent; if one or several pixels in a connected component have connectivity with one or several pixels in another connected component, these two connected components are connected, and these two connected components can form one connected component.

1.3.3 Digitizing Model

The digitizing model discussed here is used to transform spatially continuous scenes into discrete digital images, so it is a spatial quantization model.

1.3.3.1 Fundamentals
Here are some basic knowledge of the digitizing model (Marchand 2000)

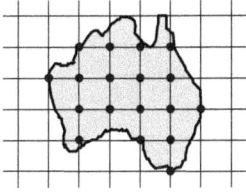

Figure 1.13: An example of a simple digitizing model.

The **preimages** and **domain** of the digitized set P are first defined:
1. Preimage: Given a set of discrete points P, a set of consecutive points S that is digitized as P is called a preimage of P;
2. Domain: The region defined by the union of all possible preimages S is called the domain of P.

There are a variety of digitizing models, in which the quantization is always a many-to-one mapping, so they are irreversible processes. Thus, the same quantized result image may be mapped from different preimages. It can also be said that the objects of different sizes or shapes may have the same discretization results.

Now consider first a simple digitizing model. A square image grid is overlaid on a continuous object S, and a pixel is represented by an intersection point p on the square grid, which is a digitized result of S if and only if $p \in S$. Figure 1.13 shows an example where S is represented by a shaded portion, a black dot represents a pixel p belonging to S, and all p make up a set P.

Here, the distance between the image grid points is also known as the sampling step, which can be denoted h. In the case of a square grid, the sampling step is defined by a real value, and $h > 0$, which defines the distance between two 4-neighbor pixels.

Example 1.4 Effect of different sampling steps.
In Figure 1.14(a), the continuous set S is digitized with a given sampling step h. Figure 1.14(b) gives the result for the same set S being digitized with another sampling step $h' = 2h$. Obviously, such an effect is also equivalent to changing the size of successive sets S by the scale h/h' and digitizing with the original sampling step h, as shown in Figure 1.14(c). ◻

An analysis of this digitizing model shows that it may lead to some inconsistencies, as illustrated in Figure 1.15:
1. A nonempty set S may be mapped to an empty digitizing set. Figure 1.15(a) gives several examples in which each nonempty set (including two thin objects) does not contain any integer points.
2. This digitizing model is not translation invariant. Figure 1.15(b) gives several examples where the same set S may be mapped to an empty, disconnected, or connected digitized set (the numbers of points in each set are 0, 2, 3, 6, respectively)

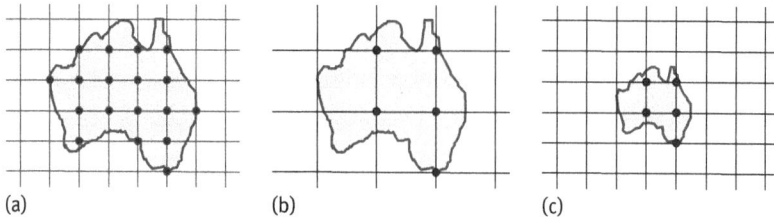

Figure 1.14: Digitizing the continuous sets with different sampling steps.

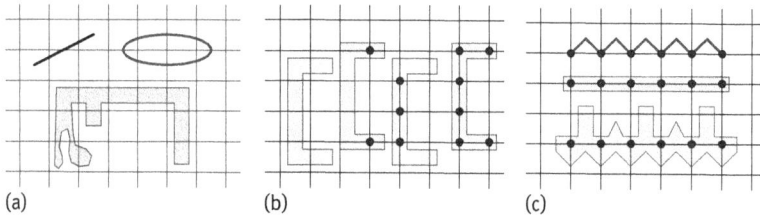

Figure 1.15: Examples of inconsistency in a digitized model.

depending on its position in the grid. In the terminology of image processing, the change of P with the translation of S is called aliasing.

3. Given a digitized set P, it is not guaranteed to accurately characterize its preimage S. Figure 1.15(c) shows a few examples where three dissimilar objects with very different shapes provide the same digitized results (points in a row).

As discussed earlier, a suitable digitizing model should have the following characteristics:

1. The digitized result for a nonempty continuous set should be nonempty.
2. The digitizing model should be as translation-invariant as possible (*i.e.*, the aliasing effect should be as small as possible).
3. Given a set P, each of its preimages should be similar under certain criteria. Strictly speaking, the domain of P should be limited and as small as possible.

Commonly used digitizing models are square box quantization (SBQ), grid-intersect quantization (GIQ) and object contour quantization (Zhang, 2005b). Only the SBQ and GIQ digitizing models are introduced below.

1.3.3.2 The Square Box Quantization

In SBQ, there is a corresponding digitization box $B_i = (x_i - 1/2, x_i + 1/2) \times (y_i - 1/2, y_i + 1/2)$ for any pixel $p_i = (x_i, y_i)$. Here, a half-open box is defined to ensure that all boxes completely cover the plane. A pixel p_i is in the digitized set P of preimage S if and only if $B_i \cap S \neq \emptyset$, that is, its corresponding digitizing cell B_i intersects S. Figure 1.16 shows the set of digitization (square box with a dot in center) obtained by SBQ from

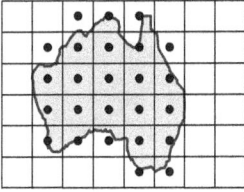

Figure 1.16: Digitizing example based on square box quantization.

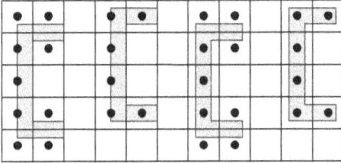

Figure 1.17: Using square box quantization can reduce aliasing.

the continuous point set S in Figure 1.13. Wherein the dotted line represents a square segmentation, which is dual with the quadratic sampling grid. It can be seen from the figure that if the square corresponding to a pixel intersects S, then this pixel will appear in the digitizing set produced by its SBQ. That is, the digitized set P obtained from a continuous point set S is the pixel set $\{p_i | B_i \cap S \neq \emptyset\}$.

The result of SBQ for a continuous line segment is a 4-digit arc. The definition of SBQ guarantees that a nonempty set S will be mapped to a nonempty discrete set P, since each real point can be guaranteed to be uniquely mapped to a discrete point. However, this does not guarantee a complete translation invariance. It can only reduce greatly the aliasing. Figure 1.17 shows several examples in which the continuous sets are the same as in Figure 1.15(b). The numbers of points in each discrete set are 9, 6, 9, 6, respectively, which are closer each other than that in Figure 1.15(b).

Finally, it should be pointed out that the following issues in SBQ have not been resolved (Marchand, 2000):

1. The number of connected components in the background may not be maintained. In particular, the holes in S may not appear in P, as shown in Figure 1.18(a), where the dark shaded area represents S (with two holes in the middle), and the black point set represents P (no holes in the middle).

2. Since $S \subset \cup_i B_i$, the number N_S of connected components in a continuous set S is not always equal to the number N_P of connected components in its corresponding discrete set P, generally $N_S \leq N_P$. Figure 1.18(b) shows two examples where the number of black points in the dark-shaded area representing S is less than the number of black dots in the shallow shaded area representing P.

3. More generally, the discrete set P of a continuous set S has no relation to the discrete set P^c of a continuous set S^c, where the superscripts denote the complement. Figure 1.18(c) shows two examples in which the dark-shaded area on the left represents S and the black-point set represents P; the dark-shaded area on the right represents S^c, and the black-point set represents P^c. Comparing these two figures, it is hardly to see the complementary relationship between P and P^c.

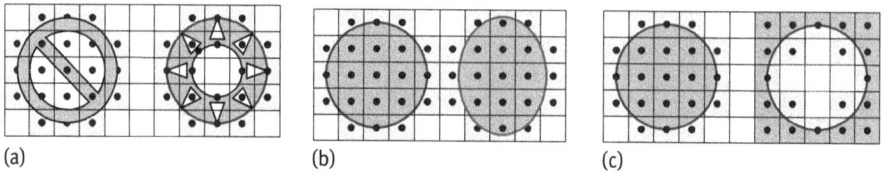

Figure 1.18: Some problems with square box quantization.

1.3.3.3 Grid-Intersect Quantization

GIQ can be defined as follows. Given a thin object C formed by continuous points, its intersection with the grid line is a real point $t = (x_t, y_t)$. This point may satisfy $x_t \in \mathbf{I}$ or $y_t \in \mathbf{I}$ (where $\mathbf{I}$ represents a set of 1-D integers) depending on that C intersects the vertical grid lines or the horizontal grid lines. This point $t \in C$ will be mapped to a grid point $p_i = (x_i, y_i)$, where $t \in (x_i - 1/2, x_i + 1/2) \times (y_i - 1/2, y_i + 1/2)$. In special cases (such as $x_t = x_i + 1/2$ or $y_t = y_i + 1/2$), the point p_i located at left or top will be taken as belonging to the discrete set P.

Figure 1.19 shows the results obtained for a curve C using the GIQ. In this example, C is clockwise (as indicated by the arrow in the figure) traced. Any short continuous line between C and the pixel marked by dot is mapped to the corresponding pixel.

It can be proved that the result of GIQ for a continuous straight line segment (α, β) is an 8-digit arc. Further, it can be defined by the intersection of (α, β) and the horizontal or vertical (depending on the slope of (α, β) line.

GIQ is often used as a theoretical model for image acquisition. The aliasing effects generated by the GIQ are shown in Figure 1.20. If the sampling step h is appropriate for the digitization of C, the resulting aliasing effect is similar to the aliasing effect produced by SBQ.

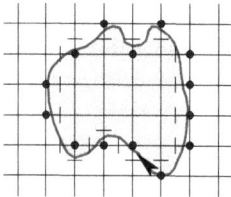

Figure 1.19: Example of grid-intersect quantization for continuous curve C.

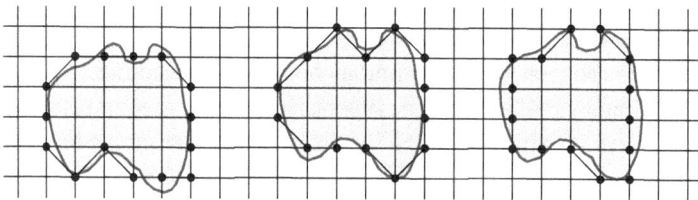

Figure 1.20: The aliasing effect of grid-intersect quantization.

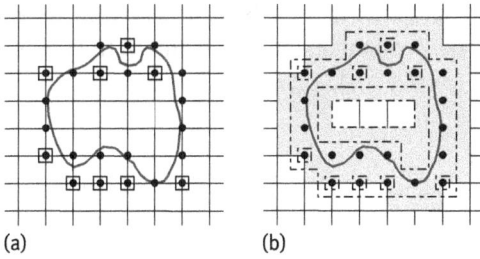

Figure 1.21: Comparison of the domains of SBQ and GIQ.

(a)   (b)

A comparison of the GIQ domain and the SBQ domain is given in the following. Figure 1.21(a) gives the result of digitizing the curve C by the SBQ method (all black dots). Pixels surrounded by small squares are those that do not appear in the GIQ result of digitizing the curve C. Clearly, GIQ reduces the number of pixels in a digitized set. On the other hand, Figure 1.21(b) shows the domain boundaries derived from the GIQ and SBQ results for corresponding C. The dotted line represents the boundary of the union of successive subsets defined by the GIQ result, and the thin line represents the boundary of the union of successive subsets defined by the SBQ result.

As shown in Figure 1.21, the area of the SBQ domain is smaller than the area of the GIQ domain. In this sense, SBQ is more accurate in characterizing a preimage of a given digitized set. However, the shape of the GIQ domain appears to be more appropriate for the description of the curve C than the shape of the SBQ domain. The explanation is that if the sampling step is chosen so that the details in C are larger than the size of the square grid, these details will intersect with the different grid lines. Further, each such intersection will be mapped to one pixel, and such pixels determine a contiguous subset as shown in Figure 1.21(a). In contrast, SBQ maps these details to a group of 4-adjacent pixels that define a wider contiguous region, which is the union of the corresponding digital boxes.

1.3.4 Digital Arcs and Chords

In Euclidean geometry, the arc is a part of the curve between two points, while the chord is the straight line connecting any two points on the conic curve. A continuous straight line segment can often be viewed as a special case of an arc.

A digitized set is a discrete set obtained by digitizing a continuous set using a digitizing model. In the study of discrete objects, they can be seen as a digitization results of continuous objects. The property of a continuous object that has been demonstrated in Euclidean geometry can be mapped to a discrete set. The digital arcs and digital chords, as well as some of their properties, are discussed below.

1.3.4.1 Digital Arcs
Given a neighborhood and the corresponding movement length, the chamfer distance between two pixels p and q with respect to this neighborhood is the length of the

shortest digital arc from p to q. Here, the **digital arc** may be defined as follows. Given a set of discrete points and the adjacency between them, the digital arc P_{pq} from point p to point q is defined as the arc $P_{pq} = \{p_i, \ i = 0, 1, \ldots, n\}$ satisfying the following conditions (Marchand, 2000):

1. $p_0 = p, p_n = q$;
2. $\forall i = 1, \ldots, n-1$, point p has exactly two adjacent points in the arc P_{pq}: p_{i-1} and p_{i+1};
3. The endpoint p_0 (or p_n) has exactly one adjacent point in the arc P_{pq}: p_1 (or p_{n-1}).

Different digital arcs (4-digit arcs or 8-digit arcs) can be defined as above, depending on the difference in adjacency (e. g., 4-adjacent or 8-adjacent).

Consider the continuous straight line segment $[\alpha, \beta]$ (the segment from α to β) in the square grid given in Figure 1.22. Using GIQ, points intersecting the grid lines between $[\alpha, \beta]$ are mapped to their nearest integer points. When there are two equal-distance closest points, the left discrete point in $[\alpha, \beta]$ can be first selected. The resulting set of discrete points $\{p_i\}_{i=0,\ldots,n}$ is called a digitized set of $[\alpha, \beta]$.

1.3.4.2 Digital Chords

Discrete linearity is first discussed below, then whether a digital arc is a **digital chord** will be discussed.

The determination of the digital chord is based on the following principle: Given a digital arc $P_{pq} = \{p_i\}_{i=0,\ldots,n}$ from $p = p_0$ to $q = p_n$, the distance between the continuous line $[p_i, p_j]$ and the sum of the segments $\cup_i[p_i, \ p_{i+1}]$ can be measured using the discrete distance function and should not exceed a certain threshold (Marchand, 2000). Figure 1.23 shows two examples where the shaded area represents the distance between P_{pq} and the continuous line segments $[p_i, p_j]$.

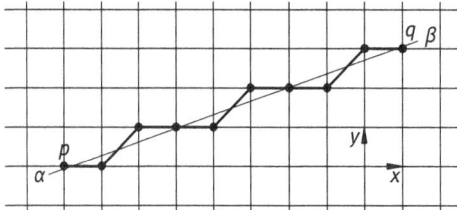

Figure 1.22: The result of grid-intersect quantization of straight line segment.

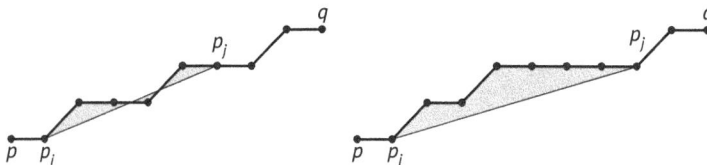

Figure 1.23: Example for judging chord properties.

It can be proved that given any two discrete points p_i and p_j in an 8-digit arc P_{pq} = $\{p_i\}_{i=0,...,n}$ and the real point p in any continuous line segment $[p_i, p_j]$, if there exists a point $p_k \in P_{pq}$ such that $d_8(p, p_k) < 1$, then P_{pq} satisfies the property of the chord.

To determine the property of the chord, a polygon that surrounds a digital arc can be first defined, which can include all the continuous segments between the discrete points on the digital arc (as shown by the shaded polygons in Figures 1.24 and 1.25). This polygon will be called a visible polygon, since any other point can be seen from any point in the polygon (i.e., the two points can be connected by a straight line in the polygon). Figures 1.24 and 1.25 show two examples of verifying that the chord properties are true and not true, respectively.

In Figure 1.24, there is always a point $p_k \in P_{pq}$ such that $d_8(p, p_k) < 1$ for all points in the shadow polygon of the digital arc P_{pq}. In Figure 1.25, there is $p \in (p_1, p_8)$ that can have $d_8(p, p_k) \geq 1$ for any $k = 0, \ldots, n$. In other words, p is located outside the visible polygon, or p_8 is not visible from p_1 in the visible polygon (and vice versa). The failure of chord property indicates that P_{pq} is not the result of digitizing a line segment.

It can be proved that in the 8-digital space, the result of the line digitization is a digital arc that satisfies the properties of the chord. Conversely, if a digital arc satisfies the property of the chord, it is the result of digitizing the line segments (Marchand, 2000).

1.4 Distance Transforms

DT is a special transform that maps a binary image into a grayscale image. DT is a global operation, but it can be obtained by local distance computation. The main concepts involved in DT are still neighborhood, adjacency, and connectivity.

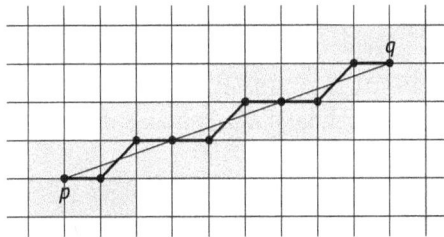

Figure 1.24: An example of verifying the correctness of the chord property.

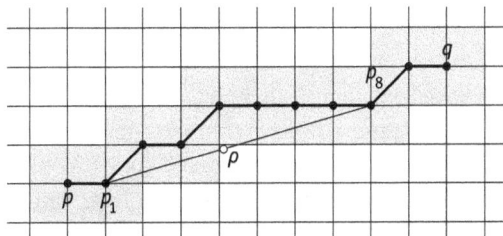

Figure 1.25: An example of the failure of the chord property.

1.4.1 Definition and Property

DT computes the distance between a point inside a region and the closest point out-side the region. In other words, the DT of a point in a region computes the nearest distance between this point and the boundary of the region. Strictly speaking, DT can be defined as follows: Given a set of points P, a subset $B \in P$, and a distance function $D(., .)$ that satisfies the metric conditions in Section 1.3.1, the distance transformation $DT(.)$ of a point $p \in P$ is

$$DT(p) = \min_{q \in B}\{D(p, q)\} \qquad (1.12)$$

The ideal measurement for distance is Euclidean distance. However, for the sake of simplicity, distance functions with integer-based arithmetic are often used, such as $D_4(., .)$ and $D_8(., .)$.

DT can generate a **distance map** of P, which is represented by a matrix $[DT(p)]$. The distance map has the same size as the original image and stores the value of $DT(p)$ for each point $p \in P$.

Given a set of points P and one of its subsets B, the distance transformation of P satisfies the following properties:
1. $DT(p) = 0$ if and only if $p \in B$.
2. Defining the point ref$(p) \in B$ such that $DT(p)$ is defined by $D[p, \text{ref}(p)]$ (ref(p) is the closest point to p in B) and the set of points $\{p|\text{ref}(p) = q\}$ forms the Voronoi cell of point q. If ref(p) is nonunique, p is on the border of the Voronoi cell.

Given a set P and its border B, the distance transformation of P has the following properties (Marchand, 2000):
1. According to the definition of $DT(.)$, $DT(p)$ is the radius of the largest disc within P that is centered at p.
2. If there is exactly one point $q \in B$ such that $DT(p) = d(p, q)$, then there exists a point $r \in P$ such that the disc of radius $DT(r)$ centered at r contains the disc of radius $DT(p)$ centered at p.
3. Conversely, if there are at least two points q and q' in B such that $DT(p) = d(p, q) = d(p, q')$, then there is no disc within P that contains the disc of radius $DT(p)$ centered at p. In that case, p is said to be the center of the maximal disc.

Example 1.5 Illustration of discrete distance map.
A distance map can be represented by a gray-level image, in which the value of each pixel is proportional to the value of the DT. Figure 1.26(a) is a binary image. If the edge of the image is regarded as the contour of the object region, Figure 1.26(b) is its corresponding gray-level image. The central pixels have higher gray-level values

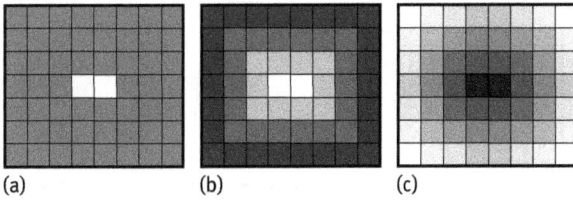

(a)   (b)   (c)

Figure 1.26: A binary image and its distance maps represented by gray-level images.

than the pixels near the boundary of the region. If the two white pixels in the center are regarded as a subset B, Figure 1.26(c) is its corresponding gray-level image. The center values are the smallest and the pixels surround have values increasing with the distance from the center.   ▣

1.4.2 Computation of Local Distances

Computing the distance using eq. (1.12) requires many calculations as it is a global operation. To solve this problem, an algorithm using only the information of the local neighborhood should be considered.

Given a set of discrete points P and one of its subsets B, D_d is the discrete distance used to compute the distance map of P. Then, for any point $p \in P - B$, there exists a point q, which is a neighbor of p (i. e., $q \in N_D(p)$), such that $DT_D(p)$, the discrete distance transformation value at p, satisfies $DT_D(p) = DT_D(q) + d_D(p, q)$. Moreover, since p and q are neighbors, $l(p, q) = d_D(p, q)$ is the length of the move between p and q. Therefore, for any point $p \notin B$, q can be described by $DT_D(q') = \min \{DT_D(p) + l(p, q'); q' \in N_D(p)\}$.

Both sequential and parallel approaches take advantage of the above property to efficiently compute discrete distance maps. A mask is defined which contains the information of local distances within the neighborhood $p \in P$. This mask is then centered at each point $p \in P$ and the local distances are propagated by summing the central value with the corresponding coefficients in the mask.

A mask $M(k, l)$ of size $n \times n$ for computing DT can be represented by an $n \times n$ matrix $M(k, l)$, in which each element represents the local distance between pixel $p = (x_p, y_p)$ and its neighbor $q = (x_p + k, y_p + l)$. Normally, the mask is centered at pixel p, and n should be an odd value.

Figure 1.27 shows several examples of such masks for the propagation of local distances p. The shaded area represents the center of the mask ($k = 0, l = 0$). The size of the mask is determined by the type of the neighborhood considered. The value of each pixel in the neighborhood of p is the length of the respective move from p. The value of the center pixel is 0. In practice, the infinity sign is replaced by a large number.

The mask of Figure 1.27(a) is based on 4-neighbors and is used to propagate D_4 distance. The mask of Figure 1.27(b) is based on 8-neighbors and is used to propagate

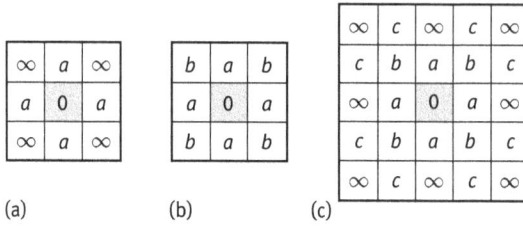

(a) (b) (c)

Figure 1.27: Masks for computing distance transform.

D_8 distance or $D_{a,b}$ ($a = 1$, $b = 1$) distance. The mask of Figure 1.27(c) is based on 16 neighbors and is used to propagate $D_{a,b,c}$ distance.

The process of calculating discrete distance maps using masks can be summarized as follows. Consider a binary image of size $W \times H$ and assume the set of its border point set B is known. The discrete distance map is a matrix, which is denoted **[DT(p)]**, of size $W \times H$ and is calculated by iteratively updating its values until a stable state is reached. The distance map is initialized as (iteration $t = 0$)

$$DT_D^{(0)}(p) = \begin{cases} 0 & \text{if } p \in B \\ \infty & \text{if } p \notin B \end{cases} \tag{1.13}$$

Then, for iterations $t > 0$, the mask $M(k, l)$ is positioned at a pixel $p = (x_p, y_p)$ and the following updating formula is used to propagate the distance values from the pixel $q = (x_p + k, y_p + l)$ onto p

$$DT_D^{(t)}(p) = \min_{k,l}\{DT_D^{(t-1)}(q) + M(k, l); \ q = (x_p + k, y_p + l)\} \tag{1.14}$$

The updating process stops when no change occurs in the distance map for the current iteration.

1.4.3 Implementation of Discrete Distance Transformation

There are two types of techniques that can be used for the implementation of discrete distance transformation: sequential implementation, and parallel implementation. Both produce the same distance map (Borgefors, 1986).

1.4.3.1 Sequential Implementation
For **sequential implementation**, the mask is divided into two symmetric submasks. Then, each of those submasks is sequentially applied to the initial distance map containing the values as defined by eq. (1.13) in a forward and backward raster scan, respectively.

Figure 1.28(a):

(dot pattern image with two circle markers)

Figure 1.28(b):

4	3	4
3	0	3
4	3	4

Figure 1.28(c):

4	3	4
3	0	

Figure 1.28(d):

	0	3
4	3	4

(a) (b) (c) (d)

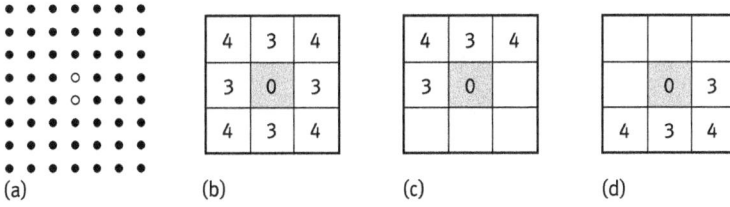

Figure 1.28: Masks for sequential distance transformation.

Example 1.6 Sequential computation of a discrete distance map.
Consider the image shown in Figure 1.28(a) and the 3×3 mask (with $a = 3$ and $b = 4$) in Figure 1.28(b). The set B is defined as the central white pixels and P is the set of all pixels. The mask is first divided into two symmetric submasks, as shown in Figure 1.28(c, d), respectively. ▣

Forward pass: The initial distance map $[DT_D(p)]^{(0)}$ is shown in Figure 1.29(a). The upper submask is positioned at each point of the initial distance map and the value of each pixel is updated by using eq. (1.14), according to the scanning order shown in Figure 1.29(b). For values corresponding to the border of the image, only coefficients of the mask that are contained in the distance map are considered. This pass results in the distance map $[DT_D(p)]^{(t')}$ shown in Figure 1.29(c).

Backward pass: In a similar way, the lower submask is positioned at each point of the distance map $[DT(p)]^{(t')}$ and the value of each pixel is updated by using eq. (1.14), according to the scanning order shown in Figure 1.30(a). This pass results in the distance map of the image shown in Figure 1.30(b).

Clearly, the complexity of this algorithm is $O(W \times H)$ since the updating formula in eq. (1.14) can be applied in constant time.

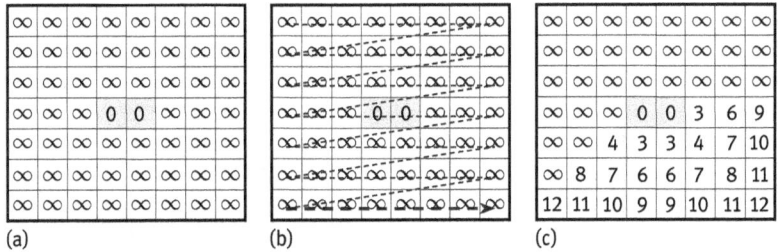

Figure 1.29(a):

∞	∞	∞	∞	∞	∞	∞	∞
∞	∞	∞	∞	∞	∞	∞	∞
∞	∞	∞	∞	∞	∞	∞	∞
∞	∞	∞	0	0	∞	∞	∞
∞	∞	∞	∞	∞	∞	∞	∞
∞	∞	∞	∞	∞	∞	∞	∞
∞	∞	∞	∞	∞	∞	∞	∞

Figure 1.29(b): (scanning order overlaid on the grid)

∞	∞	∞	∞	∞	∞	∞	∞
∞	∞	∞	∞	∞	∞	∞	∞
∞	∞	∞	∞	∞	∞	∞	∞
∞	∞	∞	0	0	∞	∞	∞
∞	∞	∞	∞	∞	∞	∞	∞
∞	∞	∞	∞	∞	∞	∞	∞
∞	∞	∞	∞	∞	∞	∞	∞

Figure 1.29(c):

∞	∞	∞	∞	∞	∞	∞	∞
∞	∞	∞	∞	∞	∞	∞	∞
∞	∞	∞	∞	∞	∞	∞	∞
∞	∞	∞	0	0	3	6	9
∞	∞	4	3	3	4	7	10
∞	8	7	6	6	7	8	11
12	11	10	9	9	10	11	12

(a) (b) (c)

Figure 1.29: Forward pass of sequential distance transformation.

(a)

12	11	10	9	9	10	11	12
11	8	7	6	6	7	8	11
10	7	4	3	3	4	7	10
9	6	3	0	0	3	6	9
10	7	4	3	3	4	7	10
11	8	7	6	6	7	8	11
12	11	10	9	9	10	11	12

(b)

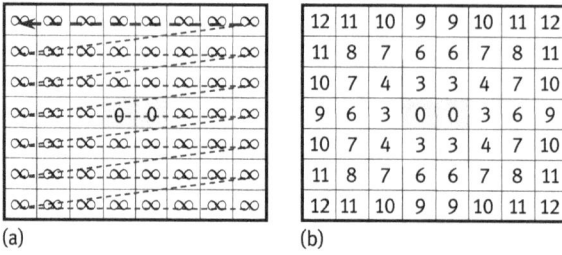

Figure 1.30: Backward pass of sequential distance transformation.

1.4.3.2 Parallel Implementation

In parallel implementation, each pixel is associated with a processor. The discrete distance map is first initialized using eq. (1.13), and the updating operation given by eq. (1.14) is applied to all pixels for each iteration. The process stops when no change occurs in the discrete distance map for the current iteration.

The **parallel implementation** can be represented by ($p = (x_p, y_p)$)

$$\mathrm{DT}^{(t)}(x_p, y_p) = \min_{k,j}\{\mathrm{DT}^{(t-1)}(x_p + k, y_p + l) + M(k, l)\} \tag{1.15}$$

where $\mathrm{DT}^{(t)}(x_p, y_p)$ is the iterative value at (x_p, y_p) for the tth iteration, k and l are the relative positions with respect to the center of mask $(0, 0)$, and $M(k, l)$ represents the mask entry value.

Example 1.7 Parallel computation of a discrete distance map.
Consider again the image shown in Figure 1.26(a). The mask used here is still the original distance mask shown in Figure 1.28(b) and the initial discrete distance map is shown in Figure 1.29(a). Figure 1.31(a, b) gives the first two temporary discrete distance maps obtained during the parallel computation. The distance map in Figure 1.31(c) is the final result.

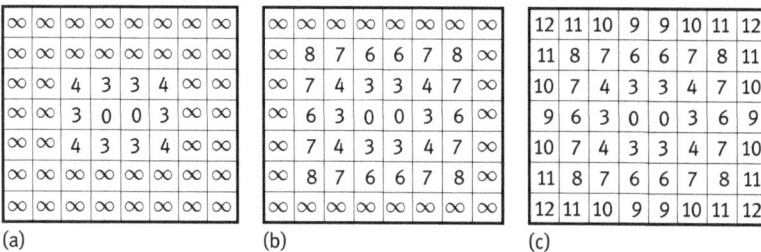

∞	∞	∞	∞	∞	∞	∞	∞
∞	∞	∞	∞	∞	∞	∞	∞
∞	∞	4	3	3	4	∞	∞
∞	∞	3	0	0	3	∞	∞
∞	∞	4	3	3	4	∞	∞
∞	∞	∞	∞	∞	∞	∞	∞
∞	∞	∞	∞	∞	∞	∞	∞

(a)

∞	∞	∞	∞	∞	∞	∞	∞
∞	8	7	6	6	7	8	∞
∞	7	4	3	3	4	7	∞
∞	6	3	0	0	3	6	∞
∞	7	4	3	3	4	7	∞
∞	8	7	6	6	7	8	∞
∞	∞	∞	∞	∞	∞	∞	∞

(b)

12	11	10	9	9	10	11	12
11	8	7	6	6	7	8	11
10	7	4	3	3	4	7	10
9	6	3	0	0	3	6	9
10	7	4	3	3	4	7	10
11	8	7	6	6	7	8	11
12	11	10	9	9	10	11	12

(c)

Figure 1.31: Temporary discrete distance maps and the final result.

1.5 Overview of the Book

This book has eight chapters. This chapter provides an overview of image analysis. The scope of image analysis is introduced along with a discussion on related topics. Some discrete distance metrics, digitization models, and the DT techniques that are specifically important for achieving the goal of image analysis (obtain the measurement data of the interest objects in image) are introduced.

Chapter 2 is titled image segmentation. This chapter introduces first the definition of segmentation and the classification of algorithms. Then, the four groups of basic techniques, as well as the extensions and generations of basic techniques are detailed. Finally, the segmentation evaluation is also discussed.

Chapter 3 is titled object representation and description. This chapter introduces first the schemes for representation classification and description classification. Then, the boundary-based representation, region-based representation, transform-based representation, the descriptors for boundary, and the descriptors for region are presented in sequence.

Chapter 4 is titled feature measurement and error analysis. This chapter discusses about direct and indirect measurements, the accuracy and precision of measurements, and the two commonly used types of connectivity. The sources of feature measurement errors are analyzed in detail. One example of error analysis procedure is also introduced.

Chapter 5 is titled texture analysis. This chapter introduces the concepts of texture and the classification of analysis approaches. The topics as statistical approaches, structural approaches, spectral approaches, texture categorization techniques, and texture segmentation techniques are discussed.

Chapter 6 is titled shape analysis. This chapter presents the definitions of shape and the tasks in shape analysis, and the different classes of 2-D shape. The main emphases are on the shape property description, technique-based descriptors, wavelet boundary descriptors, and the principle of fractal geometry for shape measurements.

Chapter 7 is titled motion analysis. This chapter provides an introduction to objective and tasks of motion analysis. The topics as motion detection, moving object detection, moving object segmentation, and moving object tracking are discussed with respective examples in details.

Chapter 8 is titled mathematical morphology. This chapter introduces both binary mathematical morphology and grayscale mathematical morphology. All are starting from basic operations, then further going to combined operations, and finally arriving at some practical algorithms.

Each chapter of this book is self-contained and has similar structure. After a general outline and an indication of the contents of each section in the chapter, the main subjects are introduced in several sections. Toward the end of every chapter, 12 exercises are provided in the Section "Problems and Questions." Some of them involve conceptual understanding, some of them require formula derivation, some of them need calculation, and some of them demand practical programming. The answers or hints for two of them are collected and presented at the end of the book to help readers to start.

The references cited in the book are listed at the end of the book. These references can be broadly divided into two categories. One category relates directly to the contents of the material described in this book, the reader can find the source of relevant definitions, formula derivations, and example explanation. References of this category are generally marked at the corresponding positions in the text. The other category helps the reader for further study, for expanding the horizons or solving specific problems in scientific research. References of this category are listed in the Section "Further Reading" at the end of each chapter, where the main contents of these references are simply pointed out to help the reader targeted to access.

To read this book, some basic knowledge is essential:
1. Mathematics: Both Linear algebra and matrix theory are important, as the image is represented by a matrix, and image processing often requires matrix manipulation. In addition, knowledge of statistics, probability theory, and stochastic modeling is also very worthwhile.
2. Computer science: Mastery of computer software technology, understanding of the computer architecture system, and application of computer programming methods are very important.
3. Electronics: Many devices involved in image processing are electronic devices, such as camera, video camera, and display screen. In addition, electronic board, FPGA, GPU, SOC, and so on are frequently used.

Some specific prerequisites for reading this book include understanding the fundamentals of signal processing and some basic knowledge of image processing (Volume I, see also Zhang 2012b).

1.6 Problems and Questions

1-1 The video sequence includes a set of 2-D images. How can the video sequence be represented in matrix form? Are there other methods of representation?
1-2 What are the relationships and differences between image analysis and image processing as well as image analysis and image understanding?
1-3* Using the database on the Internet, make a statistical investigation on the trends and characteristics of the development of image analysis in recent years.
1-4 Looking up the literature to see what new mathematical tools have been used in image analysis in recent years? What results have been achieved?
1-5 Looking up the literature to see what outstanding progress has been made in image acquisition and image display related to image analysis in recent years? What effects do they have on image analysis?

1-6 What are the functional modules included in the image analysis system? What is the relationship among each other? Which modules have been selected by this book for presentation? What relationships have they with the modules of the image processing system?

1-7* Draw the chamfer discs of $\Delta_{3,4}(15)$ and $\Delta_{3,4}(28)$, respectively.

1-8 For the digitized set obtained in Figure 1.16, the outermost pixels can be connected to form a closed contour according to the 4-connection and the 8-connection, respectively. It is required to determine whether or not they are digital arcs.

1-9 Design a nonempty set S, which may be mapped to an empty digitized set in both two digitization models. Make a schematic diagram.

1-10 With reference to Figure 1.18, discuss what would be the results if using grid-intersect quantization?

1-11 Perform the distance transformation for the images in Figure Problem 1-11 using the serial algorithm and plot the results after each step.

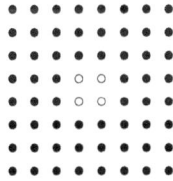

Figure Problem 1-11

1-12 Computer the distance transform map for the simple shape in Figure Problem 1-12. Explain the meaning of the values in the resulted map.

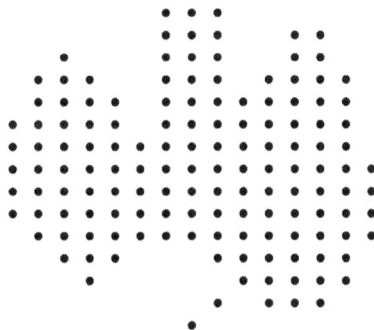

Figure Problem 1-12

1.7 Further Reading

1. **Image and Image Engineering**
   - More detailed introductions on image can be found in Volume I (Zhang, 2012b).
   - The series of survey papers for image engineering can be found in Zhang (1996a, 1996b, 1997a, 1998a, 1999a, 2000a, 2001a, 2002a, 2003a, 2004, 2005a, 2006a, 2007a, 2008a, 2009, 2010, 2011a, 2012a, 2013a, 2014, 2015a, 2016 and 2017). Some summaries on this survey series can be found in Zhang (11996d, 2002c, 2008b and 2015e).

2. **The Scope of Image Analysis**
   - More comprehensive introduction and further discussions on image analysis can be found in Joyce (1985), Lohmann (1998), Zhang (1999b), Kropatsch (2001), Umbaugh (2005), Zhang (2005b), Pratt (2007), Gonzalez (2008), Sonka (2008), Zhang (2012c) and Russ (2016).

3. **Digitization in Image Analysis**
   - Extensions to neighborhood and discrete distance, including 16 neighborhood and distance can be found in Zhang (2005b).
   - The introduction to the connectivity and the shortest path of the N_{16} space can be seen in Zhang (1999b).

4. **Distance Transforms**
   - The method of 3-D distance transformation can be seen in Ragnemalm (1993).

5. **Overview of the Book**
   - The main materials in this book are extracted from the books: Zhang (2005b), Zhang (2012c).
   - More solutions for some problems and questions can be found in Zhang (2002b).
   - More related materials can be found in Zhang (2007b) and Zhang (2013b).
   - Particular information on the analysis of face images can be found in Lin (2003), Tan (2006), Yan (2009), Li (2011a), Tan (2011), Yan (2010), Zhu (2011), Zhang (2015b).

2 Image Segmentation

Image segmentation is the first step of image analysis and is necessary for obtaining analysis results from objects in an image. In image segmentation, an image is divided into its constituent parts and those parts of interest (objects) are extracted for further analysis.

The sections of this chapter are arranged as follows:

Section 2.1 gives a formal definition of image segmentation. The image segmentation techniques are classified according to the definition. The classification methods presented here are general and essential and provide the foundation for the subsequent sections.

Section 2.2 introduces the basic knowledge of image segmentation and the basic principles and methods of various groups of segmentation techniques, such as parallel boundary, sequential boundary, parallel region, and sequential region ones.

Section 2.3 extends the basic segmentation techniques to solve the problem of image segmentation for different kinds of images and generalizes their principles to accomplish segmentation tasks that are more complicated.

Section 2.4 focuses on the evaluation of image segmentation algorithms and the comparison of different evaluation methods, which are at higher levels of image segmentation, respectively.

2.1 Definition and Classification

Research on image segmentation has attracted much attention over the years and a large number (several thousands) of algorithms have been developed (Zhang, 2006b). Descriptions of an image generally refer to the significant parts of the image. Thus, an image description requires segmenting the image into such parts. Image segmentation is one of the most critical tasks in automatic image analysis and image recognition systems because the segmentation results will affect all thenceforth tasks, such as feature extraction and object classification.

2.1.1 Definition of Segmentation

In **image segmentation**, an image is divided into different regions, each of which has certain properties (Fu, 1981). The aim of image segmentation is either to obtain a description of an image or to classify the image into meaningful classes. An example of the former is the description of an office scene. An example of the latter is the classification of an image-containing cancer cells.

Image segmentation can be defined more formally using mathematical tools and terms. To define image segmentation, the definition of the uniform predication is first

DOI 10.1515/9783110524123-002

discussed. Let S denote the grid of sample points in an image, that is, the set of points $\{(i,j)\}$, $i=1, 2, \ldots, N$, $j=1, 2, \ldots, M$, where N and M are the number of pixels in the X and Y directions, respectively. Let T be a nonempty subset of S consisting of contiguous image points.

Then, a uniform predication $P(T)$, a process that assigns the value of true or false to T, depends only on the properties related to the brightness matrix $f(i, j)$ for the points of T. Furthermore, P has the property that if Z is a nonempty subset of T, then $P(T) =$ true implies that $P(Z) =$ true.

The uniformity predication P is a function with the following characteristics:

1. P assigns the value true or false to T, depending only on properties of the values of the image elements of T.
2. $T \supset Z \wedge Z \neq 0 \wedge P(T) =$ true $\Rightarrow P(Z) =$ true; {(set Z is contained in set T) and (Z is a nonempty set) and ($P(Y)$ is true) implies always ($P(Z)$ is true)}.
3. T contains only one element: $\Rightarrow P(T) =$ true.

Now consider the **formal definition of image segmentation** (Fu, 1981). A segmentation of the grid S for a uniformity predication P is to partition S into disjoint non-empty subsets $S_1, S_2, \ldots, S_N$ such that (where S_i and S_j are adjacent):

1. $\bigcup_{i=1}^{N} S_i = S$ (the union of all S_i is (equivalent to) S}.
2. For all i and j and $i \neq j$, it has $R_i \cap R_j = \emptyset$.
3. $P(S_i) =$ true for $i = 1, 2, \ldots, N$.
4. $P(S_i \cup S_j) =$ false for $i \neq j$ (the predication of the union of sets S_i and S_j is false for i is not equal to j).
5. For $i = 1, 2, \ldots, N$, S_i is a connected component.

The first condition implies that every image point must be in a region. This means that the segmentation algorithm should not be terminated until every point is processed. The second condition implies that the regions obtained by the segmentation should not overlap with each other. In other words, a pixel cannot belong to two different regions in the segmentation result. The third condition means that the pixels in a segmented region should have certain identical properties. The fourth condition indicates that the pixels in two different regions should have some distinguishable attributes. The fifth condition implies that regions must be connected components (i. e., comprising contiguous lattice points).

2.1.2 Classification of Algorithms

Owing to the importance of image segmentation, much effort has been devoted to the segmentation process and technique development in the past few decades. This has already resulted in many (thousands) different algorithms, and the number of related algorithms is still increasing (Zhang, 2006b). A number of survey papers have been published (*e. g.*, Fu, 1981; Haralick, 1985. Sahoo, 1988; Pal, 1993; Zhang, 2008d; 2015d).

A classification of algorithms is used to partition a set of algorithms into subsets. An appropriate **algorithm classification** scheme should satisfy the following four conditions (Zhang, 1993d):
1. Every algorithm must be in a group.
2. All groups together can include all algorithms.
3. The algorithms in the same group should have common properties.
4. The algorithms in different groups should have certain distinguishable properties each other.

Classification should be performed according to some classification criteria. The first two conditions imply that the classification criteria should be suitable for classifying all different algorithms. The last two conditions imply that the classification criteria should determine the representative properties of each algorithm group. Keeping these conditions in mind, the following two criteria appear to be suitable for the classification of segmentation algorithms.

The gray-level image segmentation is generally based on one of the two basic properties of the gray-level values in images: **discontinuity** and **similarity** (Fu, 1981). Therefore, two categories of algorithms can be distinguished: boundary-based algorithms that use the discontinuity property and region-based algorithms that use the similarity property. Evidently, such a classification satisfies the above four conditions. The region-based algorithms detect the object area explicitly while the boundary-based ones detect the object contours explicitly. Moreover, the object regions and their boundaries are complementary in images.

On the other hand, according to the processing strategy, segmentation algorithms can be divided into sequential algorithms and parallel algorithms. In the former, cues from earlier processing steps are used for the consequent steps. In the latter, all decisions are made independently and simultaneously. Though parallel methods can be implemented efficiently on certain computers, the sequential methods are normally more powerful for segmenting noisy images. The parallel and sequential strategies are also complementary from the processing point of view. Clearly, such a classification, along with the previous example, satisfies the four conditions mentioned above.

There is no conflict between the above two criteria. Combining both types of categorizations, four groups of segmentation algorithms can be determined (as shown in Table 2.1):

G1: **Boundary-based parallel algorithms.**

G2: **Boundary-based sequential algorithms.**

G3: **Region-based parallel algorithms.**

G4: **Region-based sequential algorithms.**

Table 2.1: Classification of segmentation techniques.

Segmentation techniques	Parallel strategy	Sequential strategy
Boundary based	Group 1 (G1)	Group 2 (G2)
Region based	Group 3 (G3)	Group 4 (G4)

Such a classification also satisfies the above four conditions mentioned in the beginning of this section. These four groups can contain all algorithms studied by Fu (1981) and Pal (1993). The edge-detection-based algorithms are either boundary-based parallel ones or boundary-based sequential ones depending on the processing strategy used in the edge linking or following. The thresholding and pixel classification algorithms are region-based and are normally carried out in parallel. Region extraction algorithms are also region-based but often work sequentially.

2.2 Basic Technique Groups

According to Table 2.1, four groups of techniques can be identified. Some basic and typical algorithms from each group will be presented.

2.2.1 Boundary-Based Parallel Algorithms

In practice, **edge detection** has been the first step of many image segmentation procedures. Edge detection is based on the detection of the discontinuity of the edges. An edge or boundary is the place where there is a more or less abrupt change in gray-levels. Some of the motivating factors of this approach are as follows:
1. Most of the information of an image lies in the boundaries between different regions.
2. Biological visual systems make use of edge detection, but not thresholding.

By a parallel solution to the edge detection problem, it means that the decision of whether or not a set of points is on an edge is made on the basis of the gray level of the set and some set of its neighbors. The decision is independent to the other sets of points that lie on an edge. So the edge detection operator in principle can be applied simultaneously everywhere in the image.

In boundary-based parallel algorithms, two processes are involved: The edge element extraction and the edge element combination (edge point detection and grouping).

2.2.1.1 Edge Detectors
Since the edge in an image corresponds to the discontinuity in an image (e. g., the intensity surface of the underlying scene), the **differential edge detector** (DED) is

-1	0	1
-d	0	d
-1	0	1

1	d	1
0	0	0
-1	-d	-1

Figure 2.1: Masks used for gradient operators.

0	-1	0
-1	4	-1
0	-1	0

-1	-1	-1
-1	8	-1
-1	-1	-1

(a)   (b)

Figure 2.2: Masks used for Laplacian operators.

popularly employed to detect edges. Though some particularly and sophisticatedly designed edge operators have been proposed (Canny, 1986), various simple DEDs such as the Sobel, Isotropic, Prewitt, and Roberts operators are still in wide use. These operators are implemented by mask convolutions. Figure 2.1 gives the two masks used for Sobel, Isotropic, and Prewitt edge detectors. For the **Sobel detector**, $d = 2$; for the **Isotropic detector**, $d = 2^{1/2}$; for the **Prewitt detector**, $d = 1$.

These operators are first-order DEDs and are commonly called gradient operators. In addition, the **Laplacian operator**, a second-order DED is also frequently employed in edge detection. Figure 2.2(a, b) gives two commonly used Laplacian masks.

These operators are conceptually well established and computationally simple. The computation of discrete differentiation can be easily made by convolution of the digital image with the masks. In addition, these operators have great potential for real-time implementation because the image can be processed in parallel, which is important, especially in computer vision applications.

The problems with edge detection techniques are that sometimes edges detected are not at the transition from one region to another and the detected edges often have gaps at places where the transitions between regions are not abrupt enough. Therefore, the detected edges may not necessarily form a set of closed connected curves that surround the connected regions.

2.2.1.2 Hough Transform

The **Hough transform** is a global technique for edge connection. Its main advantage is that it is relatively unaffected by the gaps along boundaries and by the noise in images.

A general equation for a straight line is defined by

$$y = px + q \tag{2.1}$$

where p is the slope and q is the intercept. All lines pass through (x, y) satisfy eq. (2.1) in the XY plane. Now consider the parameter space PQ, in which a general equation for a straight line is

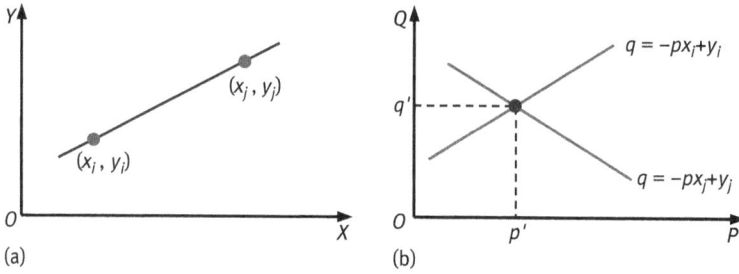

Figure 2.3: The duality of the line and the point in two corresponding spaces.

$$q = -px + y \qquad\qquad (2.2)$$

Consider Figure 2.3(a), where two points (x_i, y_i) and (x_j, y_j) are in a line, so they share the same parameter pair (p, q), which is a point in the parameter plane, as shown in Figure 2.3(b). A line in the XY plane corresponds to a point in the space PQ. Equivalently, a line in the parameter space PQ corresponds to a point in the XY plane.

The above *duality* is useful in transforming the problem of finding points lying on straight lines in the XY plane to the problem of finding a point that corresponds to those points, in the parameter space PQ. The latter problem can be solved by using an accumulative array as used in the Hough transform.

The process of the Hough transform consists of the following three steps:
1. Construct, in the parameter space, an accumulative array $A(p_{min}: p_{max}, q_{min}: q_{max})$.
2. For each point on the object (boundary), compute the parameters p and q, and then accumulate $\{A: A(p, q) = A(p, q) + 1\}$.
3. According to the maximum value in A, detect the reference point and determine the position of the object (boundary).

One example illustrating the results in different steps is presented in Figure 2.4, in which Figure 2.4(a) is an image with a circular object in the center, Figure 2.4(b) is its gradient image, Figure 2.4(c) is the image of accumulative array, and Figure 2.4(d) is the original image with detected boundary.

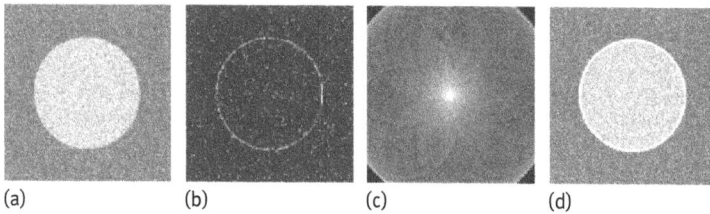

Figure 2.4: An example of circle detection using the Hough transform.

2.2.2 Boundary-Based Sequential Algorithms

In a sequential solution to the edge detection problem, the result at a point is determined upon the results of the operator at previously examined points.

2.2.2.1 Major Components

The major components of a **sequential edge detection procedure** include the following:
1. Select a good initial point. The performance of the entire procedure will depend on the choice of the initial point.
2. Determine the dependence structure, in which the results obtained at previously examined points affect both the choice of the next point to be examined and the result at the next point.
3. Determine a termination criterion to determine that the procedure is finished.

Many algorithms in this category have been proposed. A typical example is the snake, which is also called the active contour (Kass, 1988).

2.2.2.2 Graph Search and Dynamic Programming

Graph search is a global approach combining edge detection with edge linking (Gonzalez, 2002). It represents the edge segments in the form of a graph and searches in the graph for low-cost paths, which correspond to edges/boundaries, using some heuristic information and dynamic programming techniques.

A graph $G = (N, A)$ is a finite, nonempty set of nodes N, together with a set A of pairs of distinct elements (called arcs) of N. Each arc connects a node pair (n_i, n_j). A cost $c(n_i, n_j)$ can be associated with it. A sequence of nodes $n_1, n_2, \ldots, n_K$, with each node n_i being a successor of node n_{i-1} is called a path from n_1 to n_K. The cost of the entire path is

$$c(n_1, n_K) = \sum_{i=2}^{K} c(n_{i-1}, n_i) \tag{2.3}$$

The boundary between two 4-connected pixels is called an edge element. Figure 2.5 shows two examples of the edge element, one is the vertical edge element and the other is the horizontal edge element.

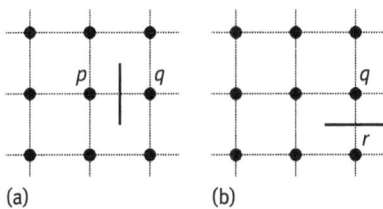

(a)   (b)   **Figure 2.5:** Example of edge elements.

One image can be considered a graph where each pixel corresponds to a node and the relation between two pixels corresponds to an arc. The problem of finding a segment of a boundary can be formulated as a search of the minimal cost path in the graph. This can be accomplished with dynamic programming. Suppose the cost associated with two pixels p and q with gray levels of $f(p)$ and $f(q)$ is

$$c(p, q) = H - [f(p) - f(q)] \tag{2.4}$$

where H is the highest gray level in the image.

The **path detection algorithm** is based on the maximization of the sum of the path's merit coefficients. Usually, they are computed using a gradient operator, which replaces each element by some quantity related to the magnitude of the slope of the underlying intensity surface. The process of the graph search for the minimal cost path will be explained using the following example.

Consider a portion of the image as shown in Figure 2.6(a), where pixels are indicated by black discs and the numbers in the parentheses represent the gray levels of pixels. Here, H in eq. (2.4) is 7. Suppose a top-down path is expected. After checking each possible edge element, the minimal cost path corresponding to the sequence of the edge elements as shown by the bold line in Figure 2.6(b) can be found.

The graph for the above search process is shown in Figure 2.7.

Example 2.1 Practical implementation of the graph search process.
The boundary of objects in noisy two-dimensional (2-D) images may be detected with a state-space search technique such as dynamic programming. A region of interest, centered on the assumed boundary, is geometrically transformed and straightened into a rectangular matrix. Along the rows (of the matrix in the transform domain), merit coefficients are calculated. Dynamic programming is used to find the optimal path in the matrix, which is transformed back to the image domain as the object boundary (Figure 2.8).

Straightening the region of interest in a 3-D image can be accomplished in several ways. In general, the region of interest is resampled along scan lines, which form the rows of the coefficient matrix. The layout order of these rows in this matrix is not

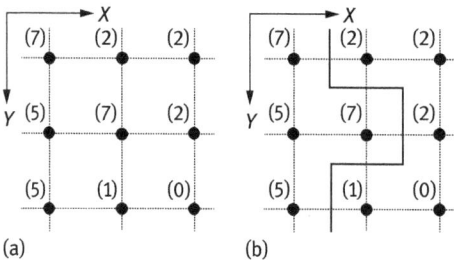

Figure 2.6: Search the sequence of edge elements.

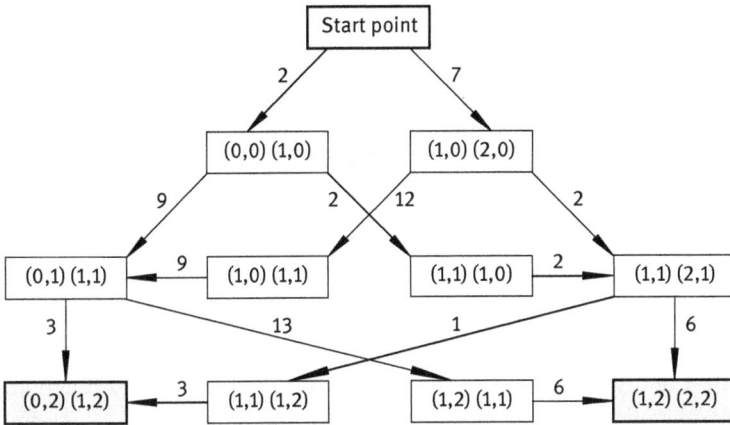

Figure 2.7: The graph for minimum-cost path searching.

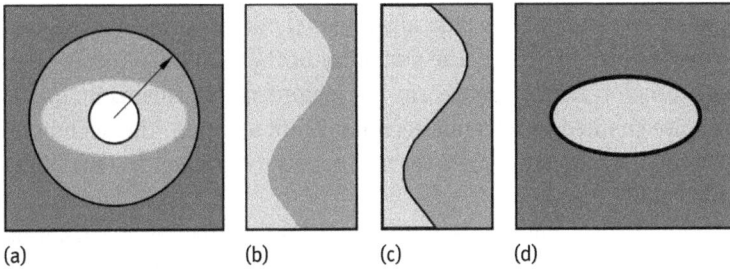

(a) (b) (c) (d)

Figure 2.8: Sketch showing the different steps of the dynamic programming technique.

important, as long as two neighboring elements in the image domain are also neighbors in the resampled domain. The way the scan lines are located in the region of interest depends on the image (*e.g.*, the scan lines are positioned perpendicular to the skeleton of the region of interest), while at the same time overlaps between scan lines are avoided. Another approach is to resample the data on a plane-by-plane basis. In this case, the number of planes in the image equal the number of planes in the resample matrix.

2.2.3 Region-based Parallel Algorithms

Thresholding or **clustering** (the latter is the multidimensional extension of the former) is a popular type of technique in the region-based parallel group. In its most general form, thresholding is described mathematically as follows:

$$g(x,y) = k \quad \text{if} \quad T_k \leq f(x,y) < T_{k+1} \quad k = 0, 1, 2, \ldots, m \tag{2.5}$$

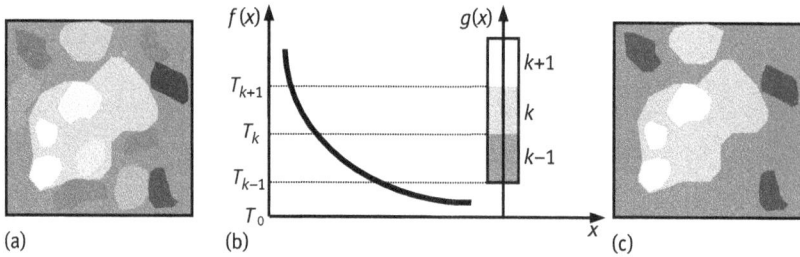

Figure 2.9: One-dimensional mapping from $f(x)$ to $S(x)$ by thresholding.

where (x, y) is the x and y coordinate of a pixel, $g(x, y)$ is the segmented function, $f(x, y)$ is the characteristic feature (e.g., gray level) function of (x, y). $T_0, \ldots, T_m$ are the threshold values, where T_0 and T_m are the minimum and maximum threshold values, respectively. Finally, m is the number of distinct labels assigned to the segmented image (Figure 2.9).

The philosophy of this approach is basically a global one because some aggregate properties of the features are used. The similarity of the feature values in each class of a segmented region form a "mode" in the feature space. This technique is more resistant to noise than edge detection techniques. In addition, this technique produces closed boundaries though sometimes it is necessary to smooth some of the noisy boundaries.

Because the number of segments is not known, an unsupervised clustering scheme may not produce the right number of segments. Besides gray-level values, the features are generally image dependent and it is not clear how these features should be defined to produce good segmentation results. Furthermore, most researchers who have used this approach generally did not use the spatial information that is inherent in an image. Although attempts have been made to utilize the spatial information, the results so far are not better than those that do not use the information.

2.2.3.1 Technique Classification

Determination of appropriate threshold values is the most important task involved in thresholding techniques. Threshold values have been determined in different techniques by using rather different criteria. Generally, the threshold T is a function of the form given by the following equation.

$$T = T[x, y, f(x, y), g(x, y)] \tag{2.6}$$

where $f(x, y)$ is the gray level of the point located at (x, y) and $g(x, y)$ denotes some local properties of this point. When T depends solely on $f(x, y)$, the technique is **point-dependent thresholding**. When T depends on both $f(x, y)$ and $g(x, y)$, the technique is **region-dependent thresholding**. If, in addition, T depends also on the spatial coordinates x and y, then the technique will be **coordinate-dependent thresholding**.

A threshold operator T can also be viewed as a test involving a function T of the form $T[x, y, N(x, y), f(x, y)]$, in which $f(x, y)$ is the characteristic feature functions of (x, y) and $N(x, y)$ denotes some local property (e. g., the average gray-level value over some neighbor pixels) of the point (x, y). Algorithms for thresholding can be divided into three types depending on the functional dependencies of the threshold operator T. When T depends only on $f(x, y)$, the threshold is called a **global threshold**. If T depends on both $f(x, y)$ and $N(x, y)$, then it is called a **local threshold**. If T depends on the coordinate values (x, y) as well as on $f(x, y)$ and $N(x, y)$, then it is called a **dynamic threshold**.

In the case of variable shading and/or variable contrast over the image, one fixed global threshold for the whole image is not adequate. Coordinate-dependent techniques, which calculate thresholds for every pixel, may be necessary. The image can be divided into several subimages. Different threshold values for those subimages are determined first and then the threshold value for each pixel is determined by the interpolation of those subimage thresholds. In a typical technique, the whole image is divided into a number of overlapping sub-images. The histogram for each subimage is modeled by one or two Gaussian distributions depending on whether it is a uni-modal or a bi-modal. The thresholds for those subimages that have bi-modal distributions are calculated using the optimal threshold technique. Those threshold values are then used in bi-linear interpolation to obtain the thresholds for every pixel.

2.2.3.2 Optimal Thresholding
Optimal thresholding is a typical technique used for segmentation. Suppose that an image contains two values combined with additive Gaussian noise. The mixture probability density function is

$$p(z) = P_1 p_1(z) + P_2 p_2(z) = \frac{P_1}{\sqrt{2\pi}\sigma_1} \exp\left[-\frac{(z - \mu_1)^2}{2\sigma_1^2}\right] + \frac{P_2}{\sqrt{2\pi}\sigma_2} \exp\left[-\frac{(z - \mu_2)^2}{2\sigma_2^2}\right] \quad (2.7)$$

where μ_1 and μ_2 are the average values of the object and the background, respectively, and σ_1 and σ_2 are the standard deviation values of the object and the background, respectively. According to the definition of probability, $P_1 + P_2 = 1$. Therefore, there are five unknown parameters in eq. (2.7).

Now look at Figure 2.10, in which $\mu_1 < \mu_2$. Suppose a threshold T is determined, all pixels with a gray-level below T are considered to belong to the background, and all pixels with a gray-level above T are considered to belong to the object. The probabilities of classifying an object pixel as a background pixel and of classifying a background pixel as an object pixel are

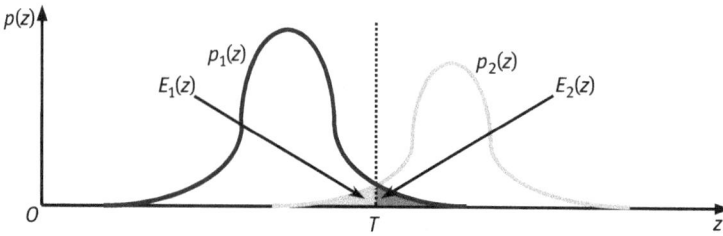

Figure 2.10: The mixture probability density function.

$$E_1(T) = \int_{-\infty}^{T} p_2(z)dz \qquad (2.8)$$

$$E_2(T) = \int_{T}^{\infty} p_1(z)dz \qquad (2.9)$$

The overall probability of error is given by eq. (2.10):

$$E(T) = P_2 E_1(T) + P_1 E_2(T) \qquad (2.10)$$

In the case where both the object and the background have the same standard deviation values, the optimal threshold is

$$T = \frac{\mu_1 + \mu_2}{2} + \frac{\sigma^2}{\mu_1 - \mu_2} \ln\left(\frac{P_2}{P_1}\right) \qquad (2.11)$$

2.2.4 Region-based Sequential Algorithms

In region-based segmentation techniques, the element of the operation is the region. The processes made on regions is either the splitting operation or the merging operation. A **split, merge, and group** (SMG) approach is described as follows, which consists of five phases executed in a sequential order.

1. Initialization phase
   The image is divided into subimages, using a quad-tree (QT) structure (see Figure 2.11).

2. Merging phase
   The homogeneity of the nodes of level L_{s+1} is evaluated to see whether all sons on level L_s are homogeneous. If a node is homogeneous, the four sons are cut from the tree and the node becomes an end-node. This process is repeated for the level until no more merges can take place or the top level is reached (a homogeneous image).

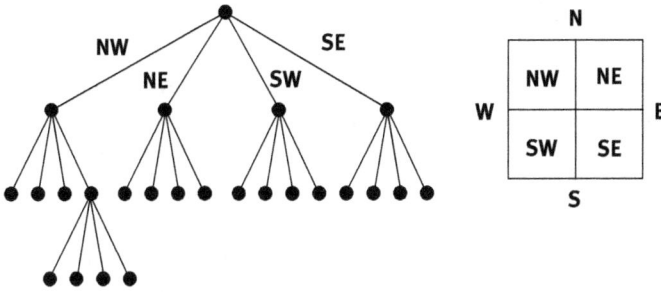

Figure 2.11: The quad-tree structure.

3. Splitting phase

   The in-homogeneous nodes on level L_s are processed. They are split into four sons that are added to the QT. The new end-nodes are then evaluated and are split again if necessary, until the quad-tree has homogeneous end-nodes only.

4. Conversion from QT to RAG (Region Adjacency Graph)

   The QT, which emphasizes hierarchical inclusion relations, has been adopted to get the best of both the splitting and the merging approaches. Outside of this structure, however, it is still possible to merge different subimages that are adjacent but cannot be merged in the QT (because of an in-homogeneous father node or differing node levels). The **region adjacency graph** (RAG), in which adjacency relations are of primary concern, allows these merges to take place.

5. Grouping phase

   The explicit neighbor relationships can be used to merge adjacent nodes that have a homogeneous union.

The path and the data structures are summarized in Figure 2.12.

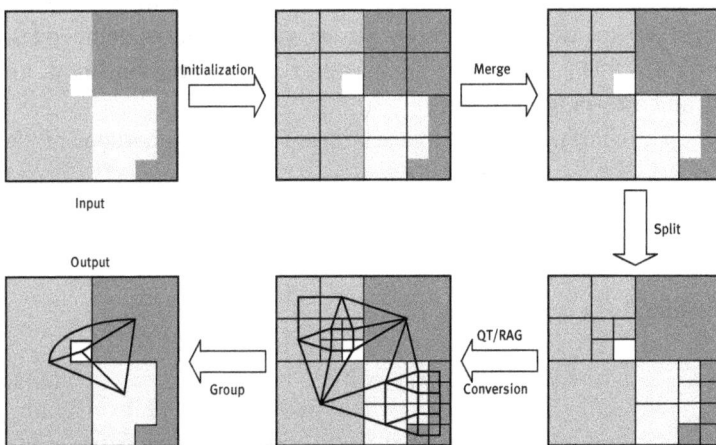

Figure 2.12: The different phases of the SMG algorithms.

2.3 Extension and Generation

In the last section, several basic algorithms for segmenting 2-D gray-level images were described. Those algorithms can be extended to segment other types of images and their principles can be generalized to complete tasks that are more complicated. Note that the distinction between the parallel and sequential strategies also extends to the known segmentation algorithms for images of higher dimensionality. The extension of parallel methods to n-D is trivial (simple), as there is no interaction between assignment decisions of different image elements. When extending sequential methods, there is a choice between applying a $(n$-1$)$-D algorithm to each (hyper-) plane of the n-D image and a true n-D segmentation algorithm. The first approach means it has to use the information in the nth dimension explicitly to achieve an n-D segmentation result while an n-D algorithm uses the information implicitly. On the other hand, it is possible that an $(n$-1$)$-D approach will be less memory expensive and faster because the extra-dimensional information can be supplied more efficiently.

2.3.1 Extending 2-D Algorithms to 3-D

A variety of imaging modalities make it possible to acquire true 3-D data. In these data, the value of a physical parameter is known for all spatial coordinates and thus it leads to a description $f(x, y, z)$. 3-D segmentation is thus required and various algorithms have been developed. Many 3-D algorithms are obtained by extending the existing 2-D algorithms. According to the classifications shown in Table 2.1, several 3-D segmentation algorithms in different groups will be discussed.

2.3.1.1 3-D Differential Edge Detectors
In 3-D space, there are three types of neighborhoods. It can also say that in the 3-D space, there are three ways to define a neighbor: voxels with joint faces, joint edges, and joint corners (Jähne, 1997). These definitions result in a **6-neighborhood**, an **18-neighborhood**, and a **26-neighborhood**.

Let $x = (x_0, \ldots, x_n)$ be some point on an image lattice. The V_1^i neighborhood of x is defined as

$$V_1^i = \{y | D_1(x, y) \leq i\} \tag{2.12}$$

and the V_∞^i neighborhood of x is defined as

$$V_\infty^i = \{y | D_\infty(x, y) \leq i\} \tag{2.13}$$

These neighbors determine the set of the lattice points within a radius of i from the center point.

Let $x = (x_0, x_1)$ be some point on a 2-D image lattice. The n-neighborhood of x, where $n = 4, 8$, is defined as

$$N_4(x) = V_1^1(x) \qquad N_8(x) = V_\infty^1(x) \tag{2.14}$$

Let $x = (x_0, x_1, x_2)$ be some point on a 3-D image lattice. The n-neighborhood of x, where $n = 6, 18, 26$, is defined as

$$N_6(x) = V_1^1(x), \qquad N_{18}(x) = V_1^2(x) \cap V_\infty^1(x), \qquad N_{26}(x) = V_\infty^1(x) \tag{2.15}$$

In digital images, differentiation is approximated by difference (e. g., by the convolution of images with discrete difference masks, normally in perpendicular directions). For 2-D detectors, two masks are used for x and y directions, respectively. For **3-D detectors**, three convolution masks for the x, y, and z directions should be designed. In the 2-D space, most of the masks used are 2×2 or 3×3. The simplest method for designing 3-D masks is just to take 2-D masks and provide them with the information for the third dimension. For example, 3×3 masks in 2-D can be replaced by $3 \times 3 \times 1$ masks in 3-D. When a local operator works on an image element, its masks also cover a number of neighboring elements. In the 3-D space, it is possible to choose the size of the neighborhoods. Table 2.2 shows some common examples of **2-D masks** and **3-D masks**.

Since most gradient operators are symmetric operators, their 3-D extensions can be obtained by the symmetric production of three respected masks. The convolution will be made with the x mask in the x–y plane, the y-mask in the y–z plane, and the z-mask in the z–x plane.

For example, the 2-D Sobel operator uses two 3×3 masks, while for the 3-D case, a straightforward extension gives the results shown in Figure 2.13(a), where three $3 \times 3 \times 1$ masks for the three respective convolution planes are depicted. This extended 3-D Sobel operator uses 18 neighboring voxels of the central voxel. For the 26-neighborhood case, three $3 \times 3 \times 3$ masks can be used. To be more accurate, the entry values of masks should be adapted according to the operators' weighting policy. According to the Sobel's weighting policy, the voxels close to the central one should have a heavier weight. Following this idea, the mask for calculating the difference

Table 2.2: Two-dimensional masks and 3-D masks for local operators.

2-D			3-D		
No. of masks	Mask size	Mask coverage	No. of masks	Mask size	Mask coverage
2	2×2	4	3	$2 \times 2 \times 1$	7
2	2×2	4	3	$2 \times 2 \times 2$	8
2	3×1	5	3	$3 \times 1 \times 1$	7
2	3×3	9	3	$3 \times 3 \times 1$	19
2	3×3	9	3	$3 \times 3 \times 3$	27

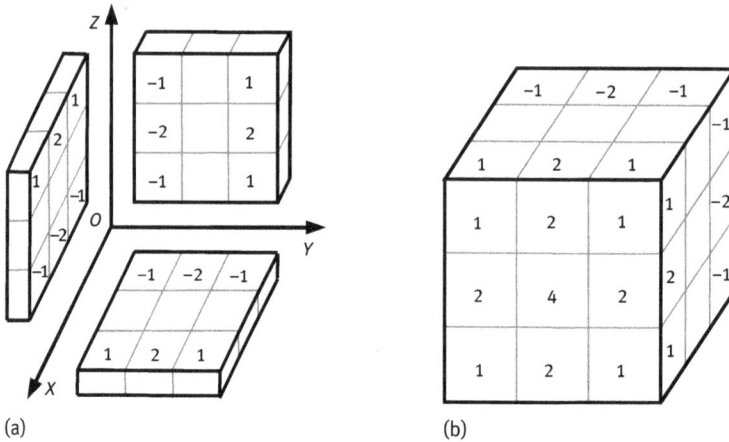

Figure 2.13: Convolution masks of 3-D Sobel operators.

along the x-axis is designed as shown in Figure 2.13(b). Other two masks are symmetric (Zhang, 1993c).

The Laplacian operator is a second-order derivative operator. In the 2-D space, the discrete analog of the Laplacian is often defined by the neighborhood that consists of the central pixel and its four horizontal and vertical neighbors. According to the linearity, this convolution mask can be decomposed into two masks, one for the x-direction and the other for the y-direction. Their responses are added to give the response of Laplacian. When extending this operator to the 3-D case, three such decomposed masks could be used. This corresponds to the 6-neighborhood example as shown in Figure 2.14(a). For the 18-neighborhood case, the mask shown in Figure 2.14(b) can be used. For the 26-neighborhood case, the mask would be like that shown in Figure 2.14(c).

2.3.1.2 3-D Thresholding Methods

Extending 2-D thresholding techniques to 3-D can be accomplished by using 3-D image concepts and by formulating the algorithms with these concepts. Instead of using the pixel in the 2-D image, the voxel is the 3-D image's primitive. The gray-level value of a voxel located at (x, y, z) will be $f(x, y, z)$ and its local properties will be represented by $g(x, y, z)$, thus the 3-D corresponding equation of eq. (2.6) is (Zhang, 1990)

$$T = T[x, y, z, f(x, y, z), g(x, y, z)] \tag{2.16}$$

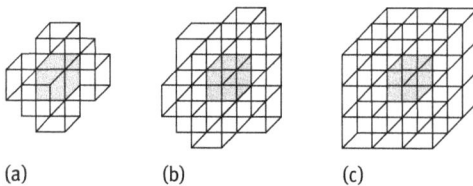

Figure 2.14: Convolution masks of 3-D Laplacian operators.

In the following, some special requirements to extend different groups of 2-D thresholding techniques will be discussed.

Point-Dependent Group: For a technique in **point-dependent thresholding** group, two process steps can be distinguished:
(i) The calculation of the gray-level histogram
(ii) The selection of the threshold values based on the histogram

When it goes from 2-D to 3-D using exclusively the individual pixel's information, the only required modification is in the calculation of the histogram from a 3-D image, which is essentially a counting and addition process. As a 3-D summation can be divided into a serial of 2-D summations, this calculation can be accomplished by using the existing 2-D programs (the direct 3-D summation is also straightforward). The subsequent threshold selection based on the resulted histogram is the same as that derived in the 1-D space. In this case, real 3-D routines are not necessary to process 3-D images.

Region-Dependent Group: In addition to the calculation of the histogram from 3-D images, 3-D local operations are also needed for the **region-dependent thresholding** group of techniques. For example, besides the gray-level histogram, the filtered gray-level image or the edge value of the image is also necessary. Those operations, as the structural information of a 3-D image has been used, should be performed in the 3-D space. In other words, truly 3-D local operations in most cases require the implementation of the 3-D versions of 2-D local operators. Moreover, as it has more dimensions in space, the variety of local operators becomes even larger. For example, one voxel in the 3-D space can have 6-, 18-, or 26-nearest or immediate neighbors, instead of 4- or 8-neighbors for a 2-D pixel.

Coordinate-Dependent Group: Two consecutive process steps can be distinguished in the **coordinate-dependent thresholding** group. One threshold value for each subimage is first calculated, then all these values are interpolated to process the whole image. The threshold value for each subimage can be calculated using either point-dependent or region-dependent techniques. Therefore, respective 3-D processing algorithms may be necessary. Moreover, 3-D interpolation is required to obtain the threshold values for all voxels in the image based on the thresholds for the subimages.

2.3.1.3 3-D Split, Merge, and Group Approach

To adapt the *SMG* algorithm in last section to 3-D images, it is necessary to change the data structures that contain the 3-D information. The RAG structure can be left unchanged. The QT structure, however, depends on the image dimensionality. The

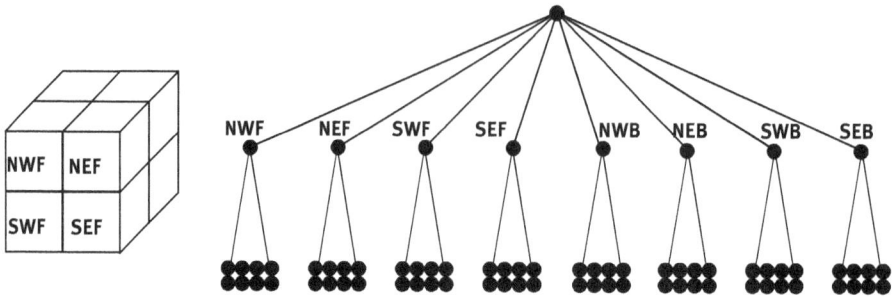

Figure 2.15: Octree structure.

3-D equivalent is called an octree. The octree structure of a 3-D image is shown in Figure 2.15. Each father in this pyramidal structure has eight sons.

The initialization, split, and merge phases are analogous to those in the 2-D case because of the similarity of the quad-tree and the octree structures. The neighbor-finding technique also functions in the same way. Only the functions used to find the path in the tree from a node to its neighbor have to be adjusted for the 3-D case.

By simply replacing the quad-tree by an octree, the 3-D algorithm will be restricted to the image containing $2^n \times 2^n \times 2^n$ voxels. In practice, 3-D images with lower z resolution are often encountered (anisotropic data). Three methods can be used to solve this problem (Strasters, 1991):

1. Adapt the image to fit into the octree structure, that is, increase the sampling density in the z direction or decrease the sampling density in the x and y directions.
2. Change the sequence of the data input, that is, input the image to the algorithm in such a way that it fits the octree structure.
3. Adapt the octree to make recursive splitting of the original image possible, that is, to allow all possible operations of the SMG algorithm on a 3-D image.

2.3.2 Generalization of Some Techniques

Many of the above-described segmentation techniques could be generalized to fit various particular requirements in real applications. Two techniques are introduced in the following as examples.

2.3.2.1 Subpixel Edge Detections

In many applications, the determination of the edge location at the pixel level is not enough. Results that are more accurate are required for registration, matching, and measurement. Some approaches to locate edge at the **subpixel** level are then proposed.

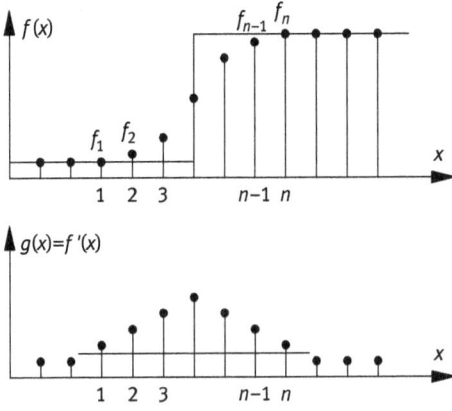

Figure 2.16: A ramp edge and its gradient.

Moment Preserving: Consider a ramp edge as shown in the upper part of Figure 2.16, where the ramp values at each position 1, 2, ..., n-1, n are indicated by $f_1, f_2, \ldots, f_{n-1}, f_n$. Edge detection is a process used to find a step edge (an idea edge), which in some sense fits best to the ramp edge. One approach is to keep the first three moments of the ramp and step edges equivalent to each other (**moment preserving**) (Tabatabai, 1984).

Consider the following formulation. The pth order of the moment ($p = 1, 2, 3$) of a signal $f(x)$ is defined by (n is the number of pixels belonging to the edge)

$$m_p = \frac{1}{n} \sum_{i=1}^{n} [f_i(x)]^p \tag{2.17}$$

It has been proven that m_p and $f(x)$ have a one-to-one correspondence. Let t be the number of pixels that have gray-level value b (background) and $n-t$ be the number of pixels whose gray-level values are o (object). To keep the first three moments of two edges equivalent, the following three equations should be satisfied

$$m_p = \frac{t}{n} b^p + \frac{n-t}{n} o^p, \qquad p = 1, 2, 3 \tag{2.18}$$

where t is

$$t = \frac{n}{2} \left[1 + s \sqrt{\frac{1}{4 + s^2}} \right] \tag{2.19}$$

in which

$$s = \frac{m_3 + 2m_1^3 - 3m_1 m_2}{\sigma^3} \quad \text{and} \quad \sigma^2 = m_2 - m_1^2 \tag{2.20}$$

The edge is located at $t + 0.5$. This value is normally a real value that provides the subpixel accuracy.

Expectation of First-Order Derivatives: This approach computes the **expectation of first-order derivatives** of the input sequence and consists of the following steps (consider 1-D case) (Hu, 1992):

(i)   For an image function $f(x)$, compute the first derivative $g(x) = |f'(x)|$.

(ii)   According to the value of $g(x)$, determine the edge region (*i. e.,* for a given threshold T determine the region $[x_i, x_j]$, where $g(x) > T$, with $1 \leq i, j \leq n$ (Figure 2.16).

(iii)   Compute the probability function $p(x)$ of $g(x)$:

$$p_k = \frac{g_k}{\sum_{i=1}^n g_i}, \qquad k = 1, 2, \ldots, n \tag{2.21}$$

(iv)   Compute the expectation value E of $g(x)$, which is taken as the location of the edge:

$$E = \sum_{k=1}^n kp_k = \sum_{k=1}^n \left(kg_k \middle/ \sum_{i=1}^n g_i \right) \tag{2.22}$$

Compared to the moment-preserving technique, the approach based on the expectation of first-order derivatives can solve the problem of multiresponse (misdetection of several edges). One comparison of these two techniques is given in Table 2.3. For some similar sequences, the subpixel-level location for the expectation technique is less sensitive to the length of the input sequence.

Using Tangent Direction Information: Both of the above methods detect the subpixel-level edges using the statistical property of the blurred boundary that has a certain width, so the edge location can be determined using the information along the normal direction to the boundary. In cases where the boundary ramp is sharp, there would be few boundary pixels used for the statistical computation, and then, the error in determining the boundary would be larger. If the shape of the object is given, then the following technique can be used, which is based on the information along the tangent direction (Zhang, 2001c). Such a technique can be split into two steps: The detection of the boundary at the pixel level and then the adjustment, according

Table 2.3: Examples of the input sequences and the detected subpixel edge locations.

No.	Input sequences	Moment	Expectation
1	0 0 0 0.5 1 1 1 1 1	3.506	3.500
2	0 0 0 0.25 1 1 1 1 1 1	3.860	3.750
3	0 0 0 0.25 1 1 1 1 1 1 1 1 1 1 1 1	3.861	3.750
4	0 0.1 0.2 0.3 0.4 0.6 0.8 1 1	4.997	4.600

to **information along the tangent direction** at the pixel-level boundary, of the edge location at the subpixel level. The first step could be a normal edge detection step. Only the second step is discussed in the following.

Consider Figure 2.17, in which the edge to be located is at the boundary of a circle. Let the function of this circle be $(x - x_c)^2 + (y - y_c)^2 = r^2$, where the center coordinates are (x_c, y_c) and the radius is r. If the boundary points along the X- and Y-axes can be found, then both (x_c, y_c) and r can be obtained, and the circle can be determined.

Figure 2.17 illustrates the principle used to find the left point along the X-axis. The pixels located inside the circle are shaded. Let x_1 be the coordinate of the left-most point along the X direction and h be the number of boundary pixels in the column with X coordinate x_1. At this column, the difference T between two cross-points that are produced by two adjacent pixels with the real circle is

$$T(h) = \sqrt{r^2 - (h - 1)^2} - \sqrt{r^2 - h^2} \qquad (2.23)$$

Suppose S is the difference between the coordinate x_1 and the real subpixel edge. In other words, S is needed to modify the pixel-level boundary to the subpixel-level boundary. It is easy to see from Figure 2.17 that S is the summation of all $T(i)$ for $i = 1, 2, \ldots, h$ and plus e

$$S = r - \sqrt{r^2 - h^2} + e < 1 \qquad (2.24)$$

where $e \in [0, \ T(h + 1)]$, whose average is $T(h + 1)/2$. Calculating $T(h + 1)$ from eq. (2.23) and putting the result into eq. (2.24) finally gives

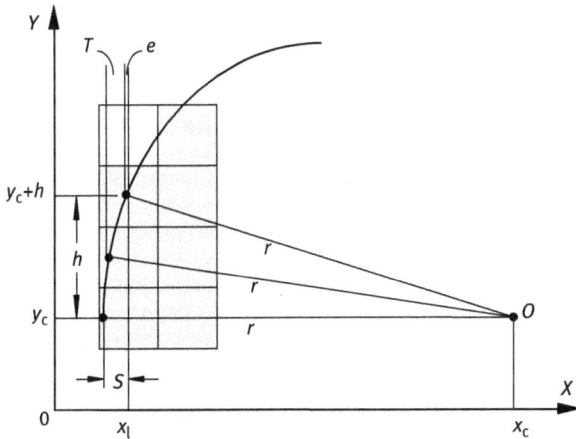

Figure 2.17: Illustrating the location of the subpixellevel edge.

$$S = r - \frac{1}{2}\left(\sqrt{r^2 - h^2} + \sqrt{r^2 - (h+1)^2} \right) \qquad (2.25)$$

There will be two kinds of errors when using S to modify the pixel-level boundary to produce the subpixel-level boundary. One error comes from the averaging of e, called $|dS_e|$. The maximal of this error is

$$|dS_e| = \frac{T(h+1)}{2} = \frac{1}{2}\left\{ \sqrt{r^2 - h^2} - \sqrt{r^2 - (h+1)^2} \right\} \qquad (2.26)$$

From Figure 2.17 and with the help of eq. (2.24), it can obtain

$$h < \sqrt{2r - 1} \qquad (2.27)$$

Expand the right side of eq. (2.27) and put the result into eq. (2.26), then eq. (2.26) becomes

$$|dS_e| \approx \frac{2h+1}{4r} \approx \frac{1}{\sqrt{2r}} \qquad (2.28)$$

Another error comes from the truncation. In the above, r is assumed to be the real radius. However, in a real situation, r can only be obtained in the first step with pixel-level accuracy. Suppose the variation of S caused by r is $|dS_r|$. Apply the derivation of eq. (2.25) to r, it has

$$\left| \frac{dS}{dr} \right| = 1 - \frac{r}{2}\left\{ 1/\sqrt{r^2 - h^2} + 1/\sqrt{r^2 - (h+1)^2} \right\} \qquad (2.29)$$

Simplifying eq. (2.29) with the help of eq. (2.27) yields (the average of dr is 0.5 pixel)

$$|dS_r| = \frac{1}{2}\left[\frac{h^2 + (h+1)^2}{4r^2} \right] < \frac{h^2 + h + 1}{4r^2} \approx \frac{1}{2r} \qquad (2.30)$$

The overall error would be the sum of $|dS_e|$ and $|dS_r|$, where the first term is the main part.

2.3.2.2 Generalization of the Hough Transform

The **generalization of the Hough transform** could solve the problem that the object has no simple analytic form but has a particular silhouette (Ballard, 1982). Suppose for a moment that the object appears in the image with a known shape, orientation, and scale. Now pick a reference point (p, q) in the silhouette and draw a line to the boundary. At the boundary point (x, y), compute the gradient angle θ (the angle between the normal at this point and the x-axis). The radius that connects the boundary point (x, y) to the reference point (p, q) is denoted r, the angle between this connection and the x-axis (vector angle) is ϕ. Both r and ϕ are functions of θ. Thus, it is possible to pre-compute the location of the reference point from boundary points (when the gradient angle is given) by

$$p = x + r(\theta) \cos[\phi(\theta)] \tag{2.31}$$
$$q = y + r(\theta) \sin[\phi(\theta)] \tag{2.32}$$

The set of all such locations, indexed by the gradient angles, comprises a table called the R-table. Figure 2.18 shows the relevant geometry and Table 2.4 shows the form of the R-table.

From the above table, it can be seen that all possible reference points can be determined given a particular θ. The R-table provides a function that connects the gradient angle θ and the coordinates of a possible reference point. The following process would be similar to the normal Hough transform.

A complete Hough transform should consider, besides translation, also the scaling and rotation. Taking all these possible variations of the boundary, the parameter space would be a 4-D space. Here, the two added parameters are the orientation parameter β (the angle between the X-axis and the major direction of the boundary) and the scaling parameter S. In this case, the accumulative array should be extended to $A(p_{min}: p_{max}, q_{min}: q_{max}, \beta_{min}: \beta_{max}, S_{min}: S_{max})$. Furthermore, eqs. (2.31) and (2.32) should be extended to eqs. (2.33) and (2.34), respectively

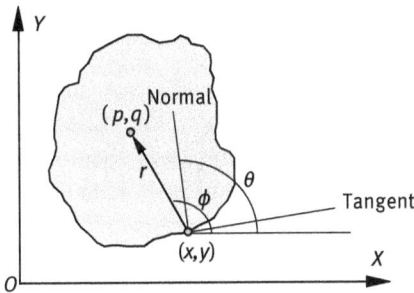

Figure 2.18: Generalized Hough transforms.

Table 2.4: An example of the R-table.

θ	$r(\theta)$	$\phi(\theta)$
θ_1	$r_1^1, r_1^2, \ldots, r_1^{N_1}$	$\phi_1^1, \phi_1^2, \ldots, \phi_1^{N_1}$
θ_2	$r_2^1, r_2^2, \ldots, r_2^{N_2}$	$\phi_2^1, \phi_2^2, \ldots, \phi_2^{N_2}$
...	...	...
θ_M	$r_M^1, r_M^2, \ldots, r_M^{N_M}$	$\phi_M^1, \phi_M^2, \ldots, \phi_M^{N_M}$

$$p = x + S \times r(\theta) \times \cos[\phi(\theta) + \beta] \tag{2.33}$$
$$q = y + S \times r(\theta) \times \sin[\phi(\theta) + \beta] \tag{2.34}$$

The accumulation becomes $A(p, q, \beta, S) = A(p, q, \beta, S) + 1$.

2.4 Segmentation Evaluation

One important fact in the development of segmentation techniques is that no general theory exists, so this development has traditionally been an *ad hoc* (made for a particular purpose) and problem-oriented process. As a result, all the developed segmentation techniques are generally application dependent. On the other hand, given a particular application, developing an appropriate segmentation algorithm is still quite a problem. Segmentation evaluation is thus an important task in segmentation research.

2.4.1 A Survey on Evaluation Methods

Many efforts have been put into segmentation evaluation and many methods have been proposed. An overview is provided below.

2.4.1.1 Classification of Evaluation Methods
Segmentation algorithms can be evaluated analytically or empirically. Accordingly, evaluation methods can be divided into two categories: analytical methods and empirical methods. The **analytical methods** directly examine and assess the segmentation algorithms themselves by analyzing their principles and properties. The **empirical methods** indirectly judge the segmentation algorithms by applying them to test images and measuring the quality of segmentation results. Various empirical methods have been proposed. Most of them can still be classified into two categories: The **empirical goodness methods** and the **empirical discrepancy methods**. In the first category, some desirable properties of segmented images, often established according to human intuition, are measured by "goodness" parameters. The performances of segmentation algorithms under investigation are judged by the values of the goodness measurements. In the second category, some references that present the ideal or expected segmentation results are found first. The actual segmentation results obtained by applying a segmentation algorithm, which is often preceded by preprocessing and/or followed by postprocessing processes, are compared to the references by counting their differences. The performances of segmentation algorithms under investigation are then assessed according to the discrepancy measurements. Following this discussion, three groups of methods can be distinguished.

A recent study on image segmentation evaluation (Zhang, 2015c) shows that most recent works are focused on the discrepancy group, while few works rely on the goodness group.

The above classification of evaluation methods can be seen more clearly in Figure 2.19, where a **general scheme for segmentation and its evaluation** is presented (Zhang, 1996c). The input image obtained by sensing is first (optionally) preprocessed to produce the segmenting image for the segmentation (in its strict sense) procedure. The segmented image can then be (optionally) postprocessed to produce the output image. Further processes such as feature extraction and measurement are done based on these output images.

In Figure 2.19, the part enclosed by the rounded square with the dash-dot line corresponds to the segmentation procedure in its narrow sense, while the part enclosed by the rounded square with the dot line corresponds to the segmentation procedure in its general form. The black arrows indicate the processing directions of the segmentation. The access points for the three groups of evaluation methods are depicted with gray arrows in Figure 2.19. Note that there is an OR condition between both arrows leading to the boxes containing the "segmented image" and the "output image" both from the **"empirical goodness method"** and the **"empirical discrepancy method."** Moreover, there is an AND condition between the arrow from the "empirical discrepancy method" to the "reference image" and the two (OR-ed) arrows going to the "segmented image" and the "output image." The analysis methods are applied to the algorithms for segmentation directly. The empirical goodness methods judge the segmented image or the

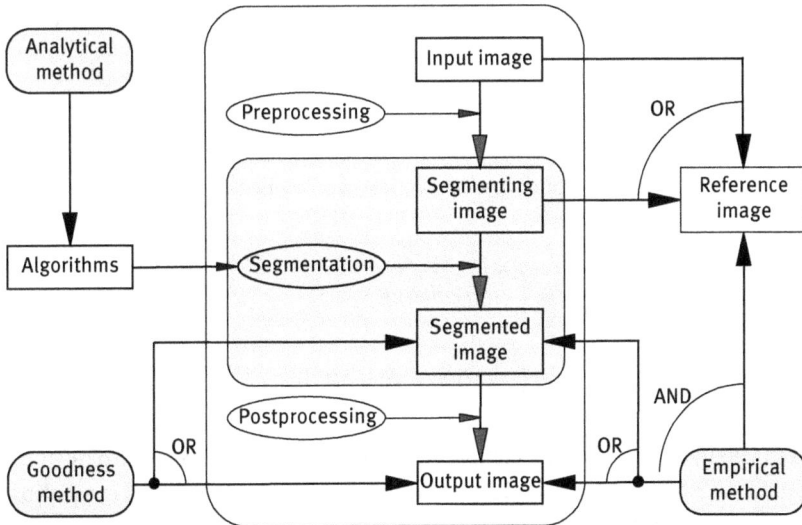

Figure 2.19: General scheme for segmentation and its evaluation.

output image to indirectly assess the performance of algorithms. For applying empirical discrepancy methods, the reference image is necessary. It can be obtained manually or automatically from the input image or the segmenting image. The empirical discrepancy methods compare the segmented image or the output image with the reference image and use their differences to assess the performances of algorithms.

In the following, some basic criteria are described according to this classification (more criteria can be found in Zhang (2001b, 2006b, and 2015c)).

2.4.1.2 Analytical Methods

Using the analytical methods to evaluate segmentation algorithms avoids the concrete implementation of these algorithms, and the results can be exempt from the influence caused by the arrangement of evaluation experiments as empirical methods do. However, not all properties of segmentation algorithms can be obtained by analytical studies. Two examples are shown as follows:

A Priori Knowledge: Such knowledge for certain segmentation algorithms is ready to be analyzed, which is mainly determined by the nature of the algorithms. However, such knowledge is usually heuristic information, and different kinds of *a priori* **knowledge** are hardly comparable. The information provided by this method is then rough and qualitative. On the other hand, not only "the amount of the relevant *a priori* knowledge that can be incorporated into the segmentation algorithm is decisive for the reliability of the segmentation methods," but it is also very important for the performance of the algorithm to know how such *a priori* knowledge has been incorporated into the algorithm (Zhang, 1991b).

Detection Probability Ratio: The analytical methods in certain cases can provide quantitative information about segmentation algorithms. Let T be the edge decision threshold, P_c be the probability of correct detection, and P_f be the probability of false detection. It can be derived that

$$P_c = \int_T^\infty p(t|edge)dt \tag{2.35}$$

$$P_f = \int_T^\infty p(t|no-edge)dt \tag{2.36}$$

The plot of P_c versus P_f in terms of T can provide a performance index of the detectors, which is called the **detection probability ratio**. Such an index is useful for evaluating the segmentation algorithms based on edge detection (Abdou, 1979).

2.4.1.3 Empirical Goodness Methods

The methods in this group evaluate the performance of algorithms by judging the quality of the segmented images. To do this, certain quality measurements should be defined. Most measurements are established according to human intuition about what conditions should be satisfied by an "ideal" segmentation (e. g., a pretty picture). In other words, the quality of segmented images is assessed by some "goodness" measurement. These methods characterize different segmentation algorithms by simply computing the goodness measurements based on the segmented image without the *a priori* knowledge of the correct segmentation. The application of these evaluation methods exempts the requirement for references so that they can be used for on-line evaluation. Different types of goodness measurements have been proposed.

Goodness Based on intra-region Uniformity: It is believed that an adequate segmentation should produce images having higher **intra-region uniformity,** which is related to the similarity of the property about the region element (Nazif, 1984). The uniformity of a feature over a region can be computed based on the variance of that feature evaluated at every pixel belonging to that region. In particular, for a gray-level image $f(x, y)$, let R_i be the ith segmented region and A_i be the area of R_i. The gray-level uniformity measure (GU) of $f(x, y)$ is given by

$$GU = \sum_i \sum_{(x,y) \in R_i} \left[f(x,y) - \frac{1}{A_i} \sum_{(x,y) \in R_i} f(x,y) \right]^2 \qquad (2.37)$$

Goodness Based on Inter-region Contrast: It is also believed that an adequate segmentation should in addition produce images having higher contrast across adjacent regions. In a simple case where a gray-level image $f(x, y)$ consists of the object with an average gray-level f_o and the background with an average gray-level f_b, a gray-level **inter-region contrast** measure (GC) can be computed by

$$GC = \frac{|f_o - f_b|}{f_o + f_b} \qquad (2.38)$$

2.4.1.4 Empirical Discrepancy Methods

In practical segmentation applications, some errors in the segmented image can be tolerated. On the other hand, if the segmenting image is complex and the algorithm used is fully automatic, the error is inevitable. The disparity between an actual segmented image and a correctly/ideally segmented image (reference image) can be used to assess the performance of algorithms. Both the actually segmented and the reference images are obtained from the same input image. The reference image is sometimes called a gold standard. A higher value of the discrepancy measurement would imply a bigger error in the actually segmented image relative to the reference image and this indicates the poorer performance of the applied segmentation algorithms.

Discrepancy Based on the Number of Misclassified Pixels: Considering the image segmentation as a pixel classification process, the percentage or the **number of pixels misclassified** is the discrepancy measurement that comes most readily to mind. One of them is called the **probability of error** (PE). For a two-class problem the PE can be calculated by

$$PE = P(O) \times P(B|O) + P(B) \times P(O|B) \tag{2.39}$$

where $P(B|O)$ is the probability of the error when classifying objects as the background, $P(O|B)$ is the probability of the error when classifying the background as objects, $P(O)$ and $P(B)$ are *a priori* probabilities of the objects and the background in images. For the multi-class problem, a general definition of PE can be found.

Discrepancy Based on the Position of Misclassified Pixels: The discrepancy measurements based only on the number of misclassified pixels do not take into account the spatial information of these pixels. It is thus possible that images segmented differently can have the same discrepancy measurement values if these measurements only count the number of misclassified pixels. To address this problem, some discrepancy measurements based on **position of mis-segmented pixels** have been proposed.

One solution is to use the distance between the misclassified pixel and the nearest pixel that actually belongs to the misclassified class. Let N be the number of misclassified pixels for the whole image and $d(i)$ be a distance metric from the ith misclassified pixel to the nearest pixel that actually is of the misclassified class. A **discrepancy measurement** (D) based on this distance is defined by Yasnoff (1977)

$$D = \sum_{i=1}^{N} d^2(i) \tag{2.40}$$

In eq. (2.4), each distance is squared. This measurement is further normalized (ND) to exempt the influence of the image size and to give it a suitable value range by Yasnoff (1977)

$$ND = 100 \times \sqrt{D}/A \tag{2.41}$$

where A is the total number of pixels in the image (*i. e.*, a measurement of area).

2.4.2 An Effective Evaluation Method

An effective approach for the objective and quantitative evaluation and comparison of segmentation algorithms is as follows (Zhang, 1994; 1998b).

2.4.2.1 Primary Conditions and General Framework
As is made evident from the discussions presented above, the following conditions should be satisfied for an evaluation and comparison procedure.

1. It should be general, that is., be suitable for all segmentation techniques and various applications. This implies that no parameters or properties of particular segmentation algorithms can be involved so that no bias with respect to some special techniques is introduced.
2. It should use **quantitative and objective criteria** for performance assessment. Here, quantitative means exactness (Zhang, 1997b), while objective refers to the segmentation goal and reality.
3. It should use images that can be reproduced by all users in different places for the purpose of algorithm testing. Moreover, these images should reflect common characteristics of real applications.

Considering these conditions, a **general evaluation framework** is designed. It consists of three related modules as shown in Figure 2.20. The information concerning segmentation goals, image contents, and imaging conditions can be selectively incorporated into this framework to generate the test images (Zhang, 1992a), which makes it suitable for various applications. This framework is also general in the sense that no internal characteristics of particular segmentation techniques are involved as only the segmentation results are taken into account in the test.

2.4.2.2 Criteria for Performance Assessment
In image analysis, the ultimate goal of image segmentation and other processes is to obtain measurements of object features. These measurements are heavily dependent on the results of these earlier processes. It is therefore obvious that the accuracy of these measurements obtained from segmented images would be an efficient

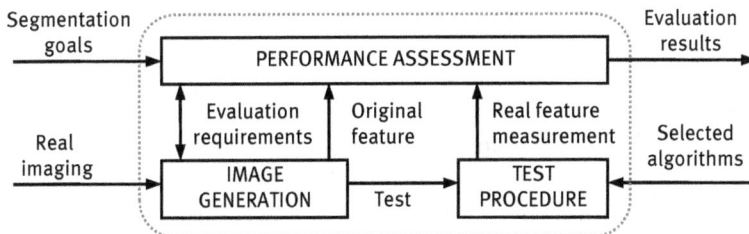

Figure 2.20: A general framework for segmentation evaluation.

index, which reveals the performance of applied segmentation algorithms. Taking into account this ultimate goal, the accuracy of these measurements is used as the criteria to judge and rank the performance of segmentation techniques, which is called **ultimate measurement accuracy** (UMA) (Zhang, 1992b).

In different analysis applications, different object features can be important. So, the UMA can represent a group of feature-dependent and goal-oriented criteria. It can be denoted by UMA_f, where f corresponds to the considered feature. The comparative nature of UMA implies that some references should be available. Depending on how the comparison is made, two UMA types can be applied, that is, the **absolute UMA** ($AUMA_f$) and the **relative UMA** ($RUMA_f$) with the following definitions

$$AUMA_f = |R_f - S_f| \tag{2.42}$$
$$RUMA_f = (|R_f - S_f|/R_f) \times 100\% \tag{2.43}$$

where R_f denotes the feature value obtained from a reference image and S_f denotes the feature value measured from a segmented image. The values of $AUMA_f$ and $RUMA_f$ are inversely proportional to the segmentation quality. The smaller the values, the better the results. From eq. (2.43), it appears that the $RUMA_f$ may be greater than 100% in certain cases. In practice, it can be limited to 100% since it does not make sense to distinguish between very inaccurate results.

Features used in eqs. (2.42) and (2.43) can be selected according to the segmentation goal. Various object features can be employed (some examples are shown in the next section) so that different situations can be covered. In addition, the combination of UMA of different object features is also possible. This combination can be, for example, accomplished by a weighted summation according to the importance of involved features in particular applications. The calculations of eqs. (2.42) and (2.43) are simple and do not require much extra effort. For a given image analysis procedure, the measurement of S_f is a task that needs to be performed anyhow. The value of R_f can be obtained by a similar process from a reference image. In fact, R_f will be produced by the image generation module and can then be used in all evaluation processes.

2.4.3 Systematic Comparison

Studies on image segmentation can be divided into three levels as shown in Table 2.5.

Most studies in segmentation concern elaborate techniques for doing segmentation tasks. Therefore, the level 0 is for segmentation algorithms. To make up the performances of these algorithms in the next level, methods for characterization are proposed and some experiments are performed to provide useful results. One more level up is to evaluate and compare the performance of these characterization methods (Zhang, 1993a).

Table 2.5: Different levels of studies in segmentation.

Level	Description	Mathematical counterpart
0	Algorithms for segmentation (2-D+3-D)	$y = f(x)$
1	Techniques for performance characterization	$y' = f'(x)$
2	Studying characterization methods	$y'' = f''(x)$

The relationships and differences between these three levels of studies could be easily explained by taking a somehow comparable example in mathematics. Given an
original function $y = f(x)$, it may need to calculate the first derivative of $f(x)$, $f'(x)$,
to assess the change rate of $f(x)$. The second derivative of $f'(x)$, $f''(x)$, can also be
calculated to judge the functionality of $f'(x)$.

2.4.3.1 Comparison of Method Groups

The three method groups for segmentation evaluation described above have their own
characteristics (Zhang, 1993a). The following criteria can be used to assess them.
1. Generality of the evaluation
2. Qualitative versus quantitative and subjective versus objective
3. Complexity of the evaluation
4. Consideration of the segmentation applications

2.4.3.2 Comparison of Some Empirical Methods

Empirical evaluation methods evaluate the performance of segmentation algorithms
by judging the quality of the segmented images or by comparing an actual segmented
image with a correctly/ideally segmented image (reference image). The performance
of these evaluation algorithms can also be compared with experiments. One typical
example is shown as follows.

Procedure

The performance of different empirical methods can be compared according to their
behavior in judging the same sequences of the segmented image. This sequence of
images can be obtained by thresholding an image with a number of ordered threshold
values. As it is known, the quality of thresholded images would be better if an appropriate threshold value is used and the quality of thresholded images would be worse
if the selected threshold values are too high or too low. In other words, if the threshold
value increases or decreases in one direction, the probability of erroneously classifying the background pixels as the object pixels goes down, but the probability of
erroneously classifying the object pixels as the background pixels goes up, or vice
verse. Since different evaluation methods use different measurements to assess this
quality, they will behave differently for the same sequence of images. By comparing
the behavior of different methods in such a case, the performance of different methods
can be revealed and ranked (Zhang, 1996c).

Compared Methods

On the basis of this idea, a comparative study of different empirical methods has been carried out. The five methods studied (and the measurements they are based on) are given as follows.

G-GU: Goodness based on the gray-level uniformity;

G-GC: Goodness based on the gray-level contrast;

D-PE: Discrepancy based on the probability of the error;

D-ND: Discrepancy based on the normalized distance;

D-AA: Discrepancy based on the absolute UMA_f using the area as a feature.

These five methods belong to five different method subgroups. They are considered for comparative study mainly because the measurements, on which these methods are based, are generally used.

Experiments and Results

The whole experiment can be divided into several steps: define test images, segment test images, apply evaluation methods, measure quality parameters, and compare evaluation results. The experiment is arranged similarly to the study of object features in the context of image segmentation evaluation.

Test images are synthetically generated (Zhang, 1993b). They are of size 256×256 with 256 gray levels. The objects are centered discs of various sizes with gray-level 144. The background is homogeneous with gray-level 112. These images are then added by independent zero-mean Gaussian noise with various standard deviations. To cope with the random nature of the noise, for each standard deviation five noise samples are generated independently and are added separately to the noise-free images in this study. Five generated test images form a test group. Figure 2.21 gives an example.

Test images are segmented by thresholding them as described earlier. A sequence of 14 threshold values is taken to segment each group of images. The five evaluation methods are then applied to the segmented images. The values of their corresponding measurements are obtained by averaging the results of the five measurements over each group. In Figure 2.22, the curves corresponding to different measurement values are plotted.

Figure 2.21: A group of test images.

Discussions

These curves can be analyzed by comparing their forms. First, as the worst segmentation results provide the highest value for all measurements, the valley values that correspond to the best segmentation results determine the margin between the two extremes. The deeper the valley, the larger the dynamic range of measurements for assessing the best and worst segmentation results. Comparing the depth of valleys, these methods can be ranked in the order of D-AA, D-PE, D-ND, G-GU, and G-GC. Note that the G-GC curve is almost unity for all segmented images, which indicates that different segmentation results can hardly be distinguished in such a case.

Secondly, for the evaluation purpose, a good method should be capable of detecting very small variations in segmented images. The sharper the curves, the higher the measure's discrimination capability to distinguish small segmentation degradation. The ranking of these five methods according to this point is the same as above. If looking at more closely, it can be seen that although D-AA and D-PE curves are parallel or even overlapped for most of the cases in Figure 2.22, the form of the D-AA curve is much sharper than that of the D-PE curve near the valley. This means that D-AA has more power than D-PE to distinguish those slightly different from the nearly best segmentation results, which is more interesting in practice. It is clear that D-AA should not be confused with D-PE. On the other hand, the flatness of the G-GC and G-GU curves around the valley show that the methods based on the goodness measurements such as GC and GU are less appropriate in segmentation evaluation.

The effectiveness of evaluation methods is largely determined by their employed image quality measurements (Zhang, 1997c). From this comparative study, it becomes evident that the evaluation methods using discrepancy measurements such as those based on the feature values of segmented objects and those based on the number of misclassified pixels are more powerful than the evaluation methods using other measurements. Moreover, as the methods compared in this study are representative of various method subgroups, it seems that the empirical discrepancy methods surpass the empirical goodness methods in evaluation.

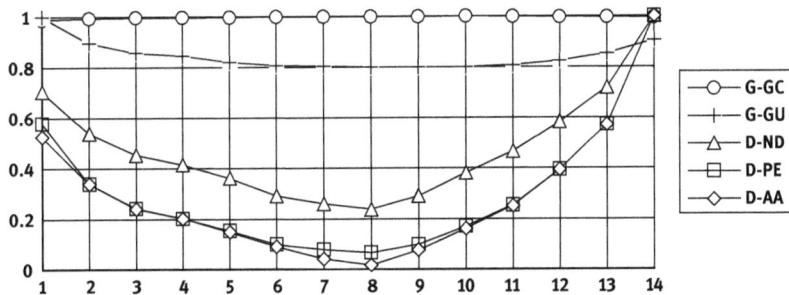

Figure 2.22: Plot of the comparison results.

2.5 Problems and Questions

2-1 Search in the literature, find some new image segmentation algorithms, classify them into four groups according to Table 2.1.

2-2 Show that the Sobel masks can be implemented by one pass of the image by using mask $[-1\ 0\ 1]$ and another pass of the image by using mask $[1\ 2\ 1]^T$.

2-3 Given an image as shown in Figure Problem 2-3, compute gradient maps by using Sobel, Isotropic, and Prewitt operators, respectively. What observations can you make from the results?

110	101	22	6	30
102	105	7	8	9
15	25	52	6	30
35	60	53	56	25
55	50	54	55	58

Figure Problem 2-3

2-4★ What is the cost value of the minimum-cost path in Figure 2.7?

2-5 An image is composed of an object whose gray-level mean is 200 and the variance is 200 on a background whose gray-level mean is 20 and the variance is 20. What is the optimal threshold value for segmenting this image? Could you propose another technique to solve this problem?

2-6 The coordinates of the three vertices of a triangle are (10, 0), (10, 10), (0, 10), respectively. Give the Hough transforms for the three edges of this triangle.

2-7 Extend the Hough transform technique to detect 2-D lines to the Hough transform technique to detect the 3-D plane.

2-8 Compute the subpixel location, using separately the moment-preserving method and the expectation of first-order derivatives method, for the input sequence, which is 0.1, 0.1, 0.1, 0.15, 0.25, 0.65, 0.75, 0.85, 0.9, 0.9.

2-9 For the shape shown in Figure Problem 2-9, construct its R-table.

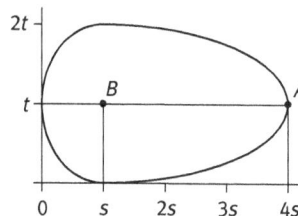

Figure Problem 2-9

2-10★ Suppose that an image is segmented by two algorithms, A and B, respectively. The results are shown in Figure Problem 2-10, respectively (bold letters belong to the objects). Use two criteria (the goodness based on intra-region uniformity and the goodness based on inter-region contrast) to judge the two-segmentation results and compare the performances of *A* and *B*.

0	0	1	0	0		0	0	1	0	0
0	0.5	1	0.5	0		0	0.5	1	0.5	0
1	1	1	1	1		1	1	1	1	1
0	0.5	1	0.5	0		0	0.5	1	0.5	0
0	0	1	0	0		0	0	1	0	0

Figure Problem 2-10

2-11   Generate an image with an irregular object in it. Add some zero-mean Gaussian noise to this image. Select the standard deviation of the Gaussian noise as the gray-level difference between the average value of the object and the average value of the background. Segment this image with any thresholding technique, and judge the segmentation result using the probability of the error.

2-12   Repeat Problem 2-11 with the absolute UMA and relative UMA defined in eqs. (2.42) and (2.43), respectively.

2.6 Further Reading

1. **Definition and Classification**
    –   Image segmentation is a basic type of image technique. Chapters on segmentation can be found in most image processing and computer vision textbooks; for example, Ballard (1982), Castleman (1996), Jähne (1997), Forsyth (2003), and Sonka (2008).
    –   Research on image segmentation has attracted a lot of attention for more than 50 years (Zhang, 2006d; 2008d; and 2015d). A large number of papers (several thousands) on segmentation research can be found in the literature. Many surveys and reviews have been published, see Zhang (2006b) for a list.
    –   Related research is far from maturation. To our knowledge, only very few books/monographs specialize in image segmentation: Medioni (2000), Zhang (2001c, 2006b).
    –   Some segmentation techniques have been designed for special purposes. For example, CBIR or CBVIR (Zhang, 2005c; 2008e).
2. **Basic Technique Groups**
    –   Edge detection is the first step in detecting the edges between different parts of an image, and thus decomposing an image into its constitutional components in 1965 (Roberts, 1965).

- The joint effect of the edge orientation and offset over the full range of significant offsets is studied to investigate fully the behavior of various differential edge operators, such as Sobel, Prewitt, and Robert's cross operators (Kitchen, 1989).

- An analytical study of the effect of noise on the performance of different edge operators can be found in the study of He (1993).

- Thresholding is the most frequently used segmentation technique. Some survey and evaluation papers are Weszka (1978), Sahoo (1988), Zhang (1990), and Marcello (2004).

- An example of color image segmentation can be found in Dai (2003, 2006).

- A comparison of several typical thresholding techniques in Xue (2012).

3. **Extension and Generation**
   - An analytical study of the magnitude and direction response of various 3-D differential edge operators can be found in Zhang (1993c).

   - The determination of the subpixel edge location is an approximation/estimation process based on statistical properties. Some error analysis for the technique using tangent direction information can be found in Zhang (2001c).

   - Other than the techniques discussed in this chapter, there are other techniques for the subpixel edge location. For example, a technique using a small mask to compute orthogonal projection has been presented in Jensen (1995).

   - An extension from object to object class in segmentation can see Xue (2011).

4. **Segmentation Evaluation**
   - Segmentation evaluation has obtained many attention in recent years, and general review of the related progress can be found in Zhang (2006c, 2008c, and 2015c).

   - An evaluation and comparison experiment for different types of segmentation algorithms can be found in Zhang (1997b).

   - In Section 2.4, the presentation and description of evaluation uses still images as examples. Many approaches discussed there can also be applied to the evaluation of video. There are also techniques specialized for video; for example, see Correia (2003).

   - The systematic study of segmentation evaluation is one level up from segmentation evaluation itself. Little effort has been spent on it. The first general overview is Zhang (1996c). Some more references can be found in Zhang (1993a, 2001c, 2008c, and 2015c).

   - The purpose of segmentation evaluation is to guide the segmentation process and to improve the performance in the segmenting images. One approach using the evaluation results to select the best algorithms for segmentation with the framework of expert system is shown in Luo (1999) and Zhang (2000c).

3 Object Representation and Description

After image segmentation, different object regions are detected in the image. Given a specific object region, or a group of object regions, it is often necessary to quantify some of its properties. This task, based on certain suitable data structures (for representation), using appropriate schemes (for description) and with a number of quantitative values (of feature measurements), constitutes one of the basic steps in object analysis and is henceforth called object characterization. This chapter concentrates on the representation and description of objects.

The sections of this chapter are arranged as follows:

Section 3.1 discusses the relation and difference between representation and description and make the technique classification for both of them.

Section 3.2 introduces some typical methods for boundary-based representation, such as chain code, boundary segment, polygon approximation, boundary signatures, and landmark point.

Section 3.3 presents several usual methods for region-based representation, such as bounding regions (including enclosures, minimally enclosing rectangles, and convex hulls), quad-trees, pyramids, and skeletons.

Section 3.4 discusses the transformation-based representation. The basic steps and characteristics of transform representation are introduced by using Fourier transform as an example. In addition, Fourier descriptor can also be derived according to the close relationship between representation and description in the transformation domain.

Section 3.5 introduces the boundary-based descriptors. In addition to some simple boundary descriptors (the length of the boundary, the diameter of the boundary, and the curvature of the boundary), two typical boundary descriptors, the shape number, and the boundary moment are also presented.

Section 3.6 describes the region-based descriptors. Based on a brief introduction of the features of the region, the center of gravity, and the density of the region (grayscale), some topological descriptors and invariant moments-based descriptors are discussed in detail.

3.1 Classification of Representation and Description

An object region can be represented in terms of its external characteristics (related to its boundaries). It can also be represented in terms of its internal characteristics (related to all the pixels compositing the region). An **external representation** is chosen when the primary focus is on shape characteristics. An **internal representation** is selected when the primary focus is on region properties.

DOI 10.1515/9783110524123-003

Some representations used in the image analysis level are listed as follows:
1. Symbolically coded edges and lines (straight line, spline fit, chain-code, end points, etc.).
2. Symbolically coded regions (convex hull, quad-tree, Euler number, spectral measures, texture measures, medial axis, etc.).
3. Symbolically coded surfaces (plane fit, generalized cylinder, generalized cone, polygon list, etc.).
4. Groups of coded edges, regions, surfaces, etc. (lists of related tokens, graph structures describing relationships, etc.).
5. Symbolically transformed edges, lines, regions, and surfaces in different spaces.
6. Symbolic projections of stored models as edges, regions, surfaces, groups of tokens, etc., corresponding to the current view.

Choosing a representation scheme is only part of the task of making the data useful to a computer. The next task is to describe the object region based on the chosen representation. Once a set of regions has been identified, the properties of the regions can be obtained via the computation of the descriptors, which could become the input of the high-level procedures that perform decision-making tasks such as recognition and inspection.

As for representation, various description schemes are possible. Many different descriptors have been proposed that provide different properties of object regions. One desired property for those descriptors is that they should be as insensitive as possible to the variations in translation, rotation, and size of object regions.

Since the regions in an image can be represented by either its boundary pixels or its internal pixels, representation schemes are often divided into boundary-based and region-based categories. In addition, except to directly represent the objects in the image with the pixels belonging to their boundaries and/or regions in the image space, transformed techniques try to represent the objects in the image with other types of representations in different kinds of spaces.

According to the above discussion, various representation methods can be broadly divided into the boundary-based, region-based, and transform-based categories, as shown in Figure 3.1(a).

The representation and description, though closely related, are distinct. Representation emphasizes more on the data structure, while description stresses more on the region properties and relationship among regions.

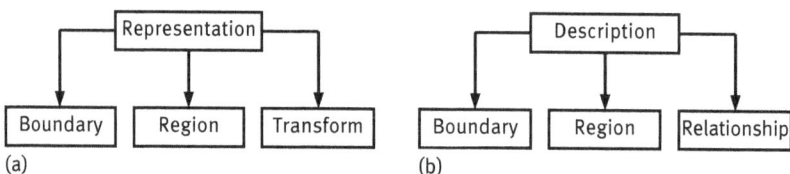

Figure 3.1: Classification of representation and description.

Object descriptions are obtained with the help of different descriptors, which provide various properties of objects. The descriptors will be broadly divided into descriptors for the boundary, descriptors for the region, and descriptors for the object relationship, as shown in Figure 3.1(b).

3.2 Boundary-Based Representation

Under the category of boundary-based representation, many techniques have been proposed.

3.2.1 Taxonomy of Boundary-Based Representation

As illustrated in Figure 3.2, one proposed **taxonomy for representation** divides the boundary-based approach into the following three classes (Costa 2001):
1. **Parametric boundaries**: The object's outline is represented as a parametric curve (e. g., with some inherent structures), which implies a sequential order along it.
2. **Set of boundary points**: The object's outline is simply represented as a set of points (e. g., a subset of the object points), without any special order among them.
3. **Curve approximation**: A set of geometric primitives (e. g., straight-line segments or splines) is fitted to the object outline, which is thus approximately represented.

Under each class, several representation techniques can be employed. Some of the frequently used techniques, such as chain codes, boundary segments, polygonal approximation, signatures, and landmarks, are briefly discussed in the following sections.

3.2.2 Chain Codes

Chain codes are used to represent a boundary by a connected sequence of straight line segments of a specified length and direction (Gonzalez 2002). The direction of

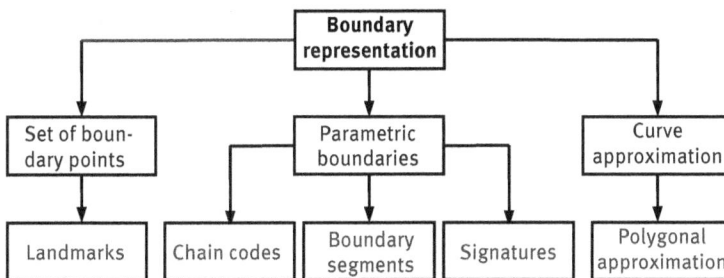

Figure 3.2: Taxonomy of boundary-based representation.

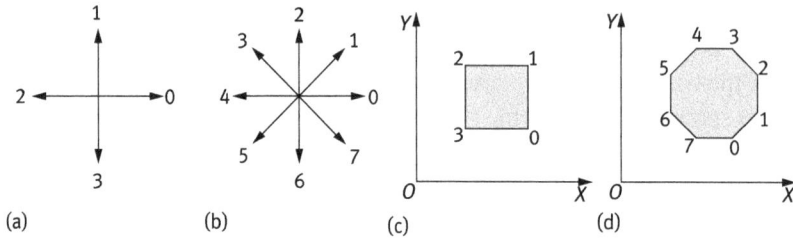

Figure 3.3: Two types of chain code.

each segment is coded by using a numbering scheme based on the pixel connectivity. The most popular schemes are 4-direction and 8-direction chain codes as shown in Figure 3.3(a, b). According to these schemes, the boundary of a region in an image can be efficiently coded using only a 2-bit or 3-bit code for a segment (note a position/location in a 256×256 image needs two 8-bit codes to represent it). Figure 3.3(c, d) gives two examples for the chains coding the boundary of the region in the image, in which one uses the scheme of Figure 3.3(a) and another uses the scheme of Figure 3.3(b).

The chain code of a boundary depends on the starting point. A normalization method is used to treat the chain code as a circular sequence of direction numbers and redefine the starting point so that the resulting sequence of numbers forms an integer of the minimum magnitude. One example is shown in Figure 3.4.

The chain code can also be normalized for rotation by using the first difference of the chain code. This difference is obtained by counting the number of direction changes (in a counterclockwise direction) that separate two adjacent elements in the code. One example is shown in Figure 3.5.

3.2.3 Boundary Segments

Decomposing a boundary into segments reduces the boundary's complexity and thus simplifies the description process (Gonzalez 2002). In cases where the boundary contains one or more significant concavities, which carry the shape information, the

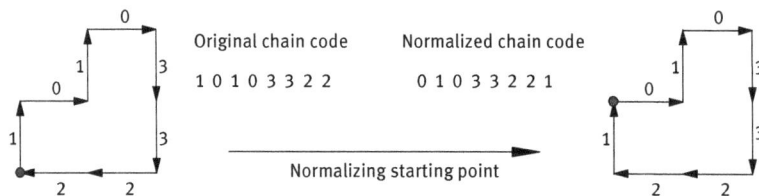

Original chain code   Normalized chain code

1 0 1 0 3 3 2 2   0 1 0 3 3 2 2 1

Normalizing starting point

Figure 3.4: Normalizing the starting point of a chain code.

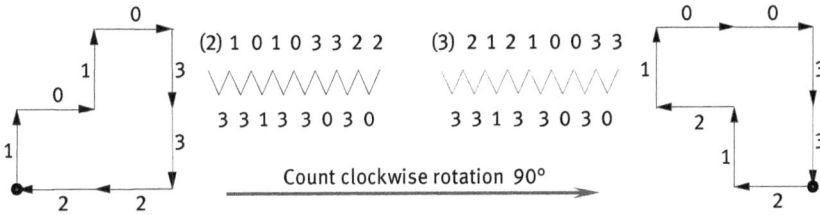

Count clockwise rotation 90°

(2) 1 0 1 0 3 3 2 2   (3) 2 1 2 1 0 0 3 3

3 3 1 3 3 0 3 0   3 3 1 3 3 0 3 0

Figure 3.5: Normalization for the rotation of a chain code.

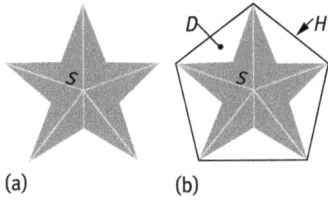

(a)   (b)

Figure 3.6: Convex hull and convex deficiency.

decomposition of the boundary can be simplified by using the convex hull (which itself is a region-based representation) as a tool. The convex hull H of an arbitrary set S is the smallest convex set containing S. The set difference $H - S$ is called the convex deficiency of the set S, which is denoted D. One example is shown in Figure 3.6. In Figure 3.6(a), a star corresponds to the set S. Figure 3.6(b) shows the convex hull H (black boundary) and the convex deficiency D (the white sectors between S and H) of S.

3.2.4 Polygonal Approximation

The goal of polygonal approximation is to capture the "essence" of the boundary shape with the fewest possible **polygonal segments** (Gonzalez 2002). Three techniques for obtaining polygons are introduced as follows.

3.2.4.1 Minimum Perimeter Polygons

Suppose that a boundary is enclosed by a set of concatenated cells as in Figure 3.7(a). It helps to visualize this enclosure as two walls corresponding to the outside and inside boundaries of the strip of the cells, while viewing the object boundary as a rubber band contained within the walls. If the rubber band is allowed to shrink, it takes the shape shown in Figure 3.7(b). If each cell encompasses only one point on the boundary, the error in each cell between the original boundary and the rubberband approximation would be at most $2^{1/2}d$, where d is the minimum possible distance between different pixels. This error can be reduced by half by forcing each cell to be centered on its corresponding pixel.

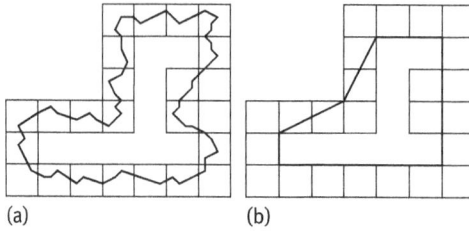

(a)　　　　　　　　(b)

Figure 3.7: Minimum perimeter polygon.

3.2.4.2 Merging Technique and Resulting Polygons

One approach is to merge points along a boundary until the least square error line of the points merged so far exceeds a preset threshold. When this condition occurs, the parameters of the line are stored, the error is set to 0, and the procedure is repeated. Merge a new point along the boundary until the error exceeds the threshold again. At the end of the procedure, the intersections of adjacent line segments form the vertices of the polygons. One example is shown in Figure 3.8.

One approach to **boundary segment splitting** is to divide a segment successively into two parts until a specified criterion is satisfied. For instance, a requirement might be that the maximum perpendicular distance from a boundary segment to the line joining its two end points does not exceed a preset threshold. If it does, the farthest point from the line becomes a vertex, which divides the initial segment into two sub-segments. One example is shown in Figure 3.9.

This approach has the advantage of seeking prominent inflection points (such as corners). For a closed boundary, the best starting points are usually the two farthest points in the boundary.

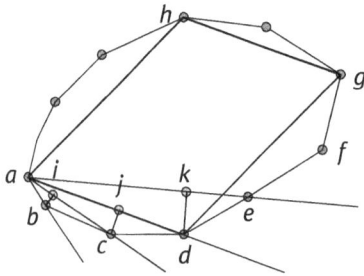

Figure 3.8: Merging technique and resulting polygon.

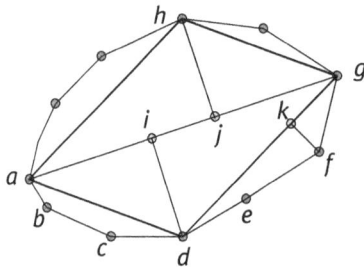

Figure 3.9: Splitting technique and resulting polygon.

3.2.5 Boundary Signatures

A **boundary signature** is a 1-D functional presentation of a boundary (Gonzalez 2002). In general, the signature is a 1-D representation of a 2-D form, or in other words, the signature describes a 2-D image using 1-D signals. The boundary signature may be generated in various ways.

3.2.5.1 Distance Versus Angle

One of the simplest ways for generating the signature of a boundary is to plot the distance from the centroid to the boundary (r) as a function of an angle (θ). Two examples of **distance versus angle signatures** are illustrated in Figure 3.10(a, b). In Figure 3.10(a), r is a constant, while in Figure 3.10(b), $r = A \sec \theta$. The generated signature is invariant to translation, but does depend on rotation and scaling. Normalization with respect to rotation can be achieved by finding a way to select the same starting point to generate the signature, regardless of the orientation of the shape. If the boundary is represented by chain codes, the technique that treats the chain code as a circular sequence of the direction numbers and redefines the starting point so that the resulting sequence of the numbers forms an integer of the minimum magnitude can be used here.

3.2.5.2 ψ-s Curve

Signatures can also be generated by traversing the boundary. For each point on the boundary, the angle between a reference line and the tangent of that point to the boundary is plotted. The resulting signature, though quite different from the $r(\theta)$ curve, would carry the information about basic shape characteristics.

One example is the ψ-s curve (**tangent angle versus arc length**). Take s as the arc length of the boundary traversed and ψ as the angle between a fixed line (here the x-axis) and a tangent to the boundary at the traversing point. The plot of ψ against s is called the ψ-s curve, which is like a continuous version of the chain-code representation (Ballard, 1982). Horizontal straight lines in the ψ-s curve correspond to straight lines on the boundary (ψ does not change). Nonhorizontal straight lines correspond to the segments of circles, since ψ is changing at a constant rate. The ψ-s curves for

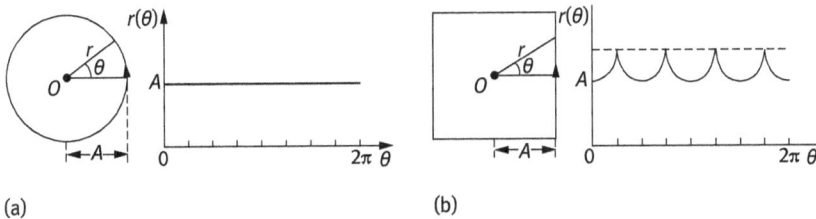

(a)

(b)

Figure 3.10: Distance versus angle signatures.

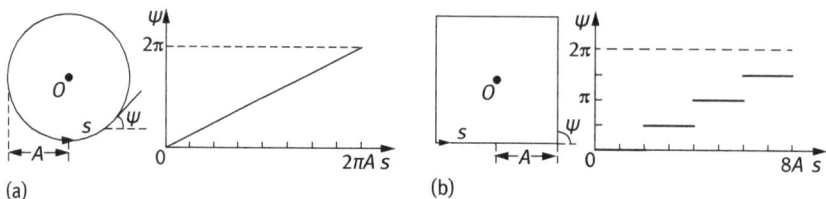

(a)

(b)

Figure 3.11: The ψ-s curves of the two forms in Figure 3.10.

the two forms in Figure 3.10 are shown in Figure 3.11. In Figure 3.11(a), ψ is increased constantly, while in Figure 3.11(b), ψ only takes four discrete values corresponding to the four edges of the square.

3.2.5.3 Slope Density Function

A variation of the above approach is the so-called **slope density function**, which can also be considered the signature of the ψ-s curve (projected) along the ψ axis. This function is simply a histogram of the tangent-angle values $h(\theta)$. As a histogram is a measurement of the concentration of values, the slope density function responds strongly to sections of the boundary with constant tangent angles and has deep valleys in sections rapidly producing various angles.

Figure 3.12 shows the slope density functions of the two forms in Figure 3.10. Note that the slope density function for the circle has the same form as the distance versus angle signatures does, but the slope density function for the square is quite different from the distance versus angle signatures for the square.

Since the tangent direction of an edge is also the direction of this edge, the slope density function of an image is also an edge direction histogram of this image (Jain, 1996). For an image with objects of regular forms, the edge direction histogram also shows some regularities (periodicity/symmetry), whereas for an image with objects of irregular forms, the edge direction histogram would show some rough form.

Two examples are shown in Figure 3.13, where Figure 3.13(a) is an image with somehow regular-formed objects (buildings) and its edge direction histogram and Figure 3.13(b) is an image with somehow irregular-formed objects (a flower) and its edge

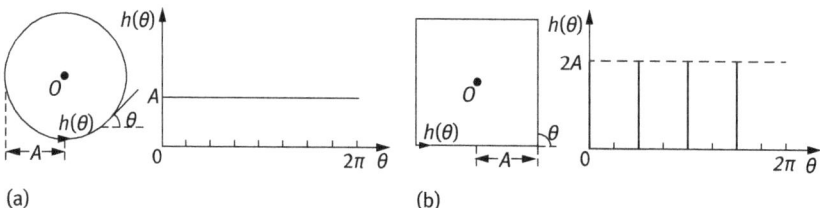

(a)

(b)

Figure 3.12: Slope density functions of the two forms in Figure 3.10.

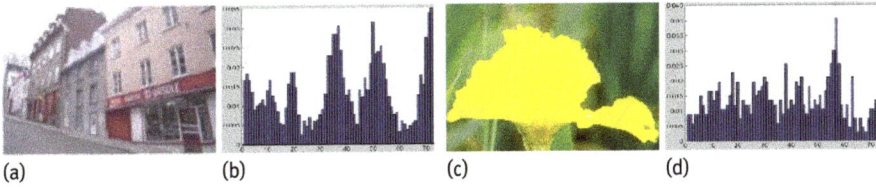

Figure 3.13: Correspondence between the form of objects and edge direction histograms.

direction histogram. The correspondence between the form of the object and the edge direction histogram is evident.

3.2.5.4 Distance Versus Arc Length

In general, contour-based signatures are created by starting from an initial point of the contour and traversing it either clockwise or counterclockwise. A simple signature illustrating this concept is to plot the distance between each contour point and the centroid as the function of the boundary points. This signature can be called the **distance versus arc length signature**. Two examples (with the two forms in Figure 3.10) are illustrated in Figure 3.14(a, b). In Figure 3.14(a), r is a constant, while in Figure 3.14(b), $r = (A^2 + s^2)^{1/2}$.

3.2.6 Landmark Points

One of the most natural and frequently used region representations are known as **landmark points** (also called critical points or salient landmarks) (Costa 2001). As an interesting way to gain insight about what is meant by salience is to imagine that you had to select the smallest number of points along a boundary so that you would be able to recover a reasonable reconstruction of the original region using these points. Figure 3.15(a) illustrates landmark points $\mathbf{S}_i = (S_{x,i}, S_{y,i})$, allowing an exact representation of a polygonal boundary. Figure 3.15(b, c) shows approximate representations of another closed boundary with a different numbers of landmark points.

It is clear that it is always possible to choose several alternative landmarks for a given boundary. For instance, more (or less) accurate representations can typically be obtained by considering more (or less) landmark points, as illustrated in

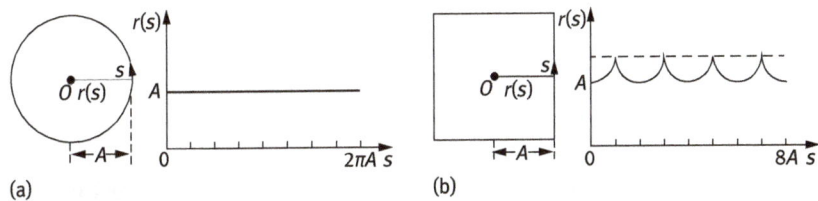

Figure 3.14: Distance versus arc length signatures.

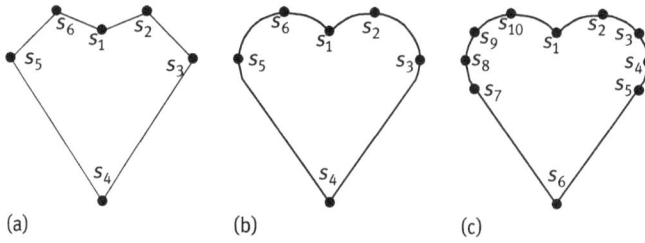

Figure 3.15: Landmark points of two regions.

Figure 3.15(b, c). In a general case, the landmark approach implies some loss of information about the shape, as the boundaries represented by landmarks becomes degenerated and is not recoverable. Consequently, landmark points define an approximated representation. However, it is always possible to obtain better representations by considering a larger number of landmarks and/or better placed landmarks. In the extreme case, if all the points of the boundary were taken as landmarks, the original boundary information would be perfectly represented. It is usually difficult to identify good landmarks in an automated manner. In many cases, landmarks are selected according to some specific constraints such as an anatomical reference point.

It is important to emphasize that proper landmark representations involve a consideration of the order of the landmark points. Several possibilities exist, which are given as follows:

1. Select a fixed reference landmark and store the coordinates of each landmark point as a feature vector, in such a way that the original shape can be recovered by joining each subsequent point. These arranged landmark points are henceforth called **ordered landmarks**. The shape S is represented as a $2n \times 1$ vector:

$$S_a = [S_{x,1}, S_{y,1}, S_{x,2}, S_{y,2}, \ldots, S_{x,n}, S_{y,n}]^{\mathrm{T}} \tag{3.1}$$

2. An alternative approach where the order of the landmark points is not taken into account is henceforth called **free landmarks**, in which the shape is represented by a set

$$S_b = \{S_{x,1}, S_{y,1}, S_{x,2}, S_{y,2}, \ldots, S_{x,n}, S_{y,n}\} \tag{3.2}$$

This does not promise exact representations. In fact, it precludes any attempt to interpolate between the landmark points. Indeed, this is a less completed representation, which is henceforth called $2n$-set and should be considered only when the order of the landmarks is not available.

3. The third interesting approach, which is referred to as the **planar-by-vector**, uses 2-D vectors corresponding to the landmark points. These vectors can be stored in an $n \times 2$ matrix S_c, which is given by

$$S_c = \begin{bmatrix} S_{x,1} & S_{y,1} \\ S_{x,1} & S_{y,2} \\ \vdots & \vdots \\ S_{x,n} & S_{y,n} \end{bmatrix} \tag{3.3}$$

4. Yet another possibility for shape representation is called the **planar-by-complex**, which is achieved by using a complex value to represent each landmark point (i. e., $S_i = S_{x,i} + jS_{y,i}$). In this representation, the whole set of the landmarks can be stored in an $n \times 1$ vector

$$S_d = [S_1, S_2, \ldots, S_n]^T \tag{3.4}$$

The above four representation schemes for a triangle with three vertices at $S_1 = (1, 1)$, $S_2 = (1, 2)$, $S_3 = (2, 1)$, which are taken as the positions of three landmarks, are summarized in Table 3.1.

Although all the above schemes allow the exact representation of the landmarks (but not necessarily the boundary), the planar-by-vector and the planar-by-complex representations are intuitively more satisfying since they do not need to map the objects into a $2n$-D space. It is observed that the concept of landmark points also provides a particularly suitable means to represent both the continuous and discrete boundaries by computers.

3.3 Region-Based Representation

Region-based representation consists of another group of representation techniques.

3.3.1 Taxonomy of Region-Based Representation

Figure 3.16 shows the subdivision of the region-based approach (Costa 2001), which includes the following components.

Table 3.1: A summary of four landmark representation schemes.

Scheme	Representation	Explanation
2n-vector	$S_a = [1, 1, 1, 2, 2, 1]^T$	S_a is a $2n \times 1$ vector of real coordinates
2n-set	$S_b = \{1, 1, 1, 2, 2, 1\}$	S_b is a set containing $2n$ real coordinates
Planar-by-vector	$S_c = \begin{bmatrix} 1 & 1 \\ 1 & 2 \\ 2 & 1 \end{bmatrix}$	S_c is an $n \times 2$ matrix containing the x- and y-real coordinates of each landmark at each of row
Planar-by-complex	$S_d = \begin{bmatrix} 1+j \\ 1+2j \\ 2+j \end{bmatrix}$	S_d is an $n \times 1$ vector of complex values, representing each of the landmarks

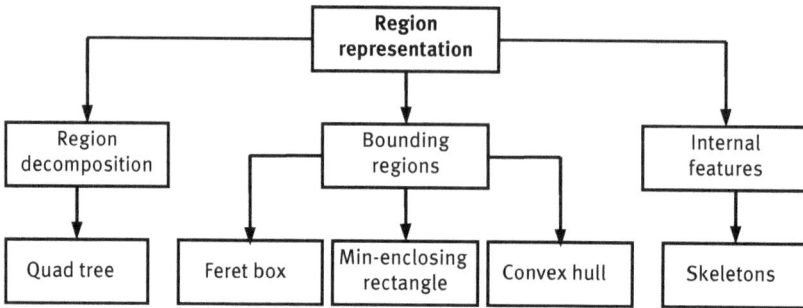

Figure 3.16: Taxonomy of region-based representation.

1. **Region decomposition**: The region is partitioned into simpler forms (*e.g.*, polygons) and represented by the set of such primitives.
2. **Bounding regions**: The region is approximated by a special predefined geometric primitive (*e.g.*, an enclosing rectangle) fitted to it.
3. **Internal features**: The region is represented by a set of features related to its internal region (*e.g.*, a skeleton).

Under each subdivision, different techniques can still be distinguished. In the following, several popularly used techniques, such as bounding regions, quad-tree, pyramid, and skeletons, are illustrated.

3.3.2 Bounding Regions

An important class of the region representation methods is based on defining a bounding region that encloses the object. Some of the most popular approaches are given in the following:

1. The **Feret box**, which is defined as the smallest rectangle (oriented according to a specific reference) that encloses the object (Levine, 1985). Usually, the rectangle is oriented with respect to the coordinate axis. An example is shown in Figure 3.17(a).
2. The **minimum enclosing rectangle** (MER), also known as the bounding box or the bounding rectangle (Castleman, 1996), which is defined as the smallest rectangle (oriented along any direction) that encloses the object. One example of MER that encloses the object as illustrated in Figure 3.17(a) is shown in Figure 3.17(b). Comparing these two figures, the difference and relationship between the Feret box and the bounding box can be perceived.
3. The **convex hull**, which has been discussed in Section 3.2.3. As an example, Figure 3.17(c) shows the convex hull of the object shown in Figure 3.17(a).

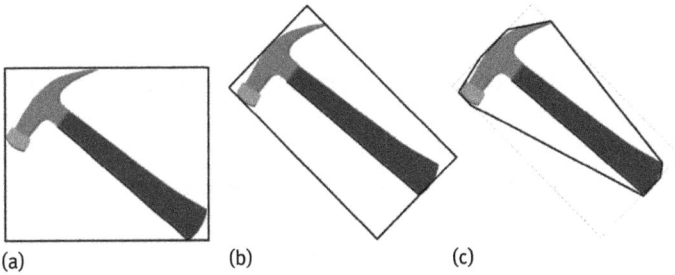

Figure 3.17: Different bounding regions of the same object.

3.3.3 Quad-Trees

A **quad-tree** is a tree in which each node has exactly four descendants, and each of the descendants has half of the node's dimensions. It is best used when the image is a square one. One illustration example is given in Figure 3.18, in which the root of the tree corresponds to the entire image and the leaves of the tree correspond to the pixels. All intermediate nodes correspond to a subdivision, which can still be subdivided. Each intermediate node has one of the three values: white, black, or gray.

A quad-tree can be used to represent a square image, as well as to represent an (irregular) object region. The total number of nodes N in the quad-tree can be calculated according to the number of levels n in the tree

$$N = \sum_{k=0}^{n} 4k = \frac{4^{n+1} - 1}{3} \approx \frac{4}{3} 4^{n} \qquad (3.5)$$

This number is one-third more than the number of pixels in the image. Since the object region delineated from the image is normally formed by connected components, the number of nodes for representing the object region can be much less than that obtained by the above equation (Figure 3.18).

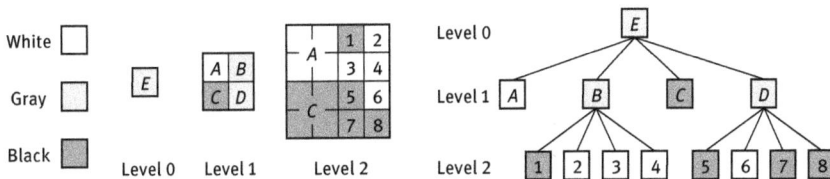

Figure 3.18: Illustrating image decomposition and quad-tree representation.

3.3.4 Pyramids

One data structure closely related to a quad-tree is the **pyramid**. The structure of a pyramid can be described by using graphs (Kropatsch 2001). A graph $G = (V, E)$ is given by a finite set of vertices, which is denoted V, a finite set of edges, which is denoted E, and a relation of incidences, which associates, with each edge e, an unordered pair $(v_1, v_2) \in V \times V$. The vertices v_1 and v_2 are called the end vertices of e.

The structure of a pyramid is determined by the neighborhood relations within the levels of the pyramid and by the "father–son" relations between adjacent levels.

A cell (if it is not at the base level) has a set of children (sons) at the level directly below it. These children are the input to the cell. At the same level, each cell has a set of neighbors (brothers/sisters). Each cell (if it is not the apex of the pyramid) has a set of parents (fathers) at the level above it.

Figure 3.19 shows a (regular) pyramid and a particular cell.

Hence, the structure of any pyramid can be described by horizontal and vertical graphs. Each (horizontal) level of a pyramid can be described by a neighborhood graph.

The (horizontal) neighborhood of a vertex $p \in V_i$ is defined by

$$H(p) = \{p\} \bigcup \{q \in V_i | (p, q) \in E_i\} \tag{3.6}$$

The vertical structure can also be described by a (bipartite) graph.

Let $R_i = \{(V_i \bigcup V_{i+1}), L_i\}$ and $L_i \subseteq (V_i \times V_{i+1})$. The set of all sons of a cell $q \in V_{i+1}$ is then defined as

$$SON(q) = \{p \in V_i | (p, q) \in L_i\} \tag{3.7}$$

In a similar manner, the set of all fathers of a cell $p \in V_i$ is defined as

$$FATHER(p) = \{q \in V_{i+1} | (p, q) \in L_i\} \tag{3.8}$$

Figure 3.19: Illustrating pyramid and cell.

Any pyramid with n levels can be described by n neighborhood graphs and $n - 1$ bipartite graphs.

Two terms are used to describe the structure of a regular pyramid, the **reduction factor** and the **reduction window**. The reduction factor r determines the rate by which the number of cells decreases from one level to the next. The reduction window (typically a square $(n \times n)$) associates with a cell in the higher level and a set of cells in the level directly below. The common notation for describing the structure of regular pyramids is $(n \times n)/r$. The classical $(2 \times 2)/4$ pyramid has no overlap since $(2 \times 2)/4 = 1$.

The quad-tree is very similar to the $(2 \times 2)/4$ pyramid. It differs mainly by the fact that a quad-tree has no links from a cell to its neighbors at the same level. Other pyramid structures with overlapped reduction windows (e. g., $(4 \times 4)/4$) are characterized by $(n \times n)/r > 1$. For completeness, if the pyramid structure satisfies $(n \times n)/r < 1$, some cells in this pyramid have no father.

Though a quad-tree can be considered a special pyramid data structure, some distinctions exist between a quad-tree and a pyramid. For example, the pyramid representation is considered a multiresolution representation while a region quad-tree is considered a variable-resolution representation (Samet 2001). Each of them has its own strengths. In particular, the region quad-tree is useful for location-based queries such as "given a location, what object is there?" On the other hand, for a feature-based query such as "given a feature, find the object's location," the region quad-tree requires that every block is to be examined to see whether the feature presents. The pyramid stores a summary in the nonleaf nodes of the data below them. Thus, the summary can be used to determine whether a searing should be performed in the leaves. If the feature is not present in a node, it cannot be present in its substructures and there is no need to search further in its descendants.

Pyramids have the following merits (Kropatsch 2001):

1. They reduce the influence of the noise by eliminating less important details in lower-resolution versions of the image.
2. They make the processing of the image independent of the resolution of the regions of interest.
3. They convert global features to local ones.
4. They reduce the computational costs using the divide-and-conquer principle.
5. They find the regions of interest for a plan-guided analysis at low cost in low-resolution images, ignoring irrelevant details.
6. They help the visual inspection of large images.
7. They increase the speed and reliability of image matching techniques by applying a coarse-to-fine strategy.

To convert a pyramid (e. g., the left of Figure 3.19) to a quad-tree (e. g., the right of Figure 3.18) simply requires a recursive search from the top to the base of the pyramid. If an array element in the pyramid is either white or black, it is used as a corresponding terminal node. Otherwise, it is used as a gray node, which is linked to the results of the recursive examination at the next level.

3.3.5 Skeletons

An important approach to representing the structural shape of a planar object region is to reduce it to a graph. This reduction may be accomplished by obtaining the skeleton of the region via a thinning (also called skeletonizing) algorithm. The *skeleton* of a region may be defined via the **medial axis transformation** (MAT). The MAT of a region R with a border B is defined as follows. First, for each point p in R, its closest neighbors in B are found. This can be achieved by using the following formula

$$d_s(p, B) = \inf\{d(p, z) | z \subset B\} \qquad (3.9)$$

If p has more than one such neighbor, it is said to belong to the medial axis (skeleton) of R. The concept of "closest" (and the resulting MAT) depends on the definition of a distance.

Although the MAT of a region yields an intuitive pleasing skeleton, direct implementation of this definition typically is expensive computationally. The implementation potentially involves calculating the distance from every interior point to every point on the boundary of a region. Algorithms with improved computation efficiency are **thinning algorithms** that iteratively delete edge points of a region subject to the constraints that the deletion of these points (1) does not remove end points, (2) does not break connectivity between the originally connected points, and (3) does not cause excessive erosion of the region.

The following thinning method consists of successive passes of two basic steps that are applied to the boundary points (pixels with value 1 and having at least one of its 8-neighbors whose value is 0) of the given region (Gonzalez 2002). For each boundary point p, its neighborhood arrangement is shown in Figure 3.20 (it is confirmed with the chain-code arrangement).

The first step is to label a boundary point p for deletion if all the following four conditions are satisfied:

(1.1) $2 \le N(p) \le 6$
(1.2) $S(p) = 1$
(1.3) $p_0 \cdot p_2 \cdot p_4 = 0$
(1.4) $p_2 \cdot p_4 \cdot p_6 = 0$

where $N(p)$ is the number of nonzero neighbors of p, and $S(p)$ is the number of 0–1 transitions in the ordered sequence $p_0, p_1, \ldots, p_7, p_0$. The point is not deleted until all border points have been processed. This delay prevents the change of data structure

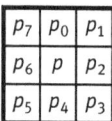

p_7	p_0	p_1
p_6	p	p_2
p_5	p_4	p_3

Figure 3.20: Neighborhood arrangement.

during the execution of the algorithm. After the labeling process has been applied to all border points, those labeled points are deleted together.

The second step labels a contour point p for deletion if all the following four conditions are satisfied:

(2.1) $2 \leq N(p) \leq 6$

(2.2) $S(p) = 1$

(2.3) $p_0 \cdot p_2 \cdot p_6 = 0$

(2.4) $p_0 \cdot p_4 \cdot p_6 = 0$

Similar to the first step, the deletion is also executed after all boundary points are labeled.

Condition (1.1) or (2.1) is violated when a contour point p only has one or seven 8-neighbors valued 1. Having only one such neighbor implies that p is the end point of a skeleton stroke and obviously it should not be deleted (Figure 3.21(a)). Deleting p, which has seven such neighbors, will cause erosion into the region (Figure 3.21(b)).

Conditions (1.3) and (2.2) are violated when applied to points on a stroke of only one pixel thick. Hence, this condition prevents disconnection of the segments of a skeleton during the thinning operation (Figure 3.21(c, d) for two different cases).

Conditions (1.4) and (1.5) are satisfied simultaneously by the minimum set of values ($p_2 = 0$ or $p_4 = 0$) or ($p_0 = 0$ and $p_6 = 0$). Thus, with reference to the neighborhood arrangement in Figure 3.20, a point that satisfies these conditions as well as the first two conditions is a east or south boundary point or a northwest corner point in the boundary (Figure 3.21(e)).

Conditions (2.3) and (2.4) are satisfied simultaneously by the minimum set of values ($p_0 = 0$ or $p_6 = 0$) or ($p_2 = 0$ and $p_4 = 0$). Thus, with reference to the neighborhood arrangement in Figure 3.20, a point that satisfies these conditions as well as the first two conditions is a north or west boundary point or a southeast corner point in the boundary (Figure 3.21(f)).

3.4 Transform-Based Representation

The underlying idea behind the transform-based approach is the application of a transform, such as the Fourier transform or wavelets, in order to represent the object in terms of the transform coefficients in the transformed spaces.

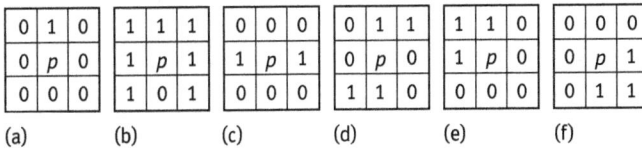

0	1	0		1	1	1		0	0	0		0	1	1		1	1	0		0	0	0
0	p	0		1	p	1		1	p	1		0	p	0		1	p	0		0	p	1
0	0	0		1	0	1		0	0	0		1	1	0		0	0	0		0	1	1
(a)			(b)			(c)			(d)			(e)			(f)					

Figure 3.21: Examples illustrating different conditions.

3.4.1 Technique Classification

Figure 3.22 shows the subdivision of the transform approach. Many transforms are proposed (only some examples are shown), which can be classified into two groups (Costa 2001):

1. Linear transform: A linear transform T is such that, given two sets of points A and B, and two scalars α and β, it has $T(\alpha A + \beta B) = \alpha T(A) + \beta T(B)$. A linear transform-based approach represents a set of points A in terms of $T(A)$, for example, the coefficients (or energies) of its transformation (e. g., Fourier).
2. Nonlinear transform: The set of points are represented in terms of their transformations, with the difference that the transforms are non-linear.

Among the linear transforms, two subgroups are distinct. One group of transforms can only be used for a mono-scale representation, such as the Fourier transform, the cosine transform, and the Z-transform. Another group of transforms can also be used for a multiscale representation, such as the Gabor transform (a special kind of short-time Fourier transform, which is also known as the windowed Fourier transform), the wavelet transform, and the scale-space approach.

3.4.2 Fourier Boundary Representation

Figure 3.23 shows a K-point digital boundary in the XY-plane.

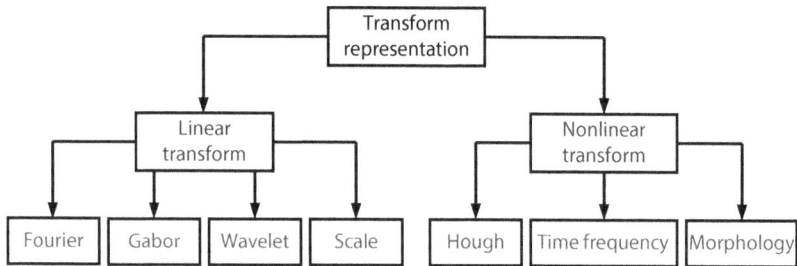

Figure 3.22: Taxonomy of transform-based representation.

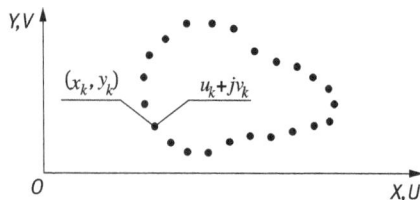

Figure 3.23: Digital boundary represented in a complex plane (overlapped with XY).

Starting from an arbitrary point, coordinate pairs $\{x_k, y_k\}$ are encountered in traversing the boundary. The boundary itself can be represented as the sequence of coordinates, and each coordinate pair can be treated as a complex number (with a real part and an imaginary part)

$$s(k) = u(k) + jv(k), \quad k = 0, 1, \ldots, N - 1 \tag{3.10}$$

The Fourier transform of $s(k)$ is

$$S(w) = \frac{1}{N} \sum_{k=0}^{N-1} s(k) \exp[-j2\pi wk/N], \quad w = 0, 1, \ldots, N - 1 \tag{3.11}$$

$S(w)$ can be called the **Fourier boundary descriptors**. This reduces a 2-D representation to a 1-D representation. The inverse Fourier transform of these coefficients restores $s(k)$, given by

$$s(k) = \sum_{w=0}^{N-1} S(w) \exp[j2\pi wk/N], \quad k = 0, 1, \ldots, N - 1 \tag{3.12}$$

In a real application, instead of using all the Fourier coefficients, only the first M coefficients are used. The result is the following approximation to $s(k)$

$$\hat{s}(k) = \sum_{w=0}^{M-1} S(w) \exp[j2\pi wk/N], \quad k = 0, 1, \ldots, N - 1 \tag{3.13}$$

Although only M terms are used to obtain each component of $\hat{s}(k)$, k still ranges from 0 to $N - 1$, which means that the approximate boundary contains the same number of points, but the number of frequency components, which are used in the reconstruction of each point, is reduced. The smaller M is, the more detail that is lost in the boundary.

One example shows a square boundary consisting of $N = 64$ points, and the results of reconstructing the original boundary using the first M coefficients is given in Figure 3.24. The following three observations can be made:

1. When the value of M is less than 8, the reconstructed boundary looks more like a circle than a square.
2. The corners only appear when M is increased to 56 or greater.
3. The edges begin to be straight only when M is increased to 61 or greater. Adding one more coefficient ($M = 62$), the reconstruction becomes almost identical to the original one.

In conclusion, a few low-order coefficients are able to capture the gross shape, but many more high-order terms are required to define accurately the sharp features such as corners and straight lines.

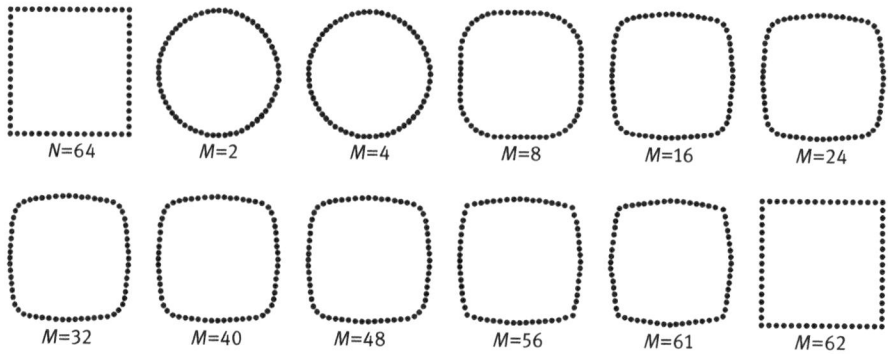

Figure 3.24: Reconstruction by the Fourier descriptor.

Table 3.2: Some basic properties of Fourier descriptors.

Transformation	Boundary	Fourier descriptor
Translation (Δx, Δy)	$s_t(k) = s(k) + \Delta x + j\Delta y$	$S_t(w) = S(w) + (\Delta x + j\Delta y)\delta(w)$
Rotation (θ)	$S_r(k) = s(k)\exp(j\theta)$	$S_r(w) = S(w)\exp(j\theta)$
Scaling (C)	$S_c(k) = C \cdot s(k)$	$S_c(w) = C \cdot S(w)$
Starting point (k_0)	$s_p(k) = s(k - k_0)$	$S_p(w) = S(w)\exp(-j2\pi k_0 w/N)$

Fourier boundary descriptors are not directly insensitive to geometric changes, such as translation, rotation, and scale, but the changes in these parameters can be related to simple transformations on the descriptors. Table 3.2 summarizes the Fourier descriptors for a boundary sequence $s(k)$ that undergoes translation, rotation, scaling, and changes in the starting point.

3.5 Descriptors for Boundary

Many techniques have been developed to describe the boundary of regions. In the following, some simple and popularly used descriptors are illustrated as examples. Descriptors that are more complicated will be discussed in specialized analysis tasks.

3.5.1 Some Straightforward Descriptors

Certain boundary descriptors can be computed and obtained easily, such as boundary length, boundary diameter, and boundary curvature.

3.5.1.1 Boundary Length
The length of a boundary (**boundary length**) is the perimeter of its enclosed region. The number of the pixels along a boundary gives a rough approximation of its length.

For a chain-coded curve with unit spacing in both directions, its length is given by the number of the vertical and horizontal components plus $2^{1/2}$ times the number of diagonal components.

The 4-connected boundary B_4 and 8-connected boundary B_8 can be defined by

$$B_4 = \{(x, y) \in R | N_8(x, y) - R \neq 0\} \tag{3.14}$$

$$B_8 = \{(x, y) \in R | N_4(x, y) - R \neq 0\} \tag{3.15}$$

The first condition shows that the boundary point belongs to the region, while the second condition shows that in the neighborhood of the boundary point there are some points that belong to the complementary of the region. Arranging all boundary points from 0 to $K-1$ (suppose the boundary has K boundary points), the length of these two boundaries can be computed by Haralick (1992)

$$\|B\| = \# \{k | (x_{k+1}, y_{k+1}) \in N_4(x_k, y_k)\} + \sqrt{2} \# \{k | (x_{k+1}, y_{k+1}) \in N_D(x_k, y_k)\} \tag{3.16}$$

where $k + 1$ is mod K. The first term corresponds to the vertical and horizontal components, while the second term corresponds to the diagonal components.

3.5.1.2 Boundary Diameter

The diameter of a boundary B is defined as

$$Dia(B) = \max_{i,j}[D(b_i, b_j)] \qquad b_i \in B, \quad b_j \in B \tag{3.17}$$

where D is a distance measurement. The value of the diameter and the orientation of a line segment connecting the two extreme points that comprise the diameter (this line is called the **major axis** of the boundary) are useful descriptors of a boundary. The **minor axis** of a boundary is defined as the line perpendicular to the major axis, of such a length that a box passing through the outer four points, where the boundary is intersected with the two axes, completely encloses the boundary. This box is called the **basic rectangle**, and the ratio of the major to the minor axis is called the **eccentricity** of the boundary, which is also a useful descriptor.

One example of the diameter of a boundary is shown in Figure 3.25. Note that the diameter can be measured using different distance metrics (such as Euclidean, city-block, chessboard, etc.) and with different results.

3.5.1.3 Boundary Curvature

The curvature of a curve is defined as the rate of the change of the slope. As the boundary is traversed in the clockwise direction, a vertex point p is said to be part of a convex segment if the change in the slope at p is nonnegative; otherwise, p is said to belong to a segment that is concave. In general, obtaining a reliable measurement of curvature at a point in a digital boundary is difficult because these boundaries tend

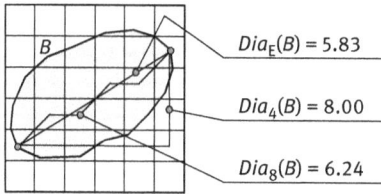

$Dia_E(B) = 5.83$

$Dia_4(B) = 8.00$

$Dia_8(B) = 6.24$

Figure 3.25: Diameter of a boundary (using different distance measurements).

to be locally "ragged." However, using the difference between the slopes of adjacent boundary segments (which have been represented as straight lines) as a descriptor of the curvature at the point, where the segments intersect, sometimes has proven to be useful.

For 3-D images, surface curvature is defined by two maximum and minimum radii. If both are positive as viewed from "inside" the object bounded by the surface, the surface is convex. If both are negative, the surface is concave, and if they have opposite signs, the surface has a saddle curvature. Figure 3.26 shows one example for each case, where Figure 3.26(a) is for convex, Figure 3.26(b) is for concave, and Figure 3.26(c) is for saddle. The total integrated curvature for any closed surface around a simply connected object (no hole, bridge, etc., exist) is always 4π.

3.5.2 Shape Numbers

Shape number is a descriptor for the boundary shape based on chain codes. According to the starting point, a chain-coded boundary has several first-order differences, in which one with the smallest value is the shape number of this boundary. For example, in Figure 3.4, the chain code for the boundary before normalization is 10103322, the difference code is 3133030, and the shape number is 03033133.

The order n of a shape number is defined as the number of digits in its representation. For a closed boundary, n is even. The value of n limits the number of different shapes of a boundary that are possible. Figure 3.27 shows all the shapes of orders 4, 6, and 8.

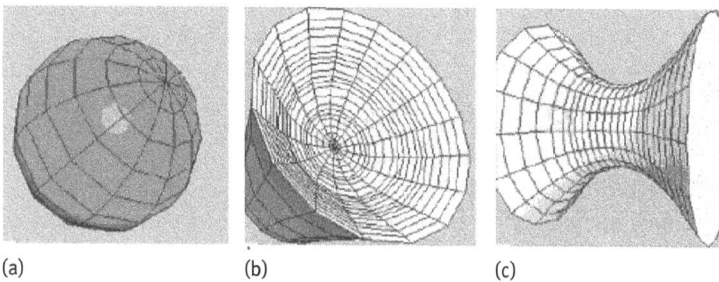

(a)   (b)   (c)

Figure 3.26: Different surface curvatures.

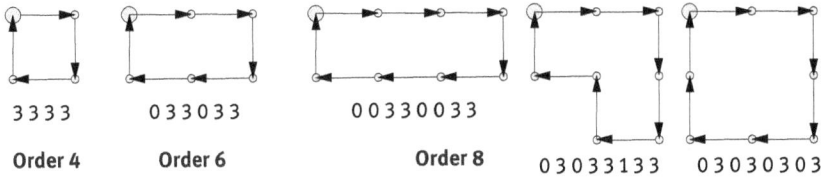

3 3 3 3   0 3 3 0 3 3   0 0 3 3 0 0 3 3

Order 4   Order 6   Order 8   0 3 0 3 3 1 3 3   0 3 0 3 0 3 0 3

Figure 3.27: All the shapes of orders 4, 6, and 8.

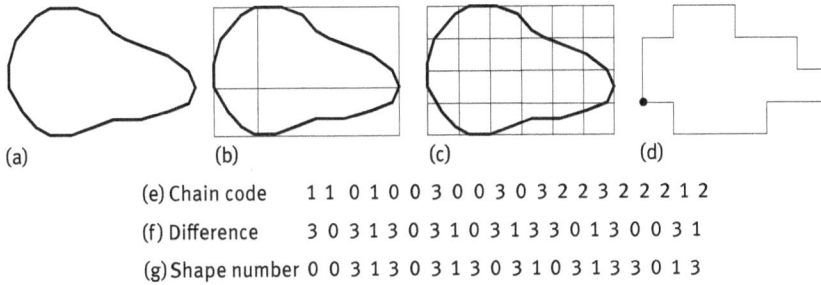

(a)   (b)   (c)   (d)

(e) Chain code   1 1 0 1 0 0 3 0 0 3 0 3 2 2 3 2 2 2 1 2

(f) Difference   3 0 3 1 3 0 3 1 0 3 1 3 3 0 1 3 0 0 3 1

(g) Shape number 0 0 3 1 3 0 3 1 3 0 3 1 0 3 1 3 3 0 1 3

Figure 3.28: Shape number computation.

In practice, for a desired shape order, the shape number of a boundary can be obtained as follows:

1. Select the rectangle whose long/short axis ratio is the closest to the given boundary, as shown in Figure 3.28(a, b).
2. Divide the selected rectangle into multiple squares (grids) according to the given order, as shown in Figure 3.28(c).
3. Select the squares such that each square has more than 50% of its area inside the boundary and a polygon that fits best to the boundary is built by these squares, as shown in Figure 3.28(d).
4. Start from the black point in Fig. 3.28(d) and obtain its chain codes, as shown in Figure 3.28(e).
5. Obtain the first-order differences of the chain codes, as shown in Figure 3.28(f).
6. Obtain the shape number of the boundary from its first-order differences, as shown in Figure 3.28(g).

3.5.3 Boundary Moments

The boundary of an object is composed of a number of segments. Each segment can be represented by a 1-D function $f(r)$, where r is an arbitrary variable. Further, the area under $f(r)$ can be normalized and counted as a histogram. For example, the segments in Figure 3.29(a) with L points can be represented as a 1-D function $f(r)$, as shown in Figure 3.29(b).

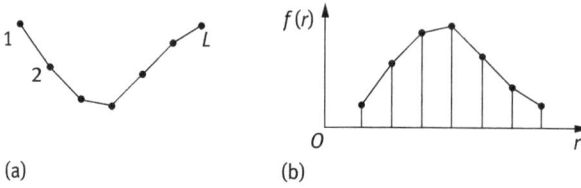

(a)   (b)

Figure 3.29: A line segment and its 1-D representation function.

The mean of $f(r)$ is given by

$$m = \sum_{i=1}^{L} r_i f(r_i)$$ (3.18)

The nth **boundary moment** of $f(r)$ about its mean is

$$\mu_n(r) = \sum_{i=1}^{L} (r_i - m)^n f(r_i)$$ (3.19)

Here, μ_n is directly related to the shape of $f(r)$. For example, μ_2 measures the spread of segments about its mean and μ_3 measures the symmetry of segments with respect to its mean. These moments are not depend on the position of segments in space, and are not sensitive to the rotation of the boundary.

3.6 Descriptors for Regions

The common properties of a region include geometric properties such as the area of the region, the centroid, and the extreme points; shape properties such as measurements of the circularity and elongation; topological properties such as the number of components and holes, and intensity properties such as mean gray tone and various texture statistics (Shapiro, 2001). Some of them are briefly discussed in the following.

3.6.1 Some Basic Descriptors

Region descriptors describe the properties of an object region in images. Some of them can be computed and obtained easily, such as the area and centroid of an object, as well as the density of an object region.

3.6.1.1 Area of Object
The area of an object region provides a rough idea about the size of the region. It can be computed by counting the number of pixels in the region:

$$A = \sum_{(x,y)\in R} 1 \qquad (3.20)$$

Given a polygon Q whose corners are discrete points, let R be the set of discrete points included in Q. Let N_B be the number of discrete points that lie exactly on the polygonal line that forms the borders of Q and let N_I be the number of interior points (i.e., non-border points) in Q. Then, $|R| = N_B + N_I$. The area of Q, denoted $A(Q)$, is the number of unit squares (possibly fractional) that are contained within Q and is given by Marchand (2000)

$$A(Q) = N_I + \frac{N_B}{2} - 1 \qquad (3.21)$$

Consider the polygon Q shown in Figure 3.30(a). The border of Q is represented as a continuous bold line. Points in Q are represented by discs (black $\bullet$ and white $\circ$). Black discs represent border points (i.e., corners and points that lie exactly on the border of Q). Empty (white) discs represent interior points. Clearly, $N_I = 71$ and $N_B = 10$. Therefore, $A(Q) = 75$.

It is necessary to distinguish the area defined by a polygon Q and the area defined by the contour of P, which is defined from the border set B. Figure 3.30(b) illustrates the resulting contour obtained when 8-connectivity is considered in P and 4-connectivity is considered in P^c. The area of P is clearly 63. One can verify that the difference between this contour-enclosed set and $A(Q)$ is exactly the area of the surface between the contour and Q (shaded area).

3.6.1.2 Centroid of Object
It is often useful to have a rough idea of where an object region is in an image. **Centroid** indicates the mass center of an object and the coordinates of a centroid are given by

$$\bar{x} = \frac{1}{A} \sum_{(x,y)\in R} x \qquad (3.22)$$

$$\bar{y} = \frac{1}{A} \sum_{(x,y)\in R} y \qquad (3.23)$$

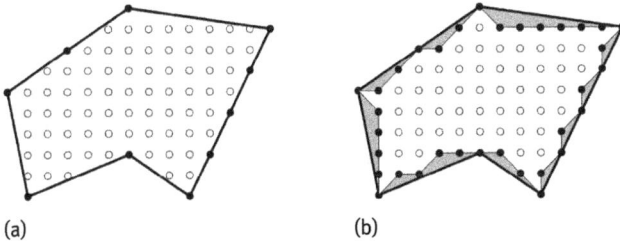

(a) (b)

Figure 3.30: Computation of the area of a polygon.

Figure 3.31: Centroid and geometric center of an irregular object.

In cases when the scale of objects is relatively small compared to the distance between objects, the object regions can be approximately represented by a point located at the centroid.

For an irregular object, its centroid and **geometric center** are often different. One example is given in Figure 3.31, where the centroid is represented by a square, the intensity weighted centroid is represented by a star, and the geometric center determined by the circumscribed circle is represented by a round point.

3.6.1.3 Region Density

Density (Densitometric) features are different from geometric features in that they must be computed based on both the original (grayscale) images and segmented images. Density is often represented in an image by gray levels. Commonly used region density features include maximum, minimum, mean, and average values as well as variance and other high order moments. They can be obtained from the gray-level histogram.

Several typical density feature descriptors are listed as follows:

1. **Transmission** (T): It is the proportion of incident light that passes through the object.

$$T = \text{light passing through object/incident light} \tag{3.24}$$

2. **Optical density** (OD): It is defined as the logarithm (base 10) of the ratio of incident light to the transmitted light (that is, the reciprocal of transmission).

$$OD = \lg(1/T) = -\lg T \tag{3.25}$$

OD is related directly to absorption, which is a complementary of transmission. Because absorption is multiplicative rather than additive, a logarithmic scale for absorption is useful to allow values of two or more absorbing items to be added simply. The value of OD goes from 0 (100% transmission) to infinity (zero transmission, for a completely opaque material).

3. **Integrated optical density** (IOD): It is a commonly used region gray-level feature. It is the summation of the pixel's OD for the whole image or a region in the image. For an $M \times N$ image $f(x, y)$, its IOD is

$$\text{IOD} = \sum_{x=0}^{M-1} \sum_{y=0}^{N-1} f(x, y) \tag{3.26}$$

Denoting the histogram of an image $H(\bullet)$ and the total number of gray-levels G,

$$\text{IOD} = \sum_{k=0}^{G-1} kH(k) \tag{3.27}$$

IOD is the weighted sum of all gray-level bins in the histogram.
4. Statistics of the above three descriptors, such as median, mean, variance, maximum, and minimum.

3.6.1.4 Invariant Moments

For a 2-D discrete region $f(x, y)$, its $(p + q)$th moment is defined as (Gonzalez 2002)

$$m_{pq} = \sum_{x} \sum_{y} x^p y^q f(x, y) \tag{3.28}$$

It has been proven that m_{pq} is uniquely determined by $f(x, y)$. On the other hand, m_{pq} also uniquely determines $f(x, y)$.

The moments around the center of gravity are called central moments. The $(p + q)$th central moment is defined by

$$\mu_{pq} = \sum_{x} \sum_{y} (x - \bar{x})^p (y - \bar{y})^q f(x, y) \tag{3.29}$$

where $\bar{x} = m_{10}/m_{00}$ and $\bar{y} = m_{01}/m_{00}$ are just the centroid coordinates defined in eqs. (3.22) and (3.23). Sample values of the first five moments for some simple images (the coordinate center is at the center of the image, pixels have unit size) shown in Figure 3.32 are given in Table 3.3 (Tomita, 1990).

The normalized central moment (normalization with respect to object area) of $f(x, y)$ is defined by

$$\eta_{pq} = \frac{\mu_{pq}}{\mu_{00}^{\gamma}} \qquad \text{where } \gamma = \frac{p+q}{2} + 1, \quad p + q = 2, 3, \dots \tag{3.30}$$

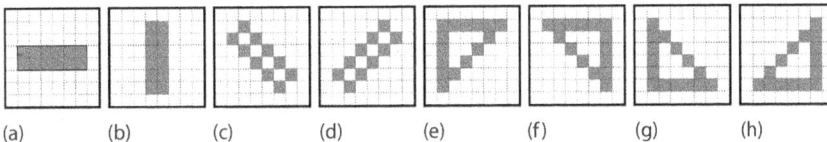

(a)   (b)   (c)   (d)   (e)   (f)   (g)   (h)

Figure 3.32: Some simple images for moment computation.

Table 3.3: The first five moments for images shown in Figure 3.32.

	(a)	(b)	(c)	(d)	(e)	(f)	(g)	(h)
μ_{20}	3	35	22.5	22.5	43.3	43.3	43.3	43.3
μ_{11}	0	0	-17.5	17.5	21.6	-21.6	-21.6	21.6
μ_{02}	35	3	22.5	22.5	43.3	43.3	43.3	43.3
μ_{21}	0	0	0	0	19.4	-19.4	19.4	-19.4
μ_{12}	0	0	0	0	-19.4	-19.4	19.4	19.4

The following seven moments that are invariants to translation, rotation, and scaling are composed by normalized second and third central moments:

$$\phi_1 = \eta_{20} + \eta_{02} \tag{3.31}$$

$$\phi_2 = (\eta_{20} - \eta_{02})^2 + 4\eta_{11}^2 \tag{3.32}$$

$$\phi_3 = (\eta_{30} - 3\eta_{12})^2 + (3\eta_{21} - \eta_{03})^2 \tag{3.33}$$

$$\phi_4 = (\eta_{30} + \eta_{12})^2 + (\eta_{21} + \eta_{03})^2 \tag{3.34}$$

$$\phi_5 = (\eta_{30} - 3\eta_{12})(\eta_{30} + \eta_{12})[(\eta_{30} + \eta_{12})^2 - 3(\eta_{21} + \eta_{03})^2] \\ + (3\eta_{21} - \eta_{03})(\eta_{21} + \eta_{03})[3(\eta_{30} + \eta_{12})^2 - (\eta_{21} + \eta_{03})^2] \tag{3.35}$$

$$\phi_6 = (\eta_{20} - \eta_{02})[(\eta_{30} + \eta_{12})^2 - (\eta_{21} + \eta_{03})^2] + 4\eta_{11}(\eta_{30} + \eta_{12})(\eta_{21} + \eta_{03}) \tag{3.36}$$

$$\phi_7 = (3\eta_{21} - \eta_{03})(\eta_{30} + \eta_{12})[(\eta_{30} + \eta_{12})^2 - 3(\eta_{21} + \eta_{03})^2] \\ + (3\eta_{12} - \eta_{30})(\eta_{21} + \eta_{03})[3(\eta_{30} + \eta_{12})^2 - (\eta_{21} + \eta_{03})^2] \tag{3.37}$$

Figure 3.33 gives one example to show the invariance of the above seven invariant moments. In Figure 3.33, Figure 3.33(a) is the original test image, Figure 3.33(b) is that image after 45° rotation counterclockwise, Figure 3.33(c) is that image reduced to half size, and Figure 3.33(d) is the mirror image of Figure 3.33(a). The seven moments defined in eqs. (3.30–3.36) are computed for all four images, and the results are

(a)   (b)   (c)   (d)

Figure 3.33: Test images for invariant moments.

Table 3.4: Some test results of invariant moments.

Invariant moment	Original image	Rotated 45°	Reduced to half	Mirror
$\phi 1$	1.510494E – 03	1.508716E – 03	1.509853E – 03	1.510494E – 03
$\phi 2$	9.760256E – 09	9.678238E – 09	9.728370E – 09	9.760237E – 09
$\phi 3$	4.418879E – 11	4.355925E – 11	4.398158E – 11	4.418888E – 11
$\phi 4$	7.146467E – 11	7.087601E – 11	7.134290E – 11	7.146379E – 11
$\phi 5$	– 3.991224E – 21	– 3.916882E – 21	– 3.973600E – 21	– 3.991150E – 21
$\phi 6$	– 6.832063E – 15	– 6.738512E – 15	– 6.813098E – 15	– 6.831952E – 15
$\phi 7$	4.453588E – 22	4.084548E – 22	4.256447E – 22	– 4.453826E – 22

listed in Table 3.4. The four values for each moment are quite close with some small differences that can be allocated to round-off in computation.

The above-discussed seven invariant moments are suitable for describing regions. They can also be used to describe the region boundary, but a modification is necessary (Yao, 1999). For a region $f(x, y)$, under a scale change, $x' = kx$, $y' = ky$, its moments are multiplied by $k^p k^q k^2$. The factor k^2 is related to the area of an object. The center moments of $f(x', y')$ become $\mu'_{pq} = \mu_{pq} k^{p+q+2}$. The normalized moments are defined as

$$\eta_{pq} = \frac{\mu_{pq}}{(\mu_{00})^\gamma} \tag{3.38}$$

In order to make the normalized moments invariant to scaling, let $\eta'_{pq} = \eta_{pq}$, and then

$$\gamma = (p + q + 2)/2 \tag{3.39}$$

For a boundary, scaling causes the change of the perimeter of the boundary. The change factor is not k^2 but k. The center moments now become $\mu'_{pq} = \mu_{pq} k^{p+q+1}$. In order to make the normalized moments invariant to scaling, let $\eta'_{pq} = \eta_{pq}$ and

$$\gamma = p + q + 1 \tag{3.40}$$

3.6.2 Topological Descriptors

Topology is the study of properties of a figure (object) that are unaffected by any deformation, as long as there is no tearing or joining of the figure (sometimes these are called **rubber-sheet distortions**). Topological properties do not depend on the notion of distance or any properties implicitly based on the concept of a distance measurement. Topological properties are useful for global descriptions of regions.

3.6.2.1 Euler Number and Euler Formula
Two topological properties that are useful for region description are the number of holes in the region and the number of connected components in the region

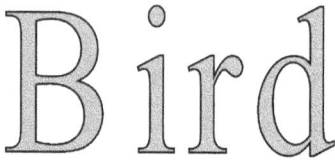

Figure 3.34: Four letters with different Euler numbers.

(Gonzalez, 2002). The number of holes H and the number of connected components C in a 2-D region can be used to define the **Euler number** E:

$$E = C - H \tag{3.41}$$

The Euler number is a topological property. The four letters shown in Figure 3.34 have Euler numbers equal to −1, 2, 1, and 0, respectively.

For an image containing C_n different connected foreground components and assuming that each foreground component C_i contains H_i holes (i. e., defines H_i extra components in the background), the Euler's number (a more generalized version) associated with the image is calculated as

$$E = C_n - \sum_{i=1}^{C_n} H_i \tag{3.42}$$

There are two Euler numbers for a binary image A, which are denoted $E_4(A)$ and $E_8(A)$ (Ritter, 2001). Each is distinguished by the connectivity used to determine the number of feature pixel components and the connectivity used to determine the number of non-feature pixel holes contained within the feature pixel connected components.

The **4-connected Euler number**, $E_4(A)$, is defined to be the number of four-connected feature pixel components minus the number of 8-connected holes, which is given by

$$E_4(A) = C_4(A) - H_8(A) \tag{3.43}$$

Here, $C_4(A)$ denotes the number of 4-connected feature components of A and $H_8(A)$ is the number of 8-connected holes within the feature components.

The **8-connected Euler number**, $E_8(A)$, is defined by

$$E_8(A) = C_8(A) - H_4(A) \tag{3.44}$$

where $C_8(A)$ denotes the number of 8-connected feature components of A and $H_4(A)$ is the number of 4-connected holes within the feature components.

Table 3.5 shows Euler numbers for some simple pixel configuration (Ritter, 2001).

Regions represented by straight-line segments (referred to as polygonal networks) have a particular simple interpretation in terms of the Euler number. Denoting the

Table 3.5: Examples of pixel configurations and Euler numbers.

No.	A	$C_4(A)$	$C_8(A)$	$H_4(A)$	$H_8(A)$	$E_4(A)$	$E_8(A)$
1	✚	1	1	0	0	1	1
2	⠶	5	1	0	0	5	1
3	◻	1	1	1	1	0	0
4	❖	4	1	1	0	4	0
5	◈	2	1	4	1	1	−3
6	▦	1	1	5	1	0	−4
7	◘	2	2	1	1	1	1

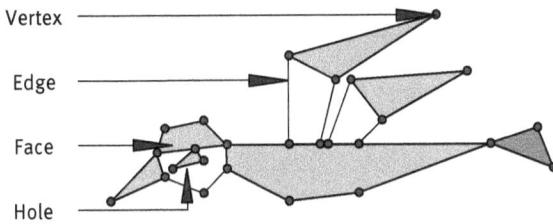

Figure 3.35: An example of a polygonal network.

number of vertices by V, the number of edges by Q, and the number of faces by F, the following relationship can be derived, which is called the **Euler formula**,

$$V - Q + F = E = C - H \tag{3.45}$$

The drawing in Figure 3.35 is an example of polygon network and has 26 vertices, 33 edges, 7 faces, 3 connected regions, and 3 holes. Therefore, its Euler number is 0.

3.6.2.2 Euler Number in 3-D
Some terms need to be defined first (Lohmann, 1998). A **cavity** is a completely enclosed component of the background. A handle has sometimes been identified by the concept of a "tunnel," which is a hole having two exits to the surface. The "number of handles" is also called the "number of tunnels," "first Betti-Number," or "genus."

The **genus** counts the maximum number of non-separating cuts it can make through the object without producing more connected components, where a "cut" must fully penetrate the object. Figure 3.36 shows examples of such cuts. There are many numbers of separating and nonseparating cuts conceivable for this object. But once you have decided on one particular nonseparating cut, no further non-separating cuts are possible since any further cut will slice the object into two. Therefore, the genus for this object is one. An object consisting of two crossed rings has a genus of 2, because there are two nonseparating cuts possible, one for each component.

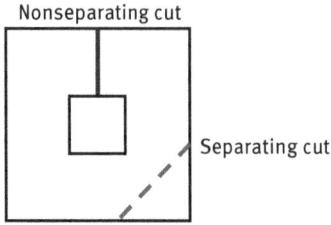

Figure 3.36: Separating and nonseparating cuts.

Now, the **3-D Euler number** is defined as the number of connected components (C) plus the number of cavities (A) minus the genus (G):

$$E = C + A - G \qquad (3.46)$$

The Euler number is clearly a global feature of an object. However, it is possible to compute both the Euler number and the genus by purely local computations (*i.e.*, by investigating small neighborhoods).

The genus of a digital object can be computed by investigating the surfaces that enclose the object and its cavities. If the object is digital, the enclosing surfaces consist of a fabric of polyhedral. Such surfaces are called **netted surfaces**. They are made by numbers of vertices (V), edges (Q), and faces (F) of voxels. The genus of a netted surface has the property given by

$$2 - 2G = V - Q + F \qquad (3.47)$$

In Figure 3.37, $F = 32$, $Q = 64$, and $V = 32$. So, $2 - 2G = 32 - 64 + 32 = 0$, that is, $G = 1$.

Sometimes, it may happen that two closed netted surfaces meet at a single edge. Such edges are counted twice – once for each surface they belong to.

Each connected component and cavity is wrapped up in a netted surface S_i. If there are M number of connected components and N number of cavities, the number of such surfaces will be $M + N$. Each of the closed surfaces encloses a single connected component or a single cavity. According to eq. (3.46), since $E_i = 1 - G_i$,

$$2 - 2G_i = 2 - 2(1 - E_i) = 2E_i \qquad (3.48)$$

$$2E = \sum_{i=1}^{M+N} 2E_i = \sum_{i=1}^{M+N} (2 - 2G_i) = \sum_{i=1}^{M+N} V_i - Q_i + F_i \qquad (3.49)$$

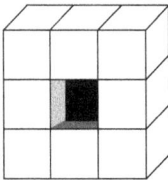

Figure 3.37: An object with one tunnel.

Let S denote the image foreground, S' denote its background, and E_k, $k = 6, 26$ denote the Euler number of a k-connected object, then the following two duality formulas exist

$$E_6(S) = E_{26}(S') \qquad E_{26}(S) = E_6(S') \qquad\qquad (3.50)$$

3.7 Problems and Questions

3-1 Provide the chain code for the object boundary in Figure Problem 3-1, as well as the starting-point normalized chain code, and the rotation normalized chain code.

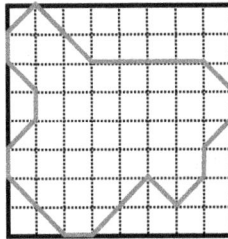

Figure Problem 3-1

3-2* By following a boundary of an object region, a signature (tangent angle signature) can be generated by traversing the boundary and taking the angle between the tangent line and a reference line as the function of the position on the boundary. Draw thus obtained signature for a square.

3-3 (1) Discuss the effect on the resulting polygon if the error threshold is set to zero in the merging method introduced in Section 3.2.4.

(2) Discuss the effect on the resulting polygon if the error threshold is set to zero in the splitting method introduced in Section 3.2.4.

3-4 Under what circumstances is the representation method based on landmark points only an approximation of the object boundary?

3-5 Draw the Feret box and the minimum enclosing rectangle of Figure 3.6(a).

3-6 Draw the quad-tree corresponding to the result in segmenting Figure Problem 3-67 by the split-merge technique.

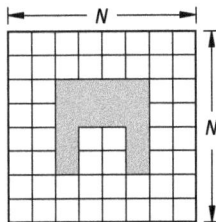

Figure Problem 3-6

3-7 For the point p in two images of Figure Problem 3-7, separately verify each condition in Section 3.3.5 for the skeleton algorithm. Will the point p be removed finally?

1	1	1
0	p	1
0	0	1

0	1	0
1	p	0
1	1	0

Figure Problem 3-7

3-8 Draw all the shapes of order 10 and find their shape number.
3-9* Find the shape number and the order of shape number for the object in Figure Problem 3-7.

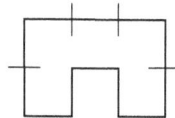

Figure Problem 3-9

3-10 Find the Euler number for each of the images in Figure 3.32 (both 4-connected and 8-connected).
3-11 Given two regions as shown in Figure Problem 3.4,

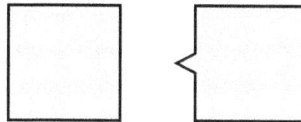

Figure Problem 3-11

(1) Which descriptor for the boundary should be used to distinguish them?

(2) Which descriptor for the region should be used to distinguish them?

3-12 Which descriptors should be used to distinguish:

(1) A square from a triangle.

(2) A square from a circle.

(3) A circle from an ellipse

3.8 Further Reading

1. **Classification of Representation and Description**
   - Specialized discussions on object representation are available in Samet (2001).
   - In addition to the descriptions for boundaries and regions, descriptions on the relations of boundary and boundary, region and region, as well as boundary and region are important. Some introductions on different relation descriptors can be found in Zhang (2003b). Discussion and application on distribution among point sets can be found in Zhang (1991a).
2. **Boundary-Based Representation**
   - More methods on boundary-based representation can be found in Otterloo (1991), Costa (2001), and Russ (2002).
3. **Region-Based Representation**
   - Topological thinning methods can be used to obtain skeleton of regions, and even skeleton of 3-D volumes (Nikolaidis, 2001).
   - More methods on region-based representation can be found in Russ (2002).
4. **Transform-Based Representation**
   - Many transforms can be used for representation. One wavelet transform-based technique for object boundary representation can be found in Zhang (2002b).
5. **Descriptors for Boundary**
   - Descriptors for object boundaries provide much information on the form of object, so are fundamental for shape analysis (Otterloo, 1991).
6. **Descriptors for Region**
   - Descriptors for regions are popularly used in many applications. More descriptors can be found, for example, in Jähne (1997) and Russ (2002).

4 Feature Measurement and Error Analysis

Object feature measurement is often the final goal of many image analysis tasks and needs to be well understood. The characteristics/properties of an object region can be quantitatively described by the values of various features. Feature measurement is often referred to as the process of obtaining quantitative and numerical description values of objects. How to determine/measure suitable features accurately and precisely is critical in image analysis.

The sections of this chapter are arranged as follows:

Section 4.1 introduces the concepts of direct and indirect measures with some examples and introduces some simple methods for combining multiple measures.

Section 4.2 provides some discussion on the two closely related but different terms: accuracy and precision, and their relationship to systematic and statistical errors.

Section 4.3 focuses on the connection ambiguity caused by using both 4-direction connectivity and 8-direction connectivity. Some contrast examples on different situations are specified.

Section 4.4 discusses the factors affecting the measurement error in detail, including the resolution of the optical lens, the sampling density at the time of imaging, the algorithm of object segmentation, and the feature computation formulas.

Section 4.5 introduces a method as an example for analyzing the measurement error, in considering the analysis of the error caused by the discrete distance instead of the Euclidean distance in the distance measurement.

4.1 Direct and Indirect Measurements

Measurements of objects can be either obtained directly by quantifying the properties (measuring the quantity) of objects or indirectly derived by combining the direct measuring results.

4.1.1 Direct Measurements

Two types of direct measurements can be distinguished:

4.1.1.1 Field Measurements
Field measurements usually are collected over a specified number of fields (e. g., images, regions in image), which is determined either by statistical considerations of the precision or by following a standard procedure.

Examples of primitive field measurements include area of features, area of features filled, area fraction, field area, field number, number of features, number of features excluded, number of intercepts, etc.

DOI 10.1515/9783110524123-004

4.1.1.2 Object-Specific Measurements

Examples of **object-specific measurements** include area, area filled, diameter (including maximum, minimum, and average), feature angle, number of features, hole count, inscribed circle (center coordinates x and y, and radius), intercept count, perimeter, position x and y, and tangent count.

4.1.2 Derived Measurements

Two types of derived measurements can be distinguished:

4.1.2.1 Field Measurements

Examples of **field measurements** include the fractal dimension of a surface, stereological parameters (such as the number of points in a volume, the area of a surface, total curvature of a curve or a surface, the density of length or surface or volume), etc.

4.1.2.2 Object-Specific Measurements

The number of derived **object-specific measurements** is unlimited in the sense that it is possible to define any combination of primitive measurements to form a new feature descriptor.

Table 4.1 lists some commonly used (direct and derived) object-specific descriptors (measured from a particular object).

4.1.3 Measurement Combinations

There are a number of possibilities to combine (already obtained) metrics to form a new (derived) metric.

Table 4.1: Commonly used object-specific descriptors.

Descriptors	Definitions
Area (A)	Number of pixels
Perimeter (P)	Length of the boundary
Longest dimension (L)	Length of the major axis
Breadth (B)	Length of the minor axis
Average diameter (D)	Average length of axes
Aspect ratio (AR)	Length of the major axis/length of the perpendicular axis
Area equivalent diameter (AD)	$(4A/\pi)^{1/2}$ (the diameter for a circular object)
Form factor (FF)	$4\pi A/P^2$ (perimeter-sensitive, always ≤ 1)
Circularity (C)	$\pi L^2/4A$ (longest dimension-sensitive, ≥ 1)
Mean intercept length	A/projected length (x or y)
Roughness (M)	$P/\pi D$ (πD refers to the perimeter of circle)
Volume of a sphere (V)	$0.75225 \times A^{2/3}$ (sphere rotated around diameter)

1. The simplest combination technique is addition. Given two metrics d_1 and d_2, then

$$d = d_1 + d_2 \qquad (4.1)$$

   is also a metric.

2. Metrics can be combined by scaling with a real-valued positive constant. Given a metric d_1 and a constant $\beta \in \mathbf{R}^+$, then

$$d = \beta d_1 \qquad (4.2)$$

   is also a metric.

3. If d_1 is a metric and $\gamma \in \mathbf{R}^+$, then

$$d = \frac{d_1}{\gamma + d_1} \qquad (4.3)$$

   is also a metric.

The operations in eqs. (4.1–4.3) may be combined. For example, if $\{d_n : n = 1, \ldots, N\}$ is a set of metrics, then $\forall \beta_n, \gamma_n \in \mathbf{R}^+$

$$d = \sum_{n=1}^{N} \frac{d_n}{\gamma_n + \beta_n d_n} \qquad (4.4)$$

is also a metric.

It should be noted that the product of two metrics is not necessarily a metric. This is because the triangle inequality may not be satisfied by the product.

4.2 Accuracy and Precision

Any measurement taken from images (*e.g.*, the size or the position of an object or its mean gray-level value) is only useful if the uncertainty of the measurement can also be estimated. The **uncertainty of the measurement** can be judged according to the concept of accuracy and/or precision.

4.2.1 Definitions

Precision (also called efficiency) is defined in terms of repeatability, describing the ability of the measurement process to duplicate the same measurement and produce the same result. **Accuracy** (also called unbiasedness) is defined as the agreement between the measurement and some objective standard taken as the "truth." In many cases, the latter requires standards that are themselves defined and measured by some national standards body.

Feature measurement can be judged according to its accuracy and/or precision. An **accurate measurement** (unbiased estimate) $\tilde{a}$ of a parameter a is the one for which $E\{\tilde{a}\} = a$. A **precise measurement** (consistent estimate) $\hat{a}$ of a parameter a based on N samples is the one which converges to a as $N \to \infty$ (under the condition that the estimator is unbiased). In real applications, a consistent estimate can be systematically different from the true value. This difference is the bias, and could be invisible as the "true value" is unknown. Take creating a public opinion poll as an example. If the questioning (sampling) is not uniformly distributed, that is, every member of the population needs to have an equal chance of being selected for the sample, there will be sampling bias. If leading questions are employed (like the imperfection of calibration of instruments), there will be systematic bias (Howard, 1998).

For scientific purposes, an unbiased estimate is the best that it can hope to achieve if a quantity cannot be directly measured. In other words, unbiasedness is the most desirable attribute that a scientific method can have. In a given experiment, the high accuracy could be achieved using a correct sampling scheme and proper measurement methods. The obtained precision depends on the object of interest and, in many cases, it could be controlled by putting more effort into refining the experiment. Note that a highly precise but inaccurate measurement is generally useless.

4.2.2 Relationships

It is possible to have a biased estimator that is "efficient" in that it converges to a stable value quickly and has a very small standard deviation, just as it is possible to have an inefficient unbiased estimator (converging slowly, but very close to the true value). For example, it is needed to estimate the position of the crosshair in the middle box as shown in Figure 4.1. Two estimation procedures are available. The first procedure yields eight estimates as shown in the left box in Figure 4.1. These estimates are not consistent, but they are unbiased if put together. They correspond to an accurate measurement. The second procedure yields the eight estimates shown in the right box in Figure 4.1. These estimates are very consistent but not accurate. They are definitely biased estimations.

Table 4.2 suggests the difference between precision and accuracy by graphical illustrations of different shooting effects at a target. For high precision, the hits are

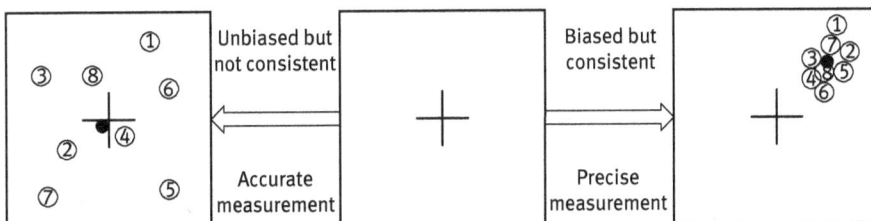

Figure 4.1: Illustrating the difference between accuracy and precision.

Table 4.2: Effects of accuracy and precision when shooting at a target.

closely clustered together. While for low precision, there is a marked scatter of hits. For high accuracy, the average of the cluster of hits tends toward the bull's eye. While for low accuracy, the hits are inaccurate or biased.

4.2.3 Statistical Error and Systematic Error

Accuracy and precision can also be discussed in relation to two important classes of errors. In Figure 4.1, the centroid of different measurements is indicated by the black circle. The **statistical error** describes the scatter of the measured value (the distribution of the individual measurements) if the same measurement is repeated over and over again. A suitable measurement of the width of the distribution gives the statistical error. From the statistical point of view, the example in the right side of Figure 4.1 is better.

However, this mean value may be further off the true value than what is given by the statistical error margins. Such a deviation is called a **systematic error**, which is indicated by the difference between the true value and the average of the measured values. A precise but inaccurate measurement is encountered when statistical error is low, but the systematic error is high (as in the example shown in the right side of Figure 4.1). On the other hand, if the statistical error is large and the systematic error is low, the individual measurements scatter widely, but their mean value is close to the true value (as in the example in the left side of Figure 4.1).

It is easy—at least in principle—to get an estimate of the statistical error by repeating the same measurement many times (Jähne, 1999). However, it is much harder to control systematic errors. They are often related to a lack of understanding of the measuring setup, and procedure. Unknown or uncontrolled parameters influencing the measuring procedure may easily lead to systematic errors. Examples of systematic errors are calibration errors and drift caused by temperature-dependent parameters in an experimental setup without temperature control.

4.3 Two Types of Connectivity

Regions and boundaries are formed with groups of pixels, and these pixels have certain connections among them. When defining a region and boundary, different types of connectivity should be used; otherwise, certain problems will arise. Some examples are shown in the following.

4.3.1 Boundary Points and Internal Points

In image measurement, **internal pixels** and **boundary pixels** should be judged with different types of connectivity to avoid ambiguity. This can be explained with the help of Figure 4.2. In Figure 4.2(a), the brighter pixels form an object region. If the internal pixels are judged according to 8-direction connectivity, then the darker pixel in Figure 4.2(b) is an internal pixel and other brighter pixels form the 4-directional boundary (as indicated by thick lines). If the boundary pixels are judged according to 8-direction connectivity, then the three darker pixels in Figure 4.2(c) are internal pixels and other brighter pixels form the 8-directional boundary (as indicated by thick lines). Figure 4.2(b, c) corresponds to our intuition. In fact, this is what it is expected and needed.

However, if the boundary pixels and internal pixels are judged using the same type of connectivity, then the two pixels marked with "?" in Figure 4.2(d, e) would have ambiguity and be judged as an internal pixel or boundary pixel. For example, if the boundary pixels are judged by 4-direction connectivity (as in Figure 4.2(b)) and the internal pixels are judged by 4-direction connectivity, too, a problem will arise. The pixels in question should be judged as internal pixels in the first view since all pixels in the neighborhood belong to the region (see the dashed lines in Figure 4.2(d)), but it should also be judged as a boundary pixel; otherwise, the boundary in Figure 4.2(b) would become unconnected. On the other hand, if the boundary pixels are judged by 8-direction connectivity (as in Figure 4.2(c)) and the internal pixels

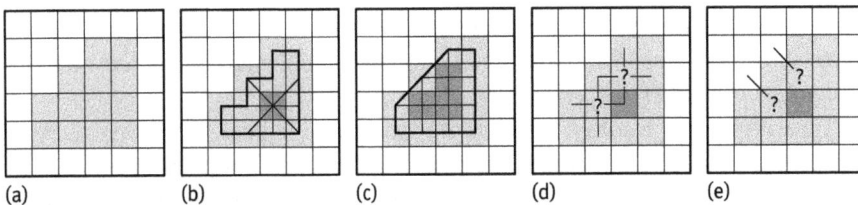

(a)   (b)   (c)   (d)   (e)

Figure 4.2: The connectivity of boundary pixels and internal pixels.

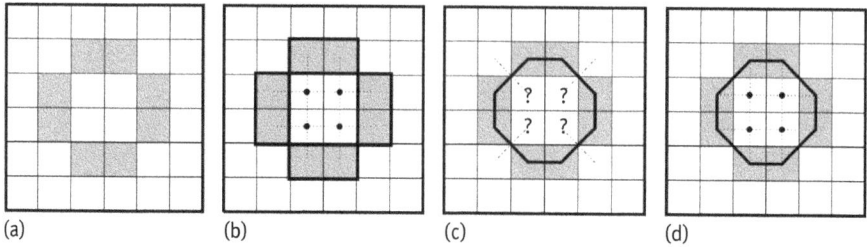

(a)   (b)   (c)   (d)

Figure 4.3: The connectivity of object pixels and hole pixels.

are judged by 8-direction connectivity, too, the ambiguity problem exists again. The pixels marked with "?" are enclosed inside the boundary as in Figure 4.2(c), so are considered internal pixels, but their 8-connected neighborhoods have pixels that do not belong to the region (see the dashed lines in Figure 4.2(e)), so they should also be considered boundary pixels.

4.3.2 Object Points and Background Points

As another example, consider Figure 4.3(a), which consists of a black object with a hole on a white background. If one uses 4-connectedness for both background and objects, then the object consists of four disconnected pieces (in contrast to intuitiveness), yet the hole is separated from the "outside" background, as shown in Figure 4.3(b). Alternatively, if one uses 8-connectedness for both background and objects, then the object is now 1-connected piece, yet the hole is now connected to the outside, as shown in Figure 4.3(c). A suitable solution for this example is to use 4-connectedness for the hole (the hole can be separated from the "outside" background) while using 8-connectedness for the object (the object is one connected piece), as shown in Figure 4.3(d).

The paradox problem as shown by the above examples is called "connectivity paradox," and it poses complications for many geometric algorithms. The solution is thus to alternatively use 4-neighbour and 8-neighbour connectivity for the boundary and internal pixels of an image.

4.3.3 Separating Connected Components

A **4-connected curve** may define more than two connected components. In the example shown in Figure 4.4(a), three 4-connected components are defined, containing the points p, q, and r, respectively.

Such a problem is resolved by using 8-connectivity as dual to 4-connectivity as shown in Figure 4.4(b). The 4-connected curve C defines an 8-connected interior component containing the points p and q and the exterior is the 8-connected component containing r. It is clear that there is no **8-connected arc** that connects p or q to any exterior point (*i.e.*, a point in the exterior 8-connected component).

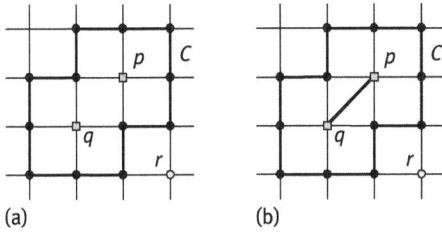

(a)   (b)

Figure 4.4: A 4-digital closed curve separates three 4-connected components.

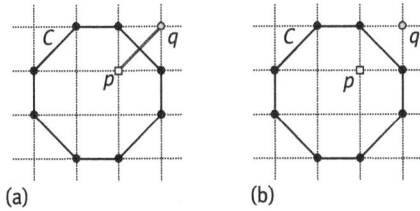

(a)   (b)

Figure 4.5: An 8-digital closed curve joins two 8-connected components.

In Figure 4.5, the second type of problem arises. The 8-connected curve C does not separate the digital plane into two 8-connected components, as shown in Figure 4.5(a). As a counter example, there exists an 8-connected arc joining two potential interior and exterior points p and q, respectively.

However, it is clear that an 8-connected curve will define two 4-connected components as its exterior and interior, as shown in Figure 4.5(b).

In conclusion, if the internal pixels were judged according to 8-connectivity, the boundary obtained would be 4-connected. While if the internal pixels are judged according to 4-connectivity, the boundary obtained would be 8-connected.

4.3.4 Open Set and Closed Set

In digital image measurements, special attention should be paid to boundary pixels. The boundary B of a region R consists of all the boundary pixels that are 4-connected or 8-connected. The other pixels of the region are called internal pixels. A boundary pixel p of a region R should satisfy two conditions: (1) p belongs to R; (2) in the neighborhood of p, there is at least one pixel that does not belong to R.

The border of a digital set is defined as follows. Given a k-connected set of points R, the complement of R, denoted R^c, define a dual connectivity relationship (denoted k'-connectivity). The border of R is the set of points B defined as the k connected set of points in R that have at least one k'-neighbor in R^c (Marchand, 2000).

Consider the digital image shown in Figure 4.6(a). Points of the 8-connected ($k = 8$) foreground F are symbolized as black circles (•) and points of the 4-connected ($k' = 4$) background F^c are symbolized as white circles (∘).

Depending on which foreground or background is chosen as an **open set**, two different border sets are defined. In Figure 4.6(b), the foreground is considered a **closed set**. Hence, it contains its border B, the points of which are surrounded by a square

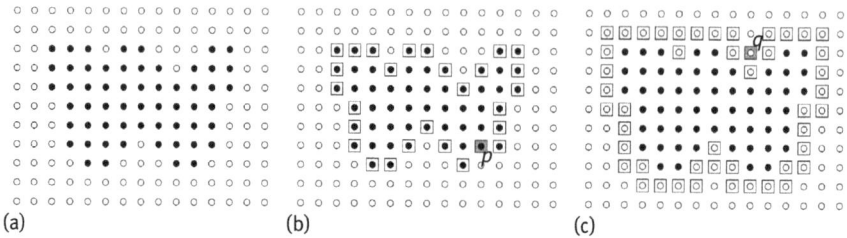

Figure 4.6: Border of a binary digital image.

box. The pixels belonging to B are black pixels with at least one white pixel among their 4-neighbors. Conversely, in Figure 4.6(c), the foreground is considered an open set. Hence, the background contains the border B. The pixels belonging to B are white pixels with at least one black pixel among their 8-neighbors.

Although the set of border points B of a connected component is a connected component with respect to the connectivity of the set it belongs to, it generally does not satisfy the conditions for being a digital closed curve. In Figure 4.6(b), the border of the foreground B is 8-connected, but the point marked with p has three 8-neighbors in B. Similarly in Figure 4.6(c), B is a 4-connected component. However, the point marked with q has three 4-neighbors in B. Therefore, in neither case B is a closed curve.

4.4 Feature Measurement Error

Images are the mapping results of the real world. However, such a mapping is a degenerated projection process. It means different real-world entities can produce the same or a similar projection. In addition, the digital image is just an approximate representation of the original analogue world/information, due to a number of factors.

4.4.1 Different Factors Influencing Measurement Accuracy

In image analysis, the measurement of features consists of starting from digitized data (image) and accurately estimating the properties of original analog entities (in the scene) that produce those digitized data. "The ability to derive accurate measures of image properties is profoundly affected by a number of issues." (Young, 1988). Along the process from scene to data (such as image acquisition, object segmentation, feature measurement), many factors will influence the accuracy of the measurements. In fact, this is an estimation process, so error is inevitable. Some important factors, which make the real data and estimated data different, are listed as follows (see Figure 4.7, in which the action points of different factors are also indicated).
1. The natural variation of the scene (such as object property and environments).
2. The image acquisition process, including digitization (sampling and quantization), calibration, etc.

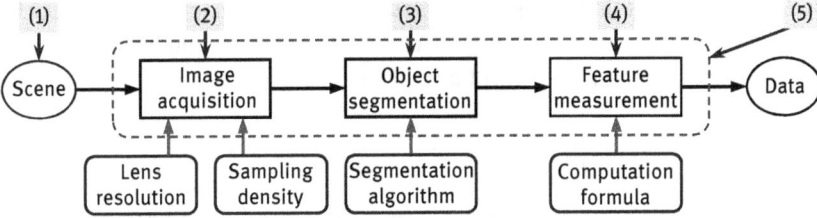

Figure 4.7: Some factors influencing measurement accuracy.

3. The various image processing and analysis procedures (such as image segmentation, feature extraction, etc.).
4. The different measurement processes and feature computation formulas (including approximation error).
5. Noise and other degradations introduced during treatments.

In the following, some factors listed in 2–4 will be discussed.

4.4.2 Influence of Optical Lens Resolution

The actual image acquisition uses optical lens. Resolution of optical lens has a major impact on the image samples. For a limited scattering optical lens, the radius of point spread function at the first zero point in the imaging plane is

$$r = \frac{1.22 \cdot \lambda}{D} d_i \tag{4.5}$$

where λ is the wavelength of light (often take $\lambda = 0.55$ µm for natural light); d_i is the distance from the lens to the imaging plane; D is the diameter of the lens. According to the Rayleigh resolution criterion, if the distance between the two points of the source image is r, then they can be distinguished.

The following gives some discussions on the use of several different imaging lens in imaging apparatus.

4.4.2.1 Normal Camera
In common case of taking picture, the distance from lens to the object $d_o \gg d_i \approx f$ (f is the focal length of the lens). Let the f-factor of the lens (the ratio of the focal length and the diameter of the lens) $n_f = f/D$; at this time, the radius r corresponding to the resolution of camera is

$$r = \frac{1.22 \cdot \lambda}{D} d_i \approx 1.22 \cdot \lambda \frac{f}{D} = 1.22 \cdot \lambda \cdot n_f \tag{4.6}$$

The above formula shows that except in a very close-up shots (macro), other shot circumstances will have better results. With the close-up shots, d_i would be much bigger than f, and the approximating effect will be relatively poor.

4.4.2.2 Telescope

If a telescope is used to observe constellations, it should be noted that the constellation is actually equivalent to point source, their image size would be many times smaller than the radius corresponding to the first zero point of point spread function of the best telescopes in the imaging plane. At this time, the constellation cannot produce the images of their own but only to copy the point spread function of telescope on the imaging plane. In this case, it is the size of the point spread function of telescope determining the resolution.

4.4.2.3 Microscope

In the optical microscope, d_i is determined by the optical tube length, which is generally between 190 and 210 mm. In contrast to common case of taking picture, except the microscope lenses with less than 10 times magnification, there are $d_i \gg d_o = f$. The numerical aperture of the lens is defined as

$$NA \approx D/2f \tag{4.7}$$

The radius r corresponding to the resolution of camera is

$$r = \frac{1.22 \cdot \lambda}{2 \cdot NA} = 0.61 \cdot \lambda/NA \tag{4.8}$$

Table 4.3 gives some theoretical resolutions of typical microscope lenses. Theoretical unit size refers to the theoretical size of a single unit in CCD. Theoretical unit number along the target diameter can be obtained by considering the field of view of the modern microscope. The diameter of the field of view of the modern microscope is 22 mm (0.9 in.), while the diameter of the field of view of the earlier microscope is 20 mm (0.8 in.).

In the following, consider the calculation of physical resolution of a CCD image. In a typical usage example, a simple CCD camera with the diagonal distance of 13 mm (0.5 inches) is used, the resulting image is 640 × 480 pixels, in which each pixel has a scale of 15.875 μm. By comparing this value with theoretical unit size in Table 4.3, it can see that the simplest camera is enough for utilization, if no lens was added between the camera and the eyepiece. Many metallography microscope allows to add a 2.5× zoom lens, which can make the camera to get unit size smaller than the above-discussed theoretical unit size.

In common cases, using an optical microscope together with a camera having resolution higher than 1,024 × 1,024 pixels could only increase the amount of data and analysis time, while cannot provide more information.

Table 4.3: Resolutions and CCD unit sizes of some typical microscope lenses.

Eyepiece magnification	F-number	Eyepiece resolution/μm	Theoretical unit size/μm	Unit number in 22 mm
4×	0.10	3.36	13.4	1,642
10×	0.25	1.34	13.4	1,642
20×	0.40	0.84	16.8	1,310
40×	0.65	0.52	20.6	1,068
60×	0.95	0.35	21.2	1,038
60× (oil)	1.40	0.24	14.4	1,528
100× (oil)	1.40	0.24	24.0	917

4.4.3 Influence of Sampling Density

There is a profound difference between image processing and image analysis, and then, the sampling theorem (a statement that can be shown to be true by reasoning) is not a proper reference for choosing a sampling density. Some points are discussed below.

4.4.3.1 Applicability of Sampling Theorem

As it is known, the Shannon sampling theorem points out that if the highest frequency component in a signal $f(x)$ is given by w_0 (if $f(x)$ has a Fourier spectrum $F(w)$, then $f(x)$ is band-limited to frequency w_0 if $F(w) = 0$ for all $|w| > w_0$), then the sampling frequency must be chosen such that $w_s > 2w_0$ (note that this is a strict inequality). It is also proved that for any signal and its associated Fourier spectrum, the signal can be limited in the time space (space-limited) or the spectrum can be limited in the frequency space (band-limited), but it cannot be limited in both the time and frequency spaces. Therefore, the applicability of the sampling theorem should be studied.

The sampling theorem is really concerned with image processing. The effects of under sampling (aliasing) and improper reconstruction techniques are usually compensated by the sensitivity (or lack of it) of the human visual system. Image analysis is not in a position to ignore these effects. The ability to derive accurate measurements of image properties is profoundly affected by a number of issues, one of which is the sampling density (Young, 1988).

The situation changes in image analysis. In general, the more pixels you can "pack" into an object, the more precise the boundary detection when measuring the object. The call for the increase in magnification to resolve small features is a better sampling requirement. Due to the misalignment of square pixels with the actual edge of an object, significant inaccuracies can occur when trying to quantify the shape of an object with only a small number of pixels. One example is shown in Figure 4.8, in which three scenarios of the effects of a minute change in position of a circular object within the pixel array and the inherent errors in size that can result are illustrated.

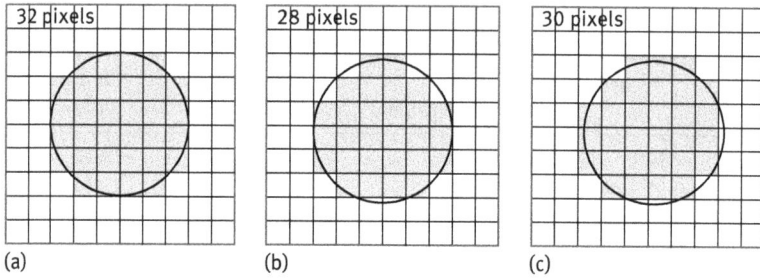

Figure 4.8: Different measurements resulting from different positioning.

The above trouble can be diminished by increasing the sampling density. However, such a problem could not be solved completely as indicated in Section 2.3. It is concluded that it is impossible to satisfy the sampling theorem in practice, except for periodic functions. As the applicability of the sampling theorem could not be guaranteed in general situations, sampling density selection should be considered for image analysis.

4.4.3.2 Selection of Sampling Density

In the structure analysis of biomedical images with a high-resolution microscope image system, the spatial resolution and the sampling density are critical. On the one hand, many of the relevant details in the stained cell and tissue images are smaller than 1 μm. On the other hand, the theoretical spatial resolution limit of a light microscope is a function of the wavelength of illumination light as well as the aperture of the optics and is approximately 0.2–0.3 μm for incoherent illumination. Theoretical resolution, d, of an optical microscope is given by the expression (ASM, 2000)

$$d = \frac{1.22 \cdot \lambda}{2 \cdot NA} \tag{4.9}$$

where λ is the wavelength of light (approximately 0.55 μm) and NA is the numerical aperture of the objective (*e.g.*, 0.10 for an objective with magnification 4×, and 1.40 for an objective with magnification 100×).

The information content of a digitized image can be described mathematically by the **modulation transfer function** (MTF, the amplitude of the Fourier transform of the point-spread function) of the complete optical and imaging system and of the spectrum of the image itself. The plot of one experimental result with cameras is shown, by the MTF in cycles per μm (vertical axis) versus the sampling density in pixel per μm (horizontal axis), in Figure 4.9 (Harms, 1984).

A reasonable choice for the sampling density necessary to detect subtle/fine cellular structures with a digital image analysis system seems to be in the range of 15–30 pixel/μm:

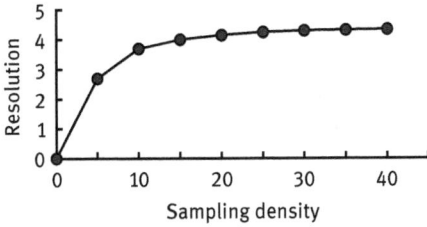

Figure 4.9: Resolution (cycle/μm) *versus* sampling density (pixel/μm).

1. The conventional sampling densities of less than 4–6 pixel/μm are sufficient in systems that locate cells and determine their size and shape as well as cellular form factors.
2. The theoretical band limit of the scanning system is approximately 4–5 cycle/μm (spatial resolution, depending on the aperture of the microscope optics and the illumination wavelength). This corresponds to a sampling density of 10–15 pixel/μm; however, the information content in the digitized image is less than that in the microscopic image as seen with human eyes.
3. At sampling densities of more than 20–30 pixel/μm, the computational costs increase with no significant increase in the information content of the image. The cutoff values of the MTF do not increase with sampling densities above 30 pixel/μm (Figure 4.9).

Based on the above observation and discussion, it can be concluded that sampling densities, which are significantly higher than what one expects from the sampling theorem, are needed for the analysis of microscopic images (Harms, 1984).

4.4.3.3 One Example in Area Measurement

In real applications, to obtain high-accuracy measurement results or to reduce the measurement error to below certain predefined values, a sampling density, which is much higher than that determined by the sampling theorem, is often used. One factor is that the selected sampling density is related to the size of the object to be measured. Take the example of measuring the area of a circular object. The relative measurement error is defined by

$$\varepsilon = \frac{|A_E - A_T|}{A_T} \times 100\% \qquad (4.10)$$

where A_E is the measured (estimated) area and A_T is the real area of the object. Based on a large number of statistical experiments, the relationship between the relative error of the area measurement and the sampling density along the diameter of the circular object is shown in Figure 4.10. The double log coordinates have been used in Figure 4.10, and in this figure, the curve of the relative error of the area measurement versus the sampling density along the diameter is almost a mono-decreasing line.

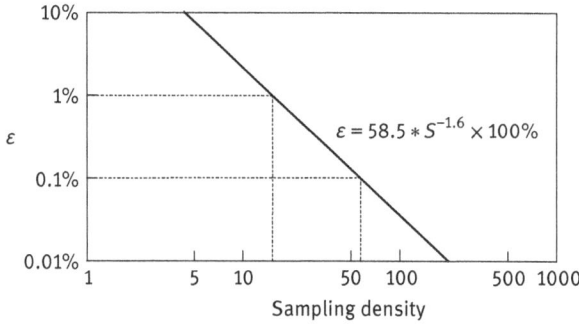

Figure 4.10: The curve of the relative error of the area measurement versus the sampling density along the diameter of a circular object.

It can be seen from Figure 4.10 that the selection of the sampling density should consider the requirement of the permitted measurement error. For example, if the relative error of the area measurement for a circular object is smaller than 1%, then 10 or more samples should be obtained along the diameter of the object. If the required relative error is smaller than 0.1%, at least 50 sample operations should be performed along the diameter of the object. Such sampling densities are often much higher than those determined by the sampling theorem. It has shown that high sampling density is needed not only for measuring the area, but also for high accuracy measurements of other features (Young, 1995).

The above discussion shows that **oversampling** is often needed for high accuracy measurements of the analogy property from digitized data. Such a sampling density is in general higher than that determined by the sampling theorem. The requirement here is different from the task of image reconstruction based on sampling, as in image processing.

4.4.4 Influence of Segmentation

Feature measurements are made for object regions in an image, so the quality of image segmentation for delineating the objects from the background certainly affects the accuracy of feature measurements (Zhang, 1995). The quality of image segmentation, in turn, is determined by the performance of the segmentation techniques applied to get segmented images.

To investigate the dependence of object feature measurements on the quality of image segmentation, a number of controlled tests are carried out. In the following, the basic testing images, segmentation procedure, and testing features are first explained (the criteria for performance assessment have been discussed in Section 2.4). Then, the experiments are performed and the result assessments are presented.

4.4.4.1 Basic Test Images

Since synthetic images have the advantage that they can easily be manipulated, that is, acquisition conditions can be precisely controlled, and experiments can be easily repeated, synthetically generated test images are used. They are produced by using the system for generating synthetic images as described in Zhang (1992a).

The basic image is 256 × 256 pixels, with 256 gray levels and composed of a centered circular disc object (diameter 128), whose gray-level value is 144, on a homogeneous background, whose gray-level value is 112. Particularities of test images generated from this basic image will be described along with the experiments.

4.4.4.2 Segmentation Procedure

A segmentation procedure that can provide gradually segmented images is useful in this task. In addition, this procedure should be relatively independent of the shape/size of objects in images. A large number of segmentation algorithms have been proposed in literature (see Chapter 2). Among them, thresholding techniques are popularly employed. The global thresholding techniques differ mainly in the way they determine the threshold values. To make the study more general, many different threshold values are applied instead of using a specific threshold technique. The goal is not to compare different thresholding techniques but to investigate the dependence of feature measurements on the threshold value, in other words, on the segmentation procedure.

The real procedure is as follows. In order to obtain a group of gradually segmented images, the test images are first multiple thresholded with a sequence of values. These threshold values are taken from the original gray levels between the background and objects. This thresholding process produces a series of labeled images. Then, one opening process is applied to each labeled image to reduce random noise effects. Finally, the biggest object is selected in each image and the holes inside the object are filled. Such a procedure is simple but effective in practice.

4.4.4.3 Testing Features

In various image analysis applications, geometric features are commonly employed (see Chapters 3 and 6). Seven geometric features are considered here. These are the area (A), perimeter (P), form factor (F), sphericity (S), eccentricity (E), normalized mean absolute curvature (N), and bending energy (B) of objects. Area and perimeter have been discussed in Section 3.5.1. Form factor, sphericity, and eccentricity will be described in Section 6.3.

The **normalized mean absolute curvature** is proportional to the average of the absolute values of the curvature function $K(m)$ of an object contour, given by

$$N \propto \sum_{m=1}^{M} |k(m)| \tag{4.11}$$

where M is the number of points on the object contour and $K(m)$ can be expressed with the chain code $C(.)$ of the object contour by $K(m) = [C(m) - C(m-1)]/\{L[C(m)] + L[C(m-1)]\}$ and $L(.)$ is the half-length of the curve segments surrounding the contour point, Bowie (1977).

Finally, the **bending energy** is proportional to the sum of the squared curvature around an object contour, Bowie (1977)

$$B \propto \sum_{m=1}^{M} K^2(m) \qquad (4.12)$$

4.4.4.4 Experiments and Results

The following four quantitative experiments have been carried out to study the dependence of RUMA (or normalized AUMA or scaled AUMA) on image segmentation. The test images will be described along with each experiment. Since the segmented images are indexed by the threshold values used for obtaining the respective images, the RUMA of features as the function of the threshold values used in the segmentation procedure is plotted. The RUMA represents the disparity between true and segmented objects. The smaller the value, the better the segmentation accuracy.

Feature Difference

The first experiment compares the normalized AUMA of these seven features for the same-segmented images. As an example, Figure 4.11 shows the results obtained from one image that contains the same object as in the basic image. The *SNR* of this image is 16. The resulted values of different features have been normalized for the purpose of comparison.

Three points can be noted from Figure 4.11. The first is that all curves have a local minimum located at the inner region between the gray levels of the object and

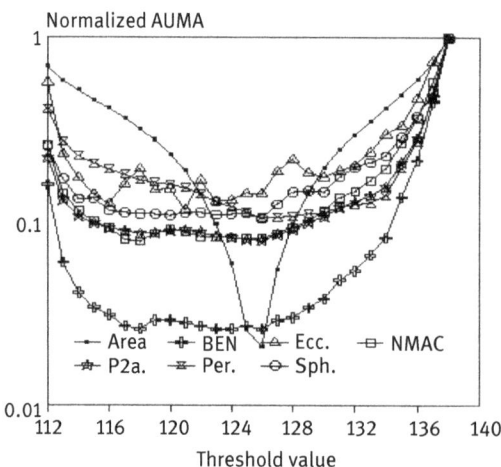

Figure 4.11: Results for feature difference.

background and have higher values at the two sides. Intuitively, this implies that the measurement accuracy of those features is related to the quality of segmented images. The second, however, is that those minima are not located at the same place. This means that there exists no unique best-segmented image with respect to all these features. The third is that the seven curves have different forms. The A curve has a deeper valley and decreases or increases around the valley quite consistently. In other words, it steadily follows the change of the threshold values. Other features are not always so sensitive to such a change. It is clear that the accurate measurement of the object area depends more on the segmentation procedure than on other features. The measurement accuracies of different features are a function of segmentation.

SNR Influence

The second experiment looks at the dependence of scaled AUMA on segmentation when images have various *SNR* levels. Four test images with different *SNR* levels are generated by adding Gaussian noise with standard deviations 4, 8, 16, and 32, and the *SNR* levels are 64, 16, 4, and 1, respectively. These values cover the range of many applications and they are compatible with other studies. These four images contain the same object as in the basic image. In Figure 4.12, the scaled AUMA curves of three features, namely A, E, and P, are presented.

The influence of noise on the results is quite different as shown in Figure 4.12. The A curves gradually shift to the left as the *SNR* level decreases, though their forms remain alike. It is thus possible, by choosing the appropriate values of algorithm parameters, to obtain similar measurement accuracy from images of different *SNR* levels. On the contrary, the E and P curves gradually move up as the *SNR* level decreases. In other words, the best measurement accuracy of E and P is associated with the *SNR* level of images. The higher the *SNR* level, the better the expected measurement accuracy. In Figure 4.12(b, c), the E curves are jagged, whereas the P curves are smoother. This implies that E is more sensitive to the variation of segmented objects due to noise. Among other features, B, N, and F curves are also smooth like the P curves, while S curves show some jags like the E curves.

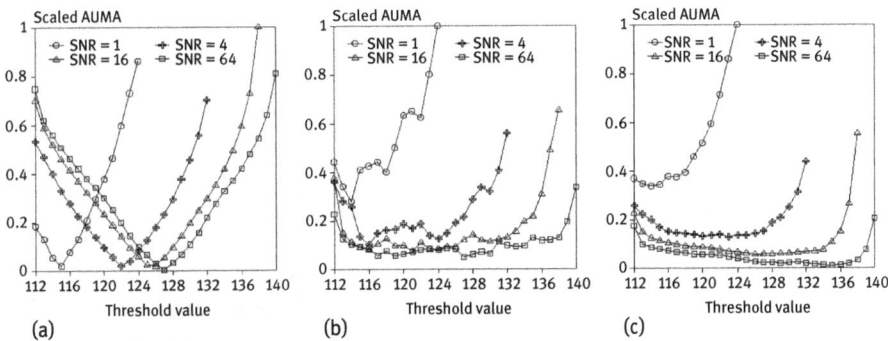

Figure 4.12: Results of SNR influence.

Influence of Object Size

In real applications, the objects of interest contained in images can have different shapes and/or sizes. The size of objects can affect the dependence of RUMA on segmentation, as shown in the third experiment. Four test images with objects of different sizes are generated. Their diameters are 128, 90, 50, and 28, respectively. The *SNR* of these images is fixed to 64 to eliminate the influence of *SNR*. The results for three features, namely *A*, *B*, and *F*, are shown in Figure 4.13.

In Figure 4.13(a, b), the measurement accuracy of *A* and *B* show an opposite tendency with respect to the change of object size. When the images to be segmented contain smaller objects, the expected measurement accuracy for *A* becomes worse while the expected measurement accuracy for *B* becomes better. Among other features, *E* and *S* exhibit a similar tendency as *A* but less significantly, while *N* curves are more comparable with *B* curves. Not all features show clear relations with object size; for example, the four *F* curves in Figure 4.13(c) are mixed. The *P* curves also show similar behavior.

Object Shape Influence

The fourth experiment is used to examine how the dependence of RUMA on segmentation is affected by object shape. Four test images containing elliptical objects of different eccentricity (*E* = 1.0, 1.5, 2.0, 2.5) are generated. Though the shapes of these four objects are quite distinct, these objects have been made similar in size (as the object size in the basic image) to avoid the influence of the object size. This was achieved by adjusting both the long and short axes of these ellipses. In Figure 4.14, the results obtained from these four images with *SNR* = 64 for three features, *A*, *N*, and *S* are given.

The difference among the four curves of the same feature in Figure 4.14 is less notable than that in Figure 4.13. In Figure 4.14(a), for example, the four *A* curves are almost overlapped with each other. This means that the influence of the object shape on the measurement accuracy is much less important than the object sizes. Other feature curves, except *B* curves, have similar behavior as *S* curves in Figure 4.14(c), while *B* curves are more like *N* curves in Figure 4.14(b).

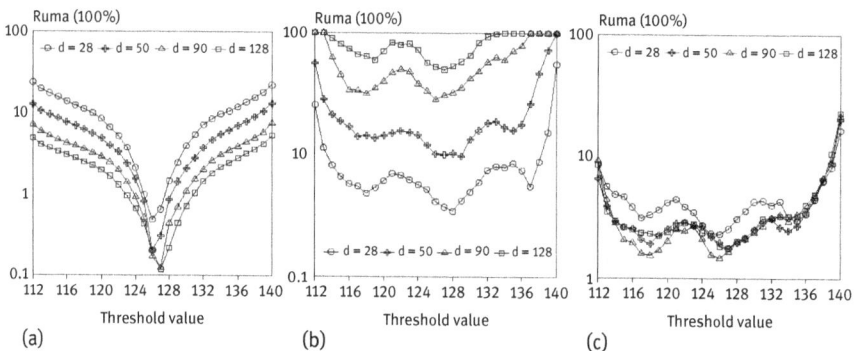

Figure 4.13: Results for object size influence.

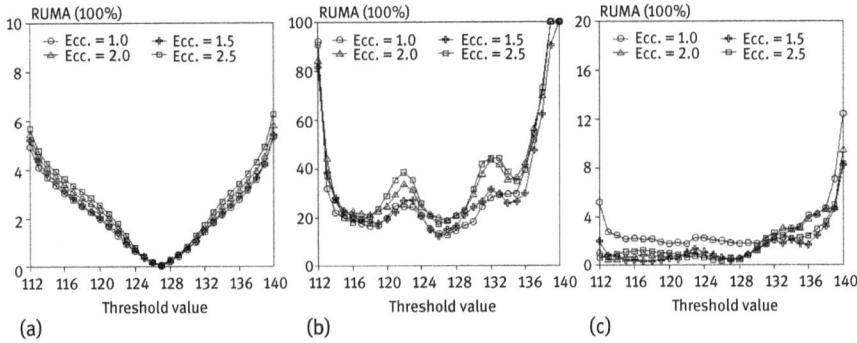

Figure 4.14: Results for object shape influence.

4.4.5 Influence of Computation Formulas

The formula for feature computing is also an important component in analyzing measurement errors. In the following, the distance measurements based on chain codes are discussed as examples.

Generally, the correct representation of simple geometric objects such as lines and circles is not clear. Lines are well defined only for angles with values of multiples of 45°, whereas for all other directions, they appear as jagged, staircase-like sequences of pixels (Jähne, 1997).

In a digital image, a straight line oriented at different angles and mapped on to a pixel grid will have different lengths if measured with the fixed link length. Suppose two points in an image are given, and a (straight) line represented by 8-directional chain code has been determined. Let N_e be the number of *even chain codes*, N_o be the number of *odd chain codes*, and N_c be the number of *corners* (i. e., the point where the chain code changes direction) in the chain codes. The total number of chain codes is N ($N = N_e + N_o$), while the length of this line can be represented using the following general formula (Dorst 1987):

$$L = A \times N_e + B \times N_o + C \times N_c \tag{4.13}$$

where A, B, and C are the weights for N_e, N_o, N_c, respectively. The length computation formula is thus dependent on these weights. Many studies have been made for these weights. Some results are summarized in Table 4.4.

Five sets of weights for A, B, and C are listed in Table 4.4. Separately introducing them into eq. (4.9) can produce five length computation formulas for lines, which are distinguished by the subscripts. For a given chain code representing a line in the image, the length computed using the five formulas are generally different. The expected measurement errors are also different; a small error is expected with formulas having the bigger labels. In Table 4.4, E represents the average mean-square difference

Table 4.4: A list of length measurement formulas.

L	A	B	C	E (%)	Explanation
L1	1	1	0	16	A short-biased estimator
L2	1.1107	1.1107	0	11	A no-bias estimator
L3	1	1.414	0	6.6	A long-biased estimator
L4	0.948	1.343	0	2.6	The error is inversely proportional to the line length
L5	0.980	1.406	−0.091	0.8	When $N = 1,000$, it becomes a no-bias estimator

between the real length and the estimated length of a line and shows the errors that may be produced by using these formulas. E is also a function of N.

Now let us have a closed look at these formulas. In 8-directional chain codes, even chain codes correspond to the codes representing horizontal or vertical straight-line segments, while odd chain codes correspond to the codes representing diagonal straight line segments. The simplest method for measuring the length of a line is to count each code in the chain code as a unit length. This results in the formula for L_1. It is expected that L_1 would be a short-biased estimator, as in an 8-directional chain code representation the even numbered codes have lengths bigger than unit. Making some scaling for these weights provides the second estimator L_2. If the distance between two adjacent pixels, along the horizontal or vertical direction, is taken as a unit, the distance between two diagonally adjacent pixels must be a $2^{1/2}$ unit. This gives the formula for L_3, which is the most popularly used formula for chain code. However, L_3 is also a biased estimator. Compared with L_1, the difference is that L_3 is a long-biased estimator. In fact, a line at different angles will have different lengths according to L_3, as the sum overestimates the value except in the case of exact $45°$ or $90°$ orientations. To compensate this effect, the weights are scaled and another estimator L_4 is obtained. The weights A and B are selected to reduce the expected error for longer lines, so L_4 would be better used for longer lines as the error is inversely proportional to the lengths of the lines. Finally, if the number of corners is considered as well as the number of codes, a more accurate estimator L_5 is obtained.

One example used to compare the five estimators, or in other words, to compare the five length computation formulas, is given in Figure 4.15. In Figure 4.15, given two points p and q, different analogue lines that join these two points are possible, only two of them (with the dotted line and dashed line, respectively) are shown as examples. The solid line provides the chain codes, and the lengths computed using the five formulas for this chain code are shown in the right of the figure. The comparison is made by taking L_5 as the reference, that is, the error produced by each formula (listed in the parentheses) is obtained by

$$\varepsilon = \frac{|L_i - L_5|}{L_5} \times 100\% \qquad i = 1, 2, 3, 4, 5 \tag{4.14}$$

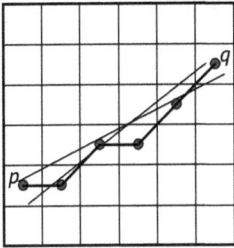

$L_1 = 5.000$ (15.326%)

$L_2 = 5.554$ (5.956%)

$L_3 = 6.242$ (5.718%)

$L_4 = 5.925$ (0.339%)

$L_5 = 5.905$ (0.0%)

Chain code: 0 1 0 1 1

Figure 4.15: Compute the chain-code length with different formulas.

4.4.6 Combined Influences

The above discussed influence factors (as well as other factors) can have some combined influence on measurements, or in other words, their effects are related Dorst (1987). For example, the influence of computation formulas on the length estimation error is related to the sampling density. With coarser sampling, various computation formulas give compatible results. While with the increase in the sampling density, the differences among the results of various computation formulas increase, too. The estimators with bigger serial numbers become even better than the estimators with smaller serial numbers. That is, the measurement errors will decrease with the increase in sampling density, and the decline is fast for estimators that show better behavior. This can be seen from Figure 4.16, in which the curves of computation error as the function of the sampling density for three estimators are plotted.

It can be seen from Figure 4.16 that the measurement errors for L_3, L_4, and L_5 decrease with the increase in sampling density. The L_4 curve declines faster than the L_3 curve and the curve L_5 declines even faster than the L_4 curve.

Finally, it should be noted that the decrease ratios of these estimators become smaller and smaller with the increase in sampling density. They finally approach their respective bounds. It has shown that these estimators all have some low bound

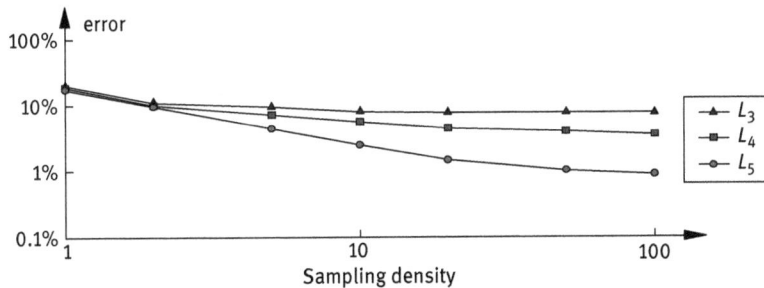

Figure 4.16: The curves of computation error versus the sampling density for three estimators.

(Table 4.3) that will not be changed by the sampling density. To get measurements with even more accuracy, some even more complicated formulas are needed (Dorst, 1987).

4.5 Error Analysis

Measurement errors arise from different sources for various reasons. Two commonly encountered measurement errors are analyzed in the following examples.

4.5.1 Upper and Lower Bounds of an 8-Digital Straight Segment

Given two discrete points p and q, the grid-intersect quantization of $[p, q]$ defines a particular straight segment between p and q. However, based on chain codes and the definition of a shift operator (below), there exists a set of inter-related digital straight segments, which can be defined between p and q, Marchand (2000).

Given a chain-code sequence $\{c_i\}_{i=1,\ldots,n}$, the **shifted chain code** is given by shift($\{c_i\}$) = $\{c_2, c_3, \ldots, c_n, c_1\}$. Given the chain code $\{c_i\}_{i=1,\ldots,n}$ of a digital straight segment P_{pq}, the shift operator can be applied successively $n - 1$ times on $\{c_i\}$ for generating $n - 1$ shifted chain codes, corresponding to different digital arcs from p to q. It is proven that any shifted chain code defines a new digital straight segment from p to q. The union of all shifted digital straight segments forms an area, which in turn defines the lower and upper bounds of the digital straight segment, as shown by the following example.

Given the two discrete points p and q illustrated in Figure 4.17(a), an 8-digital straight segment can readily be obtained as the grid-intersect quantization of the continuous segment $[p, q]$ (represented by a thick line). Its chain code is given by $\{c_i\}_{i=1,\ldots,n} = \{0, 1, 0, 0, 1, 0, 0, 1, 0\}$. Consider the eight possible shifted chain codes given in Table 4.5. These possible digital straight segments all lie within the shaded area associated with the continuous segment $[p, q]$. The upper and lower bounds of this area are called the upper and lower bounds of the digital straight segment P_{pq}. These bounds are represented as thick and dashed lines, respectively, in Figure 4.17(b).

In this example, the upper and lower bound chain codes are represented by shift (1) and shift (2), respectively.

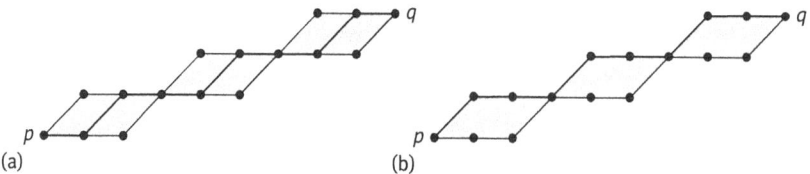

Figure 4.17: Upper and lower bounds of a digital straight segment.

Table 4.5: Shifted chain codes.

Shift	Chain code	Shift	Chain code	Shift	Chain code
	{0, 1, 0, 0, 1, 0, 0, 1, 0}	3	{0, 1, 0, 0, 1, 0, 0, 1, 0}	6	{0, 1, 0, 0, 1, 0, 0, 1, 0}
1	{1, 0, 0, 1, 0, 0, 1, 0, 0}	4	{1, 0, 0, 1, 0, 0, 1, 0, 0}	7	{1, 0, 0, 1, 0, 0, 1, 0, 0}
2	{0, 0, 1, 0, 0, 1, 0, 0, 1}	5	{0, 0, 1, 0, 0, 1, 0, 0, 1}	8	{0, 0, 1, 0, 0, 1, 0, 0, 1}

4.5.2 Approximation Errors

The relative approximation error between Euclidean distances and discrete distances depends on the value of the move lengths. In the following, the calculation of the relative error resulting from the move $d_{a,b}$ along a vertical line is discussed.

First, the definition of relative error is provided. The relative error between the values of a given discrete distance d_D and the Euclidean distance d_E between two points O and p is calculated as

$$E_D(O, p) = \frac{(1/s)d_D(O, p) - d_E(O, p)}{d_E(O, p)} = \frac{1}{s}\left[\frac{d_D(O, p)}{d_E(O, p)}\right] - 1 \qquad (4.15)$$

Parameter $s > 0$ is called the scale factor. It is used to maintain consistency between radii of the discrete and Euclidean discs. When using chamfer distances, a typical value is $s = a$.

With the aid of Figure 4.18 (only considering the first octant, that is, between lines $y = 0$ and $y = x$),

$$d_{a,b}(O, p) = (x_p - y_p)a + y_p b \qquad (4.16)$$

The error $E_{a,b}$ is measured along the line $(x = K)$ with $K > 0$, and the value of the relative error at point p is

$$E_{a,b}(O, p) = \frac{(K - y_p)a + y_p b}{s\sqrt{K^2 + y_p^2}} - 1 \qquad (4.17)$$

Typically, the graph of $E_{a,b}(O, p)$ for $y_p \in [0, K]$ is shown in Figure 4.19.

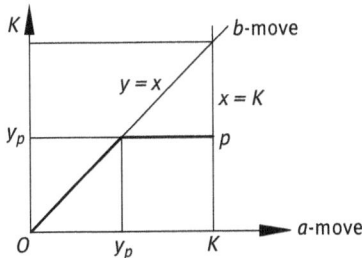

Figure 4.18: Calculation of $d_{a,b}$ in the first octant.

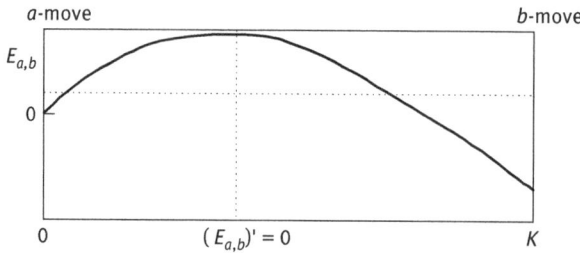

Figure 4.19: The graph of $E_{a,b}$ in the first octant.

Since $E_{a,b}$ is a convex function on $[0, K]$, its local extreme can be obtained at p such that $\partial E_{a,b}/\partial y = 0$. For all p such that $x_p = K$ and $0 \le y_p \le K$

$$\frac{\partial E_{a,b}}{\partial y}(O,p) = \frac{1}{s\sqrt{K^2 + y_p^2}}\left[(b-a) - \frac{[(K - y_p)a + y_p b]y_p}{K^2 + y_p^2}\right] \qquad (4.18)$$

In case $p = [K, (b-a)K/a]$, there is $\partial E_{a,b}/\partial y = 0$, and $E_{a,b}(O, p) = [a^2 + (b-a)^2]^{1/2}/s - 1$.

The **maximum relative error** defined as $E_{\max} = \max\ \{|E_{a,b}(O, p)|: p/x_p = K;\ 0 \le y_p \le K\}$ is then either reached at the local extreme or at a bound of the interval $y_p \in [0, K]$ (see Figure 4.19). Now, $E_{a,b}(O, p) = a/s-1$ if $p = (K, 0)$ and $E_{a,b}(O, p) = b/(2^{1/2}s) - 1$ if $p = (K, K)$. Hence,

$$E_{\max}(O, p) = \max\left\{\left|\frac{a}{s} - 1\right|,\ \left|\frac{\sqrt{a^2 + (b-a)^2}}{s} - 1\right|,\ \left|\frac{b}{\sqrt{2}s} - 1\right|\right\} \qquad (4.19)$$

Finally, although the error is calculated along a line rather than a circle, its value does not depend on the line (i. e., $E_{\max}$ does not depend on K). Hence, the value of $E_{\max}$ is valid throughout the 2-D discrete plane.

Some numerical examples for several commonly used move lengths are given in Table 4.6 (note d_4 and d_8 can be seen as particular cases of the chamfer distance in the 4- and 8-neighborhoods, respectively).

Table 4.6: Maximum relative errors for some move lengths.

Distance	Neighborhood	Move lengths and scale factor	Maximum relative error
d_4	N_4	$a = 1, s = 1$	41.43%
d_8	N_8	$a = 1, b = 1, s = 1$	29.29%
$d_{2,3}$	N_8	$a = 2, b = 3, s = 2$	11.80%
$d_{3,4}$	N_8	$a = 3, b = 4, s = 3$	5.73%

4.6 Problems and Questions

4-1 Under what conditions will eqs. (4.1–4.3) be the special cases of eq. (4.4)?

4-2 What is the applicability of a measurement with high accuracy and low precision?

4-3 In Figure Problem 4-3, the locations of the concentric circle indicate the expected measurement positions and the locations of the black squares indicate the results of real measurements.

  (1) Compute separately the average values of real measurements for two images, and compute the 4-neighbor distance between the average values and the expected values.

  (2) Compute the mean and variances of the four real measurements.

  (3) Discuss the measurement results obtained from the two images.

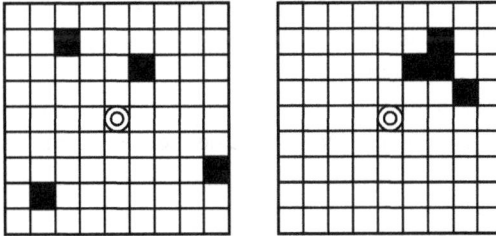

Figure Problem 4-3

4-4* A given object's area is 5. After capturing it in an image, two methods are used to estimate its area. Table Problem 4-4 gives two groups of estimated values using these two methods, respectively. Compute the means and variances of these two estimations. Compare the accuracy and the precision of these two methods.

Table Problem 4-4

First method	5.6	2.3	6.4	8.2	3.6	6.7	7.5	1.8	2.1	6.1
Second method	5.3	5.8	5.9	5.9	6.1	5.8	6.0	4.9	5.5	5.4

4-5 Find examples and explain that there is also connection confusion in 3-D images.

4-6 How do you improve the optical resolution of an image acquisition system?

4-7 Use the applicability of the sampling theorem to distinguish image processing and image analysis.

4-8 (1) Consider Figure 4.8(a). Increase the sampling rates in both the X and Y directions by a factor of two, and compute the area of the circle (suppose that the radius of the circle is 3). Compare the area measurement errors under two sampling rates.

(2) Move down the circle in Figure 4.8(a) by 1/4 pixel and then move right by 1/4 pixel. Separately compute the areas of the circle under the original sampling rates and under double increased sampling rates. Compare the area measurement errors under two sampling rates.

4-9 To remove (Gaussian) noise, a smooth filter is often employed. This process could also influence the final object measurement. Look at the literature and analyze the concrete influence.

4-10* Take a pixel in an image as the center and draw a circle with radius 2 around this pixel (suppose that the pixel size is unit). Represent this circle by an 8-connected chain code and use the five formulas for length measurement listed in Table 4.4 to compute the length of the chain code.

4-11 Consider Figure 4.8(c). Obtain the 8-connected chain code from the circle, and use the five formulas for length measurement listed in Table 4.4 to compute the length of the chain code.

4-12 Using $d_{5,7}$ as a discrete distance measurement, instead of the Euclidean distance measurement, what would be the maximum relative error produced?

4.7 Further Reading

1. **Direct and Indirect Measurements**
   - Many measurement metrics have been proposed and used in various applications. Besides those discussed in Section 4.1, more metrics can be found in Russ (2002).
2. **Accuracy and Precision**
   - More discussions on the statistical error and the system error can be found in Haralick (1993) and Jähne (1999).
3. **Two Types of Connectivity**
   - Problems caused by the existing two types of connectivity have also been discussed in many textbooks, such as Rosenfeld (1976) and Gonzalez (1987). These problems always happen in square lattices (Marchand, 2000).
4. **Feature Measurement Error**
   - More formulas for computing the length of a line can be found in Dorst (1987).
   - Many global measurements are based on local masks. Designing local masks for global feature computations is different from determining feature computation formulas (Kiryati, 1993).

- Using local masks to make the length measurement of a line in 3-D has been discussed in Lohmann (1998). Different masks designed can be seen in Borgofors (1984), Verwer (1991), Beckers (1992), and Kiryati (1993).
- Some considerations on feature measurement errors have been discussed in Zhang (1991a).

5. **Error Analysis**
   - Some other errors in feature measurement can also be analytically studied, for example, Dorst (1987).

5 Texture Analysis

Texture is a commonly used concept when describing the properties, especially those related to spatial information, of images. Texture analysis is an important branch of image analysis.

The sections of this chapter are arranged as follows:

Section 5.1 introduces the main contents of the study and application of texture, as well as the basic principles and methods of texture analysis and classification.

Section 5.2 describes the statistical methods for texture description. In addition to the commonly used gray-level co-occurrence matrix and the texture descriptors based on co-occurrence matrix, energy-based texture descriptors are introduced.

Section 5.3 discusses the structural approach to texture description. The basic method includes two components: the texture primitives and the arrangement rules. Texture mosaic is a typical method, and in recent years, the local binary pattern (LBP) has also been used.

Section 5.4 focuses on the methods of describing texture using spectrum. Fourier spectrum, Bessel–Fourier spectrum, and Gabor spectrum are discussed.

Section 5.5 introduces the idea and method of texture segmentation. Both supervised texture segmentation and the unsupervised texture segmentation are discussed. A method using wavelet for texture classification is also presented.

5.1 Concepts and Classification

Texture is commonly found in natural scenes, particularly in outdoor scenes containing both natural and man-made objects. Sand, stones, grass, leaves, and bricks all create a textured appearance in an image. Textures must be described by more than just their object classifications (Shapiro, 2001).

5.1.1 Meanings and Scale

The word "texture" has a rich range of meanings. Webster's dictionary defines **texture** as "the visual or tactile surface characteristics and appearance of something." As an illustration, consider the texture of a reptile skin. Apart from being smooth (the tactile dimension), it also has cellular and specular markings on it that form a well-defined pattern (the visual dimension).

Texture has different meanings in different application areas. In the image community, texture is a somewhat loosely defined term (Shapiro 2001) and is considered a phenomenon that is widespread, easy to recognize, and hard to define (Forsyth, 2003). In fact, a number of explanations/definitions exist. Such as "texture is characterized by tonal primitive properties as well as spatial relationships between them

DOI 10.1515/9783110524123-005

(Haralick, 1992)," "texture is a detailed structure in an image that is too fine to be resolved, yet coarse enough to produce a noticeable fluctuation in the gray levels of the neighboring cells" (Horn, 1986). A more general view (Rao, 1990) considers texture to be "the surface markings or a 2-D appearance of a surface."

Although no formal definition of texture exists, intuitively it provides measurements of properties such as smoothness, coarseness, and regularity (Gonzalez, 2002). Texture gives us information about the spatial arrangement of the color or intensities in an image (Shapiro, 2001). Many images contain regions characterized not so much by a unique value of brightness, but by a variation in brightness that is often called texture. It refers to the local variation in brightness from one pixel to the next or within a small region (Russ, 2002).

In discussing texture, one important factor is scale. Typically, whether an effect is referred to as texture or not depends on the scale at which it is viewed. A leaf that occupies most of an image is an object, but the foliage of a tree is a texture (Forsyth, 2003). For any textural surface, there exists a scale at which, when the surface is examined, it appears smooth and texture-less. Then, as resolution increases, the surface appears as a fine texture and then a coarse one. For multiple-scale textural surfaces, the cycle of smooth, fine, and coarse may repeat.

Texture can be divided into microtexture and macrotexture based on the scale (Haralick, 1992). When the gray-level primitives are small in size and the spatial interaction between gray-level primitives is constrained to be local, the resulting texture is a microtexture. The simplest example of a microtexture occurs when independent Gaussian noise is added to each pixel's value in a smooth gray-level area. As the Gaussian noise becomes more correlated, the texture becomes more of a **microtexture**. Finally, when the gray-level primitives begin to have their own distinct shape and regular organization (identifiable shape properties), the texture becomes a **macrotexture**.

In addition, from the scale point of view, gray level and texture are not independent concepts. They bear an inextricable relationship to each other very much as a particle and a wave do. Whatever exists has both particle and wave properties, and depending on the situation, either particle or wave properties may predominate. Similarly, in the image context, both gray level and texture are present, although at times one property can dominate the other, and it often tends to speak of only the gray level or only the texture. Hence, when explicitly defining gray level and texture, it is not defining two concepts but one gray-level–texture concept.

5.1.2 Research and Application Related to Texture

From the research point of view, there are four broad categories of work to be considered:
1. Texture description/characterization
   This category is related to the identification and description of 2-D texture patterns.

2. Texture segmentation

 This category is concerned with using texture as a means to perform segmenta- tion of an image, that is, to break off an image into components within which the texture is constant.

3. Texture synthesis

 This category seeks to construct large regions of texture from small example images.

4. Shape from texture

 This category involves recovering surface orientation or surface shape from image texture, or using texture as a cue to retrieve information about surface orientation and depth.

From the application point of view, texture analysis plays an important role in many areas, such as

1. Distinguish hill from mountain, forest from fruits, etc., in remote sensing applic- ations.
2. Inspect surfaces in the manufacturing of, for example, semiconductor devices.
3. Perform microanalysis for the nucleolus of cells, isotropy, or anisotropy of mater- ial, etc.
4. Recognize specimens in petrography and metallography.
5. Study metal deformation based on the orientations of its grains.
6. Visualize flow motion in biomedical engineering, oceanography, and aerodynamics.

5.1.3 Approaches for Texture Analysis

Technically, many approaches for texture analysis have been proposed. Three prin- cipal approaches used to describe the texture of a region are statistical, structural, and spectral (Gonzalez, 2002).

5.1.3.1 Statistical Techniques

They yield characterizations of textures as smooth, coarse, grainy, and so on. In statistical approaches, texture is a quantitative measurement of the arrangement of intensities in a region (Shapiro, 2001). The **statistical model** usually describes texture by statistical rules governing the distribution and relation of gray levels. This works well for many natural textures that have barely discernible primitives (Ballard, 1982). The goal of a **statistical approach** is to estimate parameters of some random pro- cess, such as fractal Brownian motions, or Markov random fields, which could have generated the original texture (Russ, 2002).

5.1.3.2 Structural Techniques

They deal with the arrangement of image primitives. In a structural approach, texture is a set of primitive texels in some regular or repeated relationship (Shapiro, 2001). **Structural approaches** try to describe a repetitive texture in terms of the primitive elements and placement rules that describe geometrical relationships between these elements (Russ, 2002).

5.1.3.3 Spectral Techniques

They are based on properties of the Fourier spectrum and are used primarily to detect global periodicity in an image by identifying high-energy, narrow peaks in the spectrum.

   Depending on the scale, different approaches have to be used. The basic interrelationships in the gray-level–texture concept are the following Haralick (1992). When a small-area patch of an image has little variation of gray-level primitives, the dominant property of that area is gray level. When a small-area patch has a wide variation of gray-level primitives, the dominant property is texture. Crucial in this distinction are the size of small-area patches, the relative sizes and types of gray-level primitives, and the number and placement or arrangement of the distinguishable primitives. As the number of distinguishable gray-level primitives decreases, the gray-level properties will predominate. In fact, when the small-area patch is only a pixel, so that there is only one discrete feature, the only property is simply the gray-level. As the number of distinguishable gray-level primitive increases within the small-area patch, the texture property will predominate. When the spatial pattern in the gray-level primitives is random and the gray-level variation between primitives is wide, a fine texture is available. When the spatial pattern becomes more definite and the gray-level regions involve more and more pixels, a coarser texture is available.

Example 5.1 Texture analysis depends on pattern and scale.
In Figure 5.1, the pattern in (a) consists of many small texture elements. It is better analyzed statistically without regard to the texture elements. The pattern in (b) consists of large texture elements. It is better analyzed structurally based on the texture

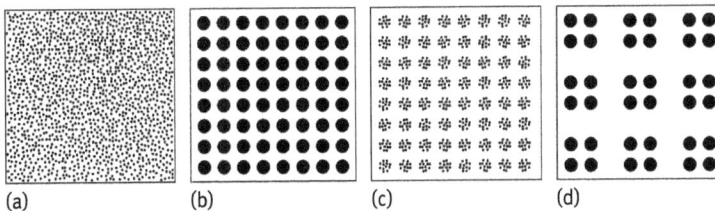

(a)   (b)   (c)   (d)

Figure 5.1: Texture analysis depends on pattern and scale.

elements. The pattern in (c) consists of many small texture elements that form local clusters. The clusters are better detected statistically by image segmentation without regard to the texture elements, and the pattern is better analyzed structurally based on the clusters. The pattern in (d) consists of large texture elements that form local clusters. The clusters are better detected structurally by grouping texture elements, and the pattern is better analyzed structurally based on the clusters.

◪

5.2 Statistical Approaches

Statistical approaches can be connected to statistical pattern recognition paradigms that divide the texture problem into two phases: training and testing. During a training phase, feature vectors from known samples are used to partition the feature space into regions representing different classes. During a testing phase, the feature-space partitions are used to classify feature vectors from unknown samples (Ballard, 1982).

5.2.1 Co-occurrence Matrix

A **co-occurrence matrix** can be computed the following way (Gonzalez, 2002). Let P be a **position operator** and C be a $K \times K$ matrix, whose element c_{ij} is the number of times that points with gray level g_i occur relative to points with gray level g_j, with $1 \le i$, $j \le K$, and K is the number of distinct gray levels in the image.

More generally, let S be a set of pixel pairs that have a spatial relationship, and the element of gray-level co-occurrence matrix is defined by Haralick (1992)

$$c(g_1, g_2) = \frac{\#\{[(x_1, y_1), (x_2, y_2)] \in S | f(x_1, y_1) = g_1 \,\&\, f(x_2, y_2) = g_2\}}{\#S} \tag{5.1}$$

The numerator on the right side of the equation is the number of pixel pairs that have a spatial relationship and have gray values g_1 and g_2, respectively. The denominator on the right side of the equation is the total number of pixel pairs. Thus, the obtained gray-level co-occurrence matrix C is normalized.

Example 5.2 Position operator and co-occurrence matrix.
Consider an image with three gray levels, $g_1 = 0$, $g_2 = 1$, and $g_3 = 2$, as follows:

$$
\begin{array}{ccccc}
0 & 0 & 0 & 1 & 2 \\
1 & 1 & 0 & 1 & 1 \\
2 & 2 & 1 & 0 & 0 \\
1 & 1 & 0 & 2 & 0 \\
0 & 0 & 1 & 0 & 1 \\
\end{array}
$$

Defining the position operator P as "one pixel to the right and one pixel below" (it is denoted by subscribes $(1, 1)$) for this image, it yields the following 3×3 co-occurrence matrix C

$$C_{(1,1)} = \begin{bmatrix} c_{11} & c_{12} & c_{13} \\ c_{21} & c_{22} & c_{23} \\ c_{31} & c_{32} & c_{33} \end{bmatrix} = \begin{bmatrix} 4 & 2 & 1 \\ 2 & 3 & 2 \\ 0 & 2 & 0 \end{bmatrix}$$

where c_{11} (top left) is the number of times that a point with level $z_1 = 0$ appears at the location one pixel below and to the right of a pixel with the same gray level, and c_{13} (top right) is the number of times that a point with level $g_1 = 0$ appears at the location one pixel below and to the right of a point with gray level $g_3 = 2$. The size of C is determined by the number of distinct gray levels in the input image. Thus, application of the concepts discussed here usually requires that intensities be re-quantized into a few gray-level bands in order to keep the size of C manageable. ◙

Example 5.3 Images and their co-occurrence matrices.
Different images can have different co-occurrence matrices as they have different texture scales. This is the basis for using a gray-level co-occurrence matrix to work out texture descriptors. Figure 5.2 shows a texture image with a small scale and several of its gray-level co-occurrence matrices. Figure 5.3 shows a texture image with a large scale and several of its gray-level co-occurrence matrices. In both groups of figures, (a) is the original image, and (b)–(e) is the gray-level cooccurrence matrices $C_{(1,0)}$, $C_{(0,1)}$, $C_{(1,-1)}$, and $C_{(1,1)}$, respectively. Comparing the two figures, it can be seen that as the gray-level changes fast in space for the image with a small scale, the entries of co-occurrence matrices are quite interspersed. On the contrary, the entries of co-occurrence matrices for the image with a large scale are concentrated around the main diagonal axis. The reason is that for the image with a large scale, the pixels (close to each other) in a pair would have similar gray-level values. In fact, the gray-level co-occurrence matrix shows the spatial information of the relative positions of different gray-level pixels. ◙

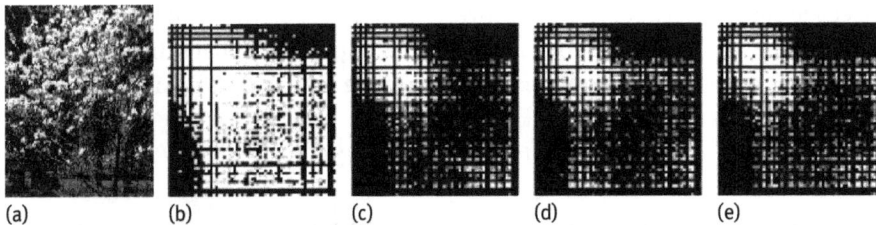

(a)　　　(b)　　　(c)　　　(d)　　　(e)

Figure 5.2: A small-scale image and its gray-level co-occurrence matrices.

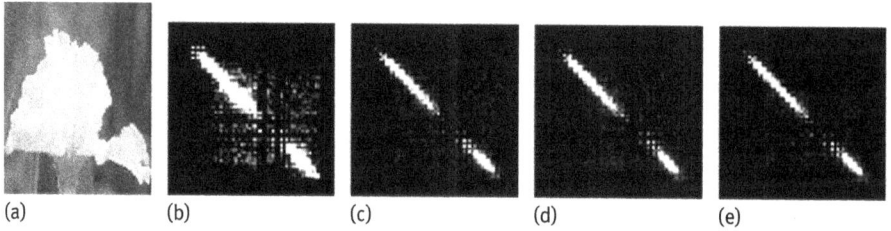

(a)   (b)   (c)   (d)   (e)

Figure 5.3: A big scale image and its gray-level co-occurrence matrices.

A simple generalization of the primitive gray-level co-occurrence approach is to consider more than two pixels at a time. This is called the generalized gray-level spatial dependence model for textures (Haralick, 1992).

Given a specific kind of spatial neighborhood and a subimage, it is possible to parametrically estimate the joint probability distribution of the gray levels over the neighborhoods in the subimage. The prime candidate distribution for the parametric estimation is a multivariate normal distribution. If $x_1, \ldots, x_N$ represent the N K-normal vectors coming from the neighborhoods in a sub-image, then the mean vector μ and covariance matrix V can be estimated by

$$\mu = \mu_0 \mathbf{1} \tag{5.2}$$

$$V = \frac{1}{N} \sum_{n=1}^{N} (x_n - \mu)(x_n - \mu)^T \tag{5.3}$$

with

$$\mu_0 = \frac{1}{N} \sum_{n=1}^{N} \mathbf{1} x_n \tag{5.4}$$

where $\mathbf{1}$ is a column vector whose components are all of the value 1.

To define the concept of generalized co-occurrence, it is necessary to decompose an image into its primitives first. Let Q be the set of all primitives on the image. Then, primitive properties, such as the mean gray level, the variance of gray levels, the region size, and the region shape, will be measured. Let T be the set of primitive properties and f be a function that assigns a property of T to each primitive in Q. Finally, it needs to specify a spatial relation between primitives, such as the distance or adjacency. Let $S \subseteq Q \times Q$ be the binary relation pairing all primitives that satisfy the spatial relation. The element of **generalized co-occurrence matrix $C_\mathbf{g}$** is defined by

$$c_g(t_1, t_2) = \frac{\#\{(q_1, q_2) \in S | f(q_1) = t_1 \,\&\, f(q_2) = t_2\}}{\#S} \tag{5.5}$$

where $c_g(t_1, t_2)$ is just the relative frequency with which two primitives occur with specified spatial relationships in the image, one primitive having the property t_1 and the other having the property t_2.

5.2.2 Descriptors Based on a Gray-Level Co-occurrence Matrix

To analyze a given C matrix in order to categorize the texture of the region over which C was computed (C has captured properties of a texture but is not directly useful for further analysis, such as comparing two textures), many derived descriptors have been defined. A few examples are shown as follows:

1. **Texture uniformity** (the highest value is achieved when the c_{ij}s are all equal, which is called the second-order moment)

$$W_U = \sum_i \sum_j c_{ij}^2 \tag{5.6}$$

2. **Texture entropy** (it is a measurement of randomness, achieving its highest value when all elements of C are maximally random)

$$W_E = -\sum_i \sum_j c_{ij} \log_2 c_{ij} \tag{5.7}$$

3. **Element difference moment** of order k (it has a relatively low value when the high values of C are near the main diagonal)

$$W_M(k) = \sum_i \sum_j (i - j)^k c_{ij} \tag{5.8}$$

It is called texture contrast when k equals 1.

4. **Inverse element difference moment** of order k (it has a relatively high value when the high values of C are near the main diagonal)

$$W_{IM}(k) = \sum_i \sum_j \frac{c_{ij}}{(i - j)^k}, \quad i \neq j \tag{5.9}$$

To avoid the problem caused by $i = j$, an alternative definition used for k equaling 1, called texture homogeneity, is given by (d is a positive constant)

$$W_H(d) = \sum_i \sum_j \frac{c_{ij}}{d + (i - j)} \tag{5.10}$$

Example 5.4 Texture images and their descriptor values.
Figure 5.4 shows five texture images. Their values of texture uniformity, entropy, contrast, and homogeneity are given in Table 5.1. It can be seen that the value of d has important influence on texture homogeneity. ◻

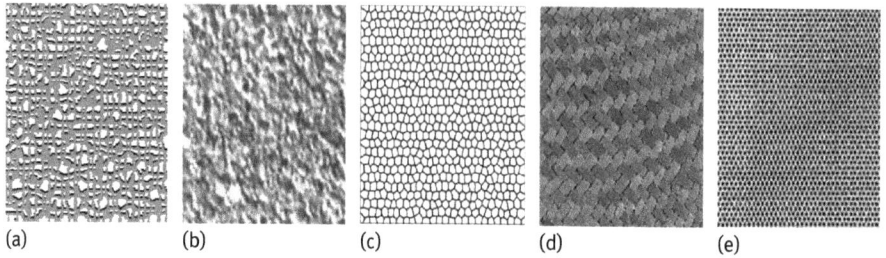

(a)   (b)   (c)   (d)   (e)

Figure 5.4: Five texture images.

Table 5.1: Different descriptor values of images in Figure 5.4.

Descriptor	(a)	(b)	(c)	(d)	(e)
W_U	0.21	5.42 E−5	0.08	0.17 E−3	1.68 E−4
W_E	0.84	4.33	2.23	3.90	4.28
W_M ($k = 1$)	74.66	54.47	101.04	24.30	76.80
$W_H(0.0001)$	4,131.05	60.53	2,820.45	155.04	144.96
$W_H(0.5)$	0.83	0.06	0.58	0.13	0.06
$W_H(3.0)$	0.14	0.04	0.15	0.07	0.03

Many other descriptors can also be defined on the basis of the gray-level co-occurrence matrix. Let

$$c_x(i) = \sum_{j=1}^{N} c_{ij}, \qquad i = 1, 2, \ldots, N \tag{5.11}$$

$$c_y(j) = \sum_{i=1}^{N} c_{ij}, \qquad j = 1, 2, \ldots, N \tag{5.12}$$

$$c_{x+y}(k) = \sum_{i=1}^{N}\sum_{j=1}^{N} c_{ij}, \qquad k = i + j = 2, 3, \ldots, 2N \tag{5.13}$$

$$c_{x-y}(k) = \sum_{i=1}^{N}\sum_{j=1}^{N} c_{ij}, \qquad k = |i - j| = 0, 1, \ldots, N - 1 \tag{5.14}$$

Then, the following 14 descriptors have been defined (Haralick, 1992; Russ 2002).
1. **Uniformity of energy** (angular second momentum)

$$W_1 = \sum_{i=1}^{N}\sum_{j=1}^{N} c_{ij}^2 \tag{5.15}$$

2. Contrast

$$W_2 = \sum_{t=0}^{N-1} t^2 \left\{ \sum_{i=1}^{N} \sum_{j=1}^{N} c_{ij} \right\} \qquad |i - j| = t \tag{5.16}$$

3. Correlation

$$W_3 = \frac{1}{\sigma_x \sigma_y} \left[\sum_{i=1}^{N} \sum_{j=1}^{N} ij c_{ij} - \mu_x \mu_y \right] \tag{5.17}$$

4. **Cluster tendency** (sum of squares variance)

$$W_4 = \sum_{i=1}^{N} \sum_{j=1}^{N} (i - \mu)^2 c_{ij} = \sum_{i=1}^{N} (i - \mu)^2 c_x(i) \tag{5.18}$$

5. Inverse difference moment (Homogeneity)

$$W_5 = \sum_{i=1}^{N} \sum_{j=1}^{N} \frac{1}{1 + (i - j)^2} c_{ij} \tag{5.19}$$

6. Sum average

$$W_6 = \sum_{i=2}^{2N} i c_{x+y}(i) \tag{5.20}$$

7. Sum variance

$$W_7 = \sum_{i=2}^{2N} (i - W_6)^2 c_{x+y}(i) \tag{5.21}$$

8. Sum entropy

$$W_8 = -\sum_{i=2}^{2N} c_{x+y}(i) \log \left[c_{x+y}(i) \right] \tag{5.22}$$

9. Entropy

$$W_9 = -\sum_{i=1}^{N} \sum_{j=1}^{N} c_{ij} \log \left[c_{ij} \right] \tag{5.23}$$

10. Difference variance

$$W_{10} = \sum_{i=2}^{2N} (i - d)^2 c_{x-y}(i) \tag{5.24}$$

where $d = \sum_{i=2}^{2N} i c_{x-y}(i)$

11. Difference entropy

$$W_{11} = -\sum_{i=2}^{N-1} c_{x-y}(i) \log\left[c_{x-y}(i)\right]$$ (5.25)

12. Information measurements of correlation 1

$$W_{12} = \frac{W_9 - E_1}{\max(E_x, E_y)}$$ (5.26)

where $E_1 = -\sum_{i=1}^{N}\sum_{j=1}^{N} c_{ij} \log\left[c_x(i)c_y(j)\right], \quad E_x = -\sum_{i=1}^{N} c_x(i) \log\left[c_x(i)\right], E_y = -\sum_{j=1}^{N} c_y(j)$

$\log\left[c_y(j)\right]$

13. Information measurements of correlation 2

$$W_{13} = \sqrt{1 - \exp\left[-2(E_2 - W_9)\right]}$$ (5.27)

where $E_2 = -\sum_{i=1}^{N}\sum_{j=1}^{N} c_x(i)c_y(j) \log\left[c_x(i)c_y(j)\right]$

14. Maximum correlation coefficient

$$W_{14} = \text{second biggest eigenvalue of } R \quad R(i,j) = \sum_{k=1}^{N} \frac{c_{ik}c_{jk}}{c_x(i)c_y(j)}$$ (5.28)

5.2.3 Law's Texture Energy Measurements

This technique is based on the **texture energy** and uses local masks to measure the amount of variation within a fixed-size window. The image is first convolved with a variety of kernels. If $f(x, y)$ is the input image and $M_1, M_2, \ldots, M_N$ are the kernels (masks), the images $g_n = f^* M_n, n = 1, 2, \ldots, N$ are computed. Then, each convolved image is processed with a nonlinear operator to determine the total textural energy in each pixel's neighborhood. When the neighborhood is $k \times k$, the energy image corresponding to the nth kernel is defined by

$$T_n(x, y) = \frac{1}{k \times k} \sum_{i=-(k-1)/2}^{(k-1)/2} \sum_{j=-(k-1)/2}^{(k-1)/2} |g_n(x + i, y + j)|$$ (5.29)

Associated with each pixel position (x, y), is a textural feature vector $[T_1(x, y)T_2(x, y)\ldots T_N(x, y)]^T$.

The textural energy approach is very much in the spirit of the transform approach, but it uses smaller windows or neighborhood support (applying a discrete orthogonal transform, such as DCT, locally to each pixel's neighborhood is also usable).

The commonly used kernels have supports for 3×3, 5×5, and 7×7 neighborhoods. Their 1-D forms are illustrated in the following formulas (L: level, E: edge, S: shape, W: wave, R: ripple, O: oscillation).

$$
\begin{aligned}
L_3 &= \begin{bmatrix} 1 & 2 & 1 \end{bmatrix} \\
E_3 &= \begin{bmatrix} -1 & 0 & 1 \end{bmatrix} \\
S_3 &= \begin{bmatrix} -1 & 2 & -1 \end{bmatrix}
\end{aligned}
\tag{5.30}
$$

$$
\begin{aligned}
L_5 &= \begin{bmatrix} 1 & 4 & 6 & 4 & 1 \end{bmatrix} \\
E_5 &= \begin{bmatrix} -1 & -2 & 0 & 2 & 1 \end{bmatrix} \\
S_5 &= \begin{bmatrix} -1 & 0 & 2 & 0 & -1 \end{bmatrix} \\
W_5 &= \begin{bmatrix} -1 & 2 & 0 & -2 & 1 \end{bmatrix} \\
R_5 &= \begin{bmatrix} 1 & -4 & 6 & -4 & 1 \end{bmatrix}
\end{aligned}
\tag{5.31}
$$

$$
\begin{aligned}
L_7 &= \begin{bmatrix} 1 & 6 & 15 & 20 & 15 & 6 & 1 \end{bmatrix} \\
E_7 &= \begin{bmatrix} -1 & -4 & -5 & 0 & 5 & 4 & 1 \end{bmatrix} \\
S_7 &= \begin{bmatrix} -1 & -2 & 1 & 4 & 1 & -2 & -1 \end{bmatrix} \\
W_7 &= \begin{bmatrix} -1 & 0 & 3 & 0 & -3 & 0 & 1 \end{bmatrix} \\
R_7 &= \begin{bmatrix} 1 & -2 & -1 & 4 & -1 & -2 & 1 \end{bmatrix} \\
O_7 &= \begin{bmatrix} -1 & 6 & -15 & 20 & -15 & 6 & -1 \end{bmatrix}
\end{aligned}
\tag{5.32}
$$

In many practical cases, a set of nine 5×5 convolution masks are used to compute the texture energy. To obtain these nine masks, the vectors L_5, E_5, S_5, and R_5 are used. L_5 gives a center-weighted local average, E_5 detects edges, S_5 detects spots, and R_5 detects ripples. The 2-D convolution masks are obtained by computing the outer product of each pair of vectors. For example, the mask $E_5^T L_5$ is computed as the product of E_5^T and L_5 as follows:

$$
\begin{bmatrix} -1 \\ -2 \\ 0 \\ 2 \\ 1 \end{bmatrix} \times \begin{bmatrix} 1 & 4 & 6 & 4 & 1 \end{bmatrix} = \begin{bmatrix} -1 & -4 & -6 & -4 & -1 \\ -2 & -8 & -12 & -8 & -2 \\ 0 & 0 & 0 & 0 & 0 \\ 2 & 8 & 12 & 8 & 2 \\ 1 & 4 & 6 & 4 & 1 \end{bmatrix}
\tag{5.33}
$$

The first step is to remove the effect of illumination by moving a small window around the image, and subtracting the local average from each pixel, to produce a preprocessed image, in which the average intensity of each neighborhood is close to zero. The size of the window depends on the class of imagery. After the preprocessing, each of the 16 5×5 masks is applied to the preprocessed image, producing 16 filtered images. Let $F_k(i, j)$ be the result of filtering the image with the kth mask at pixel (i, j). Then, the **texture energy map** E_k for filter k is defined by

$$E_k(r, c) = \sum_{i=-2}^{c+2} \sum_{i=r-2}^{r+2} |F_k(i, j)| \tag{5.34}$$

Each texture energy map is a full image, representing the application of the kth mask to the input image.

Once the 16 energy maps are produced, certain symmetric pairs are combined to produce the nine final maps, replacing each pair with its average. For example, $E_5^T L_5$ measures horizontal edge content, and $L_5^T E_5$ measures vertical edge content. The average of these two maps measures total edge content. The nine resultant energy maps are $L_5^T E_5/E_5^T L_5$, $L_5^T S_5/S_5^T L_5$, $L_5^T R_5/R_5^T L_5$, $E_5^T E_5$, $E_5^T S_5/S_5^T E_5$, $E_5^T R_5/R_5^T E_5$, $S_5^T S_5$, $S_5^T R_5/R_5^T S_5$, $R_5^T R_5$.

5.3 Structural Approaches

The structural model regards the texture elements as repeating patterns and describes such patterns in terms of generating rules. Formally, these rules can be viewed as grammars. This model is best for describing textures where there is much regularity in the placement of elements and the texture is imaged at a high resolution (Ballard, 1982).

5.3.1 Two Basic Components

To characterize a texture, the gray-level primitive properties must be characterized as well as the spatial relationships between them. This implies that texture-structure is really a two-layered structure. The first layer specifies the local properties that manifest themselves in gray-level primitives. The second layer specifies the organization among the gray-level primitives. Structural approaches for texture description are characterized by two components: Texture elements and arrangement rules.

5.3.1.1 Texture Elements
A **texture element** is a connected component of an image, which can be characterized by a group of properties. The properties of texture regions depend on the property and number of texture elements. However, there is no commonly recognized set of texture elements. The simplest texture element is pixel, whose property is gray level. A little more complex texture element is a group of connected pixels with the same gray levels. Such a texture element can be described with the size, orientation, shape, and/or average gray level.

5.3.1.2 Arrangement Rules
To describe texture, after obtaining the texture elements, it is needed to specify the **arrangement rules**. These rules are often defined by certain formal language/grammar.

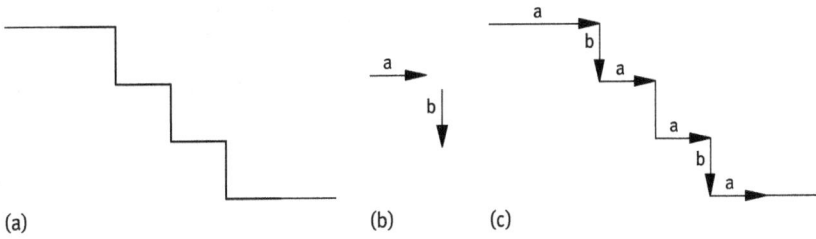

Figure 5.5: Generating the staircase structure using rules.

First, look at a simple example. Consider the staircase pattern shown in Figure 5.5(a). Two primitive elements a and b are shown in Figure 5.5(b) and three rules can be defined.

1. $S \rightarrow aA$ indicates that the variable S, which is also a starting symbol, may be replaced by aA.
2. $A \rightarrow bS$ indicates that the variable A may be replaced by bS.
3. $A \rightarrow b$ indicates that the variable A may be replaced by constant b.

Now, if applying the first rule, it will get aA. If applying the second rule to A, it leads back to the first rule and the procedure can be repeated. However, if applying the third rule to A, the procedure will be terminated. The application of the three rules can generate the structure as shown in Figure 5.5(c).

The above method can be easily extended to describe more complex texture patterns. Suppose there are the following eight rewriting rules (a represents a pattern, b means shifting downwards, and c means shifting to the left).

1. $S \rightarrow aA$ indicates that the variable S, which is also a starting symbol, may be replaced by aA.
2. $S \rightarrow bA$ indicates that the variable S, which is also a starting symbol, may be replaced by bA.
3. $S \rightarrow cA$ indicates that the variable S, which is also a starting symbol, may be replaced by cA.
4. $A \rightarrow aS$ indicates that the variable A may be replaced by aS.
5. $A \rightarrow bS$ indicates that the variable A may be replaced by bS.
6. $A \rightarrow cS$ indicates that the variable A may be replaced by cS.
7. $A \rightarrow c$ indicates that the variable A may be replaced by constant c.
8. $A \rightarrow a$ indicates that the variable A may be replaced by constant a.

Then, different 2-D patterns will be generated.

For example, a is a disk-like pattern, as shown in Figure 5.6(a), then four consecutive applications of rules (1) followed by rule (8) will generate Figure 5.6(b). While the consecutive applications of rules (1), (4), (1), (5), (3), (6), (2), (4), and (8) will generate Figure 5.6(c).

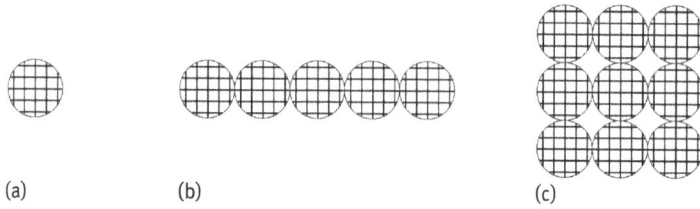

Figure 5.6: Two-dimensional pattern generation.

5.3.2 Typical Structural Methods

Based on the above-explained principles, many structural methods for texture analysis have been developed. Two of them are described in the following.

5.3.2.1 Textures Tessellation

Highly patterned textures tessellate the plane in an ordered way, and thus, the different ways in which this can be done must be understood. In a regular **tessellation**, the polygons surrounding a vertex all have the same number of sides, as shown by Figure 5.7, where Figure 5.7(a) is for triangular, Figure 5.7(b) is for rectangular, and Figure 5.7(c) is for hexagonal (Ballard, 1982).

Semiregular tessellations have two kinds of polygons (differing in the number of sides) surrounding a vertex. Four semiregular tessellations of the plane are shown in Figure 5.8.

These tessellations are conveniently described by listing, in order, the number of sides of the polygons surrounding each vertex. Thus, a hexagonal tessellation is

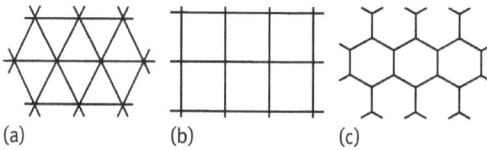

Figure 5.7: Different tessellations of the image plane.

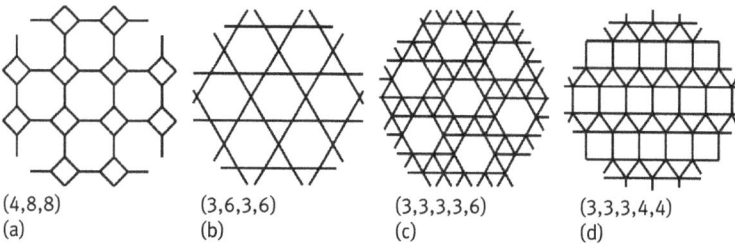

Figure 5.8: Semiregular tessellations.

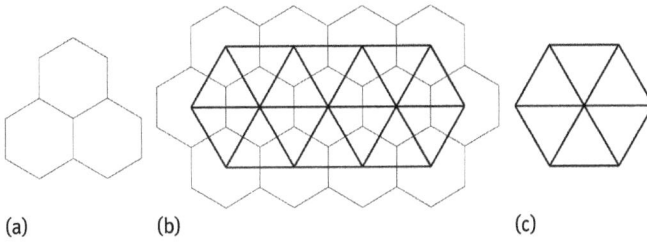

(a)   (b)   (c)

Figure 5.9: The primitive placement tessellation as the dual of the primitive tessellation.

described by (6, 6, 6) and every vertex in the tessellation of Figure 5.9 can be denoted by the list (3, 12, 12). It is important to note that the tessellations of interest are those that describe the placement of primitives rather than the primitives themselves. When the primitives define a tessellation, the tessellation describing the primitive placement will be the dual of this graph as shown in Figure 5.9. Figure 5.9(a, c) shows primitive tessellation and placement tessellation, respectively. Figure 5.9(b) is the result of their combination.

5.3.2.2 Voronoi Polygon

Suppose that a polygon has a set of already-extracted texture elements and that each one can be represented by a meaningful point, such as its centroid (Shapiro 2001). Let S be the set of these points. For any pair of points P and Q in S, the **perpendicular bisector** of the line joining them can be constructed. This perpendicular bisector divides the plane into two half planes, one of which is the set of points that are closer to P and the other of which is the set of points that are closer to Q. Let $H^Q(P)$ be the half plane that is closer to P with respect to the perpendicular bisector of P and Q. This process can be repeated for each point Q in S. The **Voronoi polygon** of P is the polygonal region consisting of all points that are closer to P than to any other point of S and is defined by

$$V(P) \bigcap_{Q \in S, Q \neq P} H^Q(P) \qquad (5.35)$$

Figure 5.10 illustrates the Voronoi polygons for a set of circular texture elements. This pattern produces hexagonal polygons for internal texture elements, of which texture elements at the image boundary have various shapes.

5.3.3 Local Binary Mode

LBP is a texture analysis operator. It is a texture metric defined with the help of local neighborhood. It belongs to the group of estimation method based on point samples, with the advantages of scale invariance, rotation invariance, and low computational complexity.

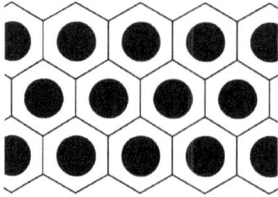

Figure 5.10: The Voronoi tessellation of a set of circular texels.

5.3.3.1 Space LBP

The original LBP operator performs thresholding on the pixels in a 3×3 neighborhood in order, treats the result as a binary number and labels it for the center pixel. An example of a basic LBP operator is shown in Figure 5.11, where the left side is a texture image from which a 3×3 neighborhood is taken. The (count-clockwise) order of the pixels in the neighborhood is denoted by the numbers in parentheses, their values are represented by the following window. The result obtained by using 50 as the threshold value is a binary image. The binary number is 10111001 and the decimal value is 185. The histograms obtained from 256 different labels (for an 8-bit grayscale image) can be further used as the texture descriptors for image regions.

The basic LBP operator can be extended by using neighborhoods of different sizes. Neighborhoods can be circular, and for noninteger coordinate locations, the bilinear interpolation can be used to compute pixel values to eliminate restrictions on the neighborhood radius and the number of pixels in the neighborhood. Let (P, R) denote a rounded pixel neighborhood, where P represents the number of pixels in the neighborhood and R represents the radius of the circle. Figure 5.12 shows several examples of circular neighborhoods with their (P, R).

Another extension to the basic LBP operator is the uniform pattern. The pixels in a neighborhood are considered in a circular order, and if it contains up to two transitions from 0 to 1 or from 1 to 0, then the binary pattern is considered as uniform. The pattern 00000000 (0 transition) and pattern 11111001 (2 transitions) are uniform; while the pattern 10111001 (4 transitions) and pattern 10101010 (7 transitions) are not uniform. They have no obvious texture structure and can be regarded as noise. In the calculation of LBP labels, a separate label is used for each uniform pattern, and for all nonuniform models, the same label is commonly used. In this way, the antinoise ability can be enhanced. For example, when using the neighborhood $(8, R)$, there are

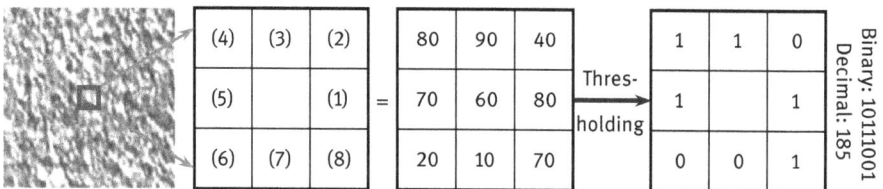

Figure 5.11: Basic LBP operator.

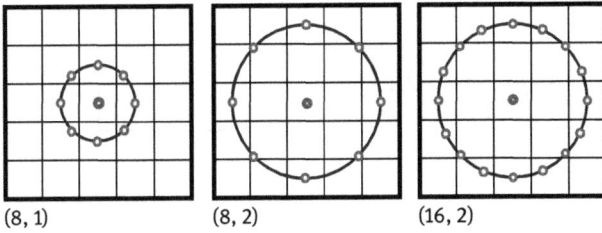

(8, 1)   (8, 2)   (16, 2)

Figure 5.12: A set of neighborhoods corresponding to different (P, R).

256 patterns, 58 of which are uniform, so there are totally 59 labels. To sum up, $\text{LBP}_{P,R}^{(u)}$ can be used to represent such an LBP operator with uniform patterns.

Different local primitives can be obtained according to the labels of LBP, which correspond to different local texture structures. Figure 5.13 shows some (meaningful) examples where a hollow dot represents 1 and a solid dot represents 0.

Once the image $f_L(x, y)$ labeled with the LBP label is calculated, the LBP histogram can be defined as

$$H_i = \sum_{x,y} I\{f_L(x, y) = i\}, \quad i = 0, \ldots, n - 1 \tag{5.36}$$

where n is the number of distinct labels given by the LBP operator, $I(z)$ is defined by

$$I(z) = \begin{cases} 1 & z \to \text{true} \\ 0 & z \to \text{false} \end{cases} \tag{5.37}$$

5.3.3.2 Time-Space LBP

The original LBP operator can be extended to the time-space representation to perform **dynamic texture analysis** (DTA), which is the **volume local binary pattern** (VLBP) operator. It can be used for analyzing the dynamic changes of texture in 3-D(X, Y, T) space, including both motion and appearance. In the 3-D(X, Y, T) space, three sets of planes can be considered: XY, XT, YT. The three classes of LBP labels obtained are XY-LBP, XT-LBP, and YT-LBP, respectively. The first category contains spatial information, the latter two categories both contain time-space information. Three LBP histograms can be obtained from three categories of LBP labels, which

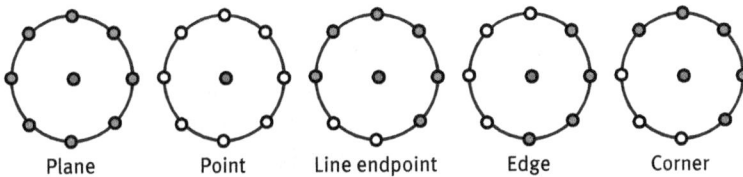

  Plane   Point   Line endpoint   Edge   Corner

Figure 5.13: Local primitives obtained with the LBP labels.

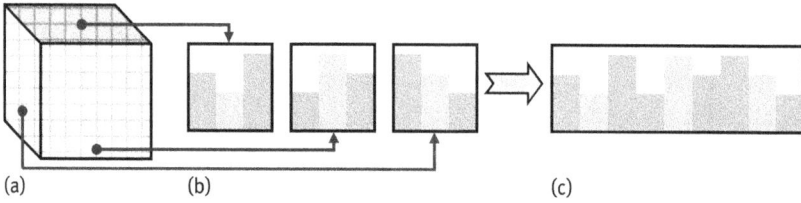

Figure 5.14: Histogram representation of local binary pattern for a 3-D volume.

can also be spanned together into a unified histogram. One example is shown in Figure 5.14. Figure 5.14(a) shows three planes of the dynamic texture. Figure 5.14(b) gives the LBP histogram for each plane. Figure 5.14(c) is the feature histogram after the stitching of three histograms.

Given a dynamic texture of $X \times Y \times T : (x_c \in \{0, \dots, X-1\}, y_c \in \{0, \dots, Y-1\}, t_c \in \{0, \dots, T-1\})$, the dynamic texture histogram can be written as:

$$H_{i,j} = \sum_{x,y,t} I\{f_j(x,y,t) = i\}, i = 0, \cdots, n_j - 1; \quad j = 0, 1, 2 \tag{5.38}$$

where n_j is the number of different labels produced by the LBP operator on the jth plane $(j = 0 : XY, j = 1 : XT$ and $j = 2 : YT)$, and $f_j(x, y, t)$ denotes the LBP code of the upper center pixel (x, y, t) on the jth plane.

For dynamic textures, it is not necessary to set the range of the time axis to be the same as the spatial axis. In other words, the distances between the temporal and spatial sampling points may be different in the XT and YT planes. More generally, the distances between the sample points on the XY, XT, and YT planes may be all different.

When it is necessary to compare dynamic textures that differ both in spatial and temporal scales, the histogram needs to be normalized to obtain a consistent description

$$N_{i,j} = \frac{H_{i,j}}{\sum_{k=0}^{n_j-1} H_{k,j}} \tag{5.39}$$

In this normalized histogram, the description of the dynamic texture can be obtained efficiently based on LBP labels obtained from three different planes. The labels obtained from the XY plane contain information about the appearance, and the labels obtained from the XT and YT planes include the symbiotic statistics of motions in the horizontal and vertical directions. This histogram constitutes a global description for the dynamic texture with spatial and temporal characteristics. Because it has nothing to do with the absolute gray value, and only with the relative relationship between the local gray levels, so it will be stable when the illumination changes. The disadvantage of LBP features does come from the ignoring of the absolute gray level and the no distinction between the relative strength. Therefore, the noise may change the weak relative relation and change its texture.

5.4 Spectral Approaches

Spectral approaches for texture analysis can be based on various spectra. Three of them, the Fourier spectrum, Bessel–Fourier spectrum, and Gabor spectrum are introduced below.

5.4.1 Fourier Spectrum

The following three features of the Fourier spectrum are useful for texture descriptions (Gonzalez, 2002):
1. Prominent peaks in the spectrum give the principal direction of the texture patterns.
2. The location of the peaks in the frequency plane gives the fundamental spatial period of the pattern.
3. Eliminating any periodic component via filtering leaves nonperiodic image elements, which can then be described by statistical techniques.

Expressing the spectrum in polar coordinates can yield a function $S(r, \theta)$, where S is the **spectrum function** and r and θ are the variables in this coordinate system. For each direction θ, $S(r, \theta)$ may be considered a 1-D function $S_\theta(r)$. Similarly, for each frequency r, $S_r(\theta)$ is a 1-D function. Analyzing $S_\theta(r)$ for a fixed value of θ yields the behavior of the spectrum (such as the presence of peaks) along a radial direction from the origin, whereas analyzing $S_r(\theta)$ for a fixed value of r yields the behavior along a circle centered on the origin.

A more global description is obtained by integrating these functions:

$$S(r) = \sum_{\theta=0}^{\pi} S_\theta(r) \tag{5.40}$$

$$S(\theta) = \sum_{r=1}^{R_0} S_r(\theta) \tag{5.41}$$

where R_0 is the radius of a circle centered at the origin.

Figure 5.15 shows two texture regions and the plotted $S(\theta)$, and the orientation of the texture pattern can be quickly determined from their spectrum curves.

If a texture is spatially periodic or directional, its power spectrum will tend to have peaks for the corresponding spatial frequencies. These peaks can form the basis of the features as a pattern recognition discriminator. One way to define features

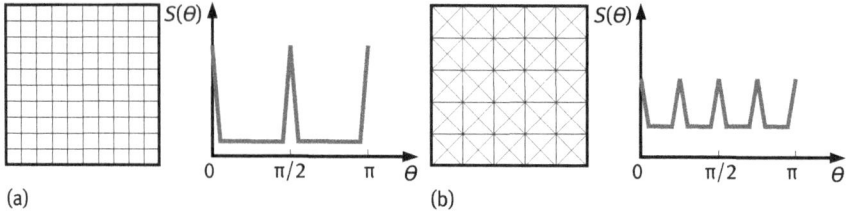

Figure 5.15: Two texture regions and their spectrum curves.

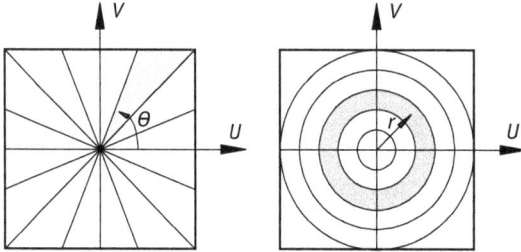

Figure 5.16: Partitioning the Fourier domain into bins.

is to partition the Fourier space into bins. Two kinds of bins, radial and angular (corresponding to the wedge filter and the ring filter, respectively) bins, are commonly used, as shown in Figure 5.16.

Radial features are given by ($|F|^2$ is the Fourier power spectrum)

$$R(r_1, r_2) = \sum\sum |F|^2(u, v) \tag{5.42}$$

where the limits are defined by

$$r_1^2 \leq u^2 + v^2 < r_2^2$$
$$0 \leq u, v < n - 1 \tag{5.43}$$

Radial features are correlated with the texture coarseness. A smooth texture will have high values of $R(r_1, r_2)$ for small radii, whereas a coarse and grainy texture will tend to have relatively higher values for larger radii.

Angular orientation features are given by

$$A(\theta_1, \theta_2) = \sum\sum |F|^2(u, v) \tag{5.44}$$

where the limits are defined by

$$\theta_1 \leq \tan^{-1}(v/u) < \theta_2$$
$$0 < u, \ v \leq n - 1 \tag{5.45}$$

Angular orientation features exploit the sensitivity of the power spectrum to the directionality of the texture. If a texture has many lines or edges in a given direction θ, $|F|^2$ will tend to have values clustered around the direction in the frequency space $\theta + \pi/2$.

5.4.2 Bessel–Fourier Spectrum

Variational calculus has been used to determine the **Bessel–Fourier boundary function** (Beddow, 1997)

$$G(R, \theta) = \sum_{m=0}^{\infty} \sum_{n=0}^{\infty} (A_{m,n} \cos m\theta + B_{m,n} \sin m\theta) J_m \left(Z_{m,n} \frac{R}{R_v} \right) \quad (5.46)$$

where angle θ; $A_{m,n}$, $B_{m,n}$ are the Bessel–Fourier coefficients, J_m is the first kind mth order Bessel function, $Z_{m,n}$ is the zero root of the Bessel function, and R_v is the radius of the field of view.

The following descriptors for texture can be derived from eq. (5.46):

1. Bessel–Fourier coefficients.
2. Moments of the gray-level distribution (gray-level histogram).
3. **Partial rotational symmetry**: The texture is composed of discrete gray levels. An R-fold symmetry operation can be performed by comparing the gray-level value at $G(R, \theta)$ with that at $G(R, \theta + \Delta\theta)$. This provides the index C_{rt}, the partial rotational symmetry index of the texture, as

$$C_{rt} = \frac{\sum_{m=0}^{\infty} \sum_{n=0}^{\infty} (H_{m,n} R^2 \cos m(2\pi/R)) J_m^2 (Z_{m,n})}{\sum_{m=0}^{\infty} \sum_{n=0}^{\infty} (H_{m,n} R^2) J_m^2 (Z_{m,n})} \quad (5.47)$$

where $R = 1, 2, \ldots, H_{m,n} R^2 = A_{m,n} R^2 + B_{m,n} R^2$.

4. **Partial translational symmetry**: A partial translational symmetry is recognized when the gray-level values are compared along the radius. So, $G(R, \theta)$ is compared with $G(R + \Delta R, \theta)$. The partial translational symmetry index of the texture is $0 < C_{tt} < 1$ and is defined as

$$C_{tt} = \frac{\sum_{m=0}^{\infty} \sum_{n=0}^{\infty} H_{m,n}^2 J_m^2 (Z_{m,n}) - [A_{m,n} A_{m-1,n} + B_{m,n} B_{m-1,n}] J_m^2 (Z_{m-1,n}) \frac{\Delta R}{2R_v}}{2 \sum_{m=0}^{\infty} \sum_{n=0}^{\infty} H_{m,n}^2 J_m^2 (Z_{m,n})} \quad (5.48)$$

5. **Coarseness**: The coarseness, F_{crs}, is defined as the difference between the gray-level values of four neighboring points around a specific point (x, y). Analysis has shown that the coarseness is related directly to the partial rotational and translational symmetries as given by

$$F_{crs} = 4 - 2(C_{rt} + C_{tt}) \quad (5.49)$$

6. **Contrast:** In the case of a distribution in which the values of the variables are concentrated near the mean, the distribution is said to have a large kurtosis. The contrast, F_{con}, is defined as the kurtosis by

$$F_{con} = \mu^4/\sigma^4 \qquad (5.50)$$

where μ^4 is the fourth moment of the gray-level distribution about the mean and σ^2 is the variance.

7. **Roughness:** The roughness, F_{rou}, is related to the contrast and coarseness as follows,

$$F_{rou} = F_{crs} + F_{con} \qquad (5.51)$$

8. **Regularity:** The regularity, F_{reg}, is a function of the variation of the textural elements over the image. It is defined by

$$F_{reg} = \sum_{t=1}^{m} C_{rt} + \sum_{t=1}^{n} C_{tt} \qquad (5.52)$$

An image with high rotational and high translational symmetries will have high regularity.

5.4.3 Gabor Spectrum

The Fourier transform uses the entire image to compute the coefficients, while using a set of 2 D Gabor filters can decompose an image into a sequence of frequency bands (**Gabor spectrum**). The impulse response of the complex 2-D Gabor filter is the product of the Gaussian low-pass filter with a complex exponential (Theodoridis, 2003) as given by

$$G(x, y) = L'(x, y) \exp[j(w_x x + w_y y)] \qquad (5.53)$$

where

$$L'(x, y) = \frac{1}{\lambda\sigma^2} L\left(\frac{x'}{\sigma}, \frac{y'}{\sigma}\right) \qquad (5.54)$$

$$L(x, y) = \frac{1}{2\pi} \exp\left(-\frac{x^2 + y^2}{2}\right) \qquad (5.55)$$

and

$$x' = x\cos\theta + y\sin\theta \qquad (5.56)$$
$$y' = -x\sin\theta + y\cos\theta \qquad (5.57)$$

It is thus clear that $L'(x, y)$ is a version of the Gaussian $L(x, y)$ that is spatially scaled and rotated by θ. The parameter σ is the spatial scaling parameter, which controls the width of the filter impulse response, and λ defines the aspect ratio of the filter that determines the directionality of the filter, that is no longer circularly symmetric.

The orientation angle θ is computed by

$$\theta = \arctan(w_y/w_x) \tag{5.58}$$

By varying the free parameters σ, λ, θ, $w = (w_x^2 + w_y^2)^{1/2}$, the filters of arbitrary orientations and bandwidth characteristics are obtained.

Gabor filters come in pairs: One recovers the **symmetric components** in a particular direction, given by

$$G_s(x, y) = \cos(k_x x + k_y y) \exp\left(-\frac{x^2 + y^2}{2\sigma^2}\right) \tag{5.59}$$

and the other recovers the **antisymmetric components**, given by

$$G_a(x, y) = \sin(k_x x + k_y y) \exp\left(-\frac{x^2 + y^2}{2\sigma^2}\right) \tag{5.60}$$

where k_x and k_y give the spatial frequency with the strongest response, and σ is referred to as the scale of the filter.

The magnitude distributions of the above two Gabor filters are drawn in Figure 5.17, where $k_x = k_y$. Figure 5.17(a) is for the symmetric filter, while Figure 5.17(b) is for the antisymmetric filter.

By rotating and scaling the above two Gabor filters, it is possible to obtain a set of filters with different orientations and bandwidths to compute the Gabor spectrum, as shown in Figure 5.18. In Figure 5.18, each ellipse covers the range determined by the half-peak value of the filter. Along the circle, the orientation angles of two adjacent

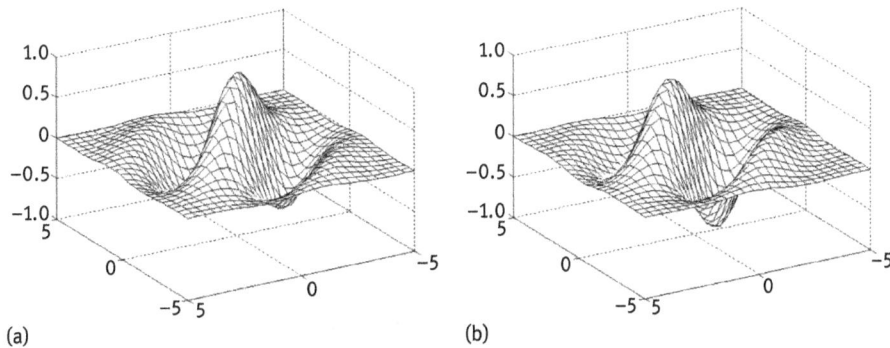

(a)   (b)

Figure 5.17: The magnitude distributions of symmetric and antisymmetric filters.

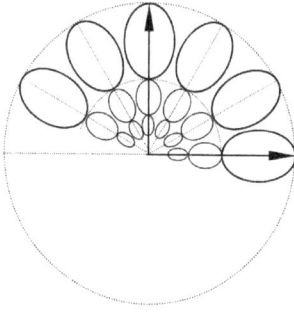

Figure 5.18: A set of filters obtained by rotating and scaling the two Gabor filters.

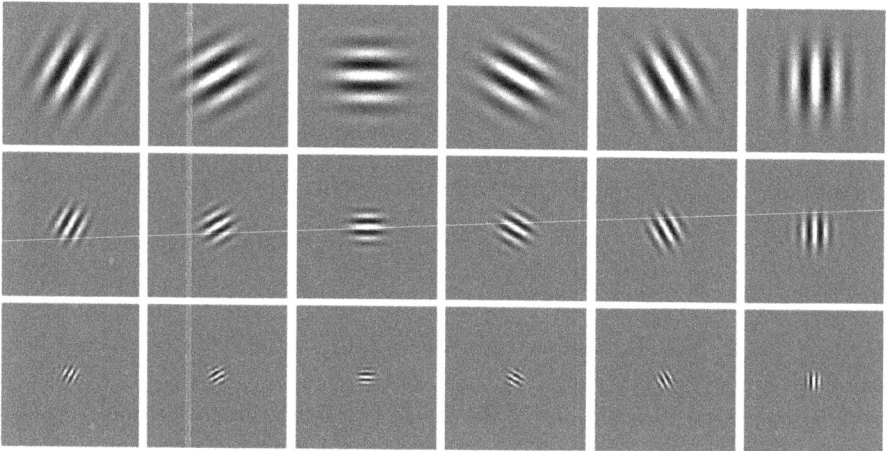

Figure 5.19: A set of filters obtained by rotating and scaling the symmetric filter.

filters are different by 30°, and along the radius, the scales of two adjacent filters are different by 2 (their ranges of the half-peak values are touched).

The gray-level images representing the above two filters are shown in Figures 5.19 and 5.20, respectively. Figure 5.19 is obtained by rotating and scaling the symmetric filter. Figure 5.20 is obtained by rotating and scaling the antisymmetric filter.

5.5 Texture Segmentation

Texture segmentation is used to segment an image into regions, where each region has a homogeneous texture and each pair of adjacent regions is differently textured (Haralick, 1992).

One example is shown in Figure 5.21. In the left image, two letters can be read because they have a different gray-level value compared to that of the background. In the right image, two letters can still be read because they have a different texture compared to that of the background, though the average gray-level values of the two parts are the same.

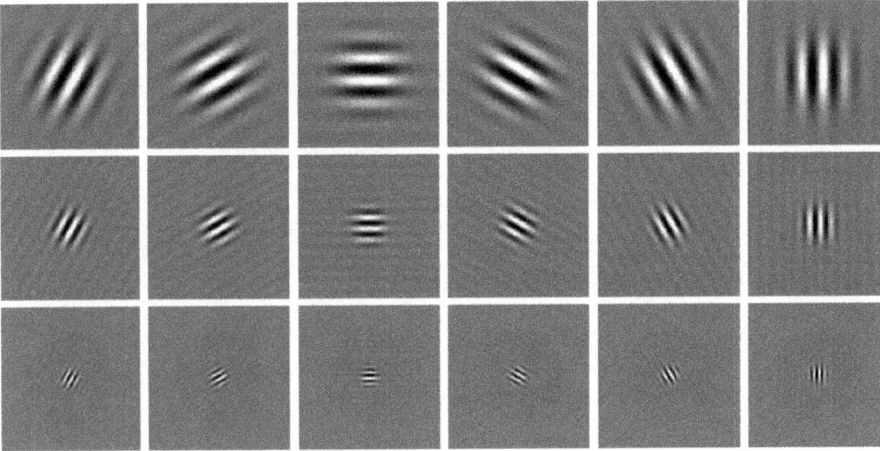

Figure 5.20: A set of filters obtained by rotating and scaling the antisymmetric filter.

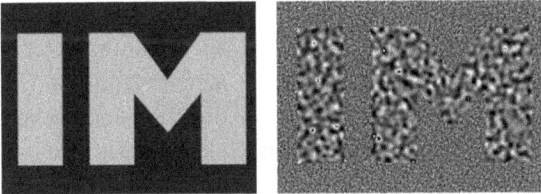

Figure 5.21: Foreground and background separation with different properties.

Any texture measurement that provides a value or a vector of values at each pixel, describing the texture in a neighborhood of that pixel, can be used to segment the image into regions of similar textures.

Techniques for texture segmentation can be divided into supervised ones and unsupervised ones.

5.5.1 Supervised Texture Segmentation

Based on the above classification technique, an algorithm for texture segmentation can be developed. It consists of two steps, namely the presegmentation step and the postsegmentation step. These two steps correspond to the two stages of classification, which are the training and the classification.

5.5.1.1 Presegmentation

The presegmentation step makes a preliminary clustering based on extracted features. The features used are the texture energy, which are computed from a $(2n + 1) \times (2n + 1)$ window in the original image $f(x, y)$ by

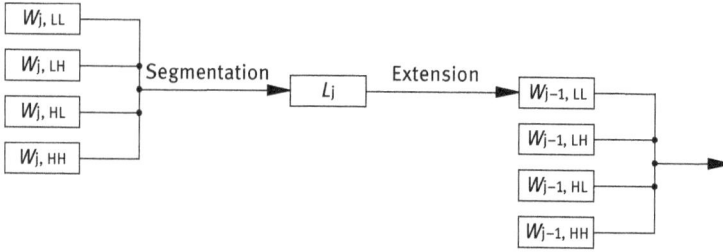

Figure 5.22: The layer structure used in presegmentation.

$$e(x, y) = \frac{1}{(2n + 1)^2} \sum_{k=-n}^{n} \sum_{l=-n}^{n} |f(k, l) - m(x, y)| \tag{5.61}$$

where $m(x, y)$ are the average of $f(x, y)$ in the window.

Considering the multiresolution structure of the wavelet decomposition, the **pre-segmentation** step takes the reverse direction as the wavelet decomposition. It starts from a high level and proceeds to the low level, as shown by Figure 5.22. By combining the *LL*, *HL*, *LH*, and *HH* channels (Section 4.6) in a higher level, a 4-D feature vector is constructed, and the image in this level is classified using the K-means algorithm. The segmented image is extended (enlarged) both in the horizontal direction and the vertical direction by 2 and is used for a lower level segmentation. This process is repeated until the finest image resolution is obtained.

5.5.1.2 Postsegmentation

During the presegmentation, sequences of segmented images in different scales have been obtained. Extending all small images to the size of the biggest image $L_1(x, y)$, which has the size of $M/2$ by $M/2$, the feature vector space can be represented by

$$E(i, j) = [E_1(i, j), E_2(i, j), \cdots, E_P(i, j)]^T \tag{5.62}$$

where P is the total number of channels.

Suppose the total number of texture classes in the image is Q. For a particular texture q, its P-D feature vector is

$$F_q(i, j) = [F_{q1}(i, j), E_{q2}(i, j), \cdots, E_{qP}(i, j)]^T \tag{5.63}$$

where $F_{qP}(i, j) = E_q(i, j)$ and $\{(i, j)\} = \{(x, y)|L_1(x, y) = q; x, y = 1, 2, \ldots, M/2\}$. Suppose the number of pixels satisfying $L_1(x, y) = q$ is U, the variance vector of $F_q(i, j)$ is $s_q^2 = [s_{q1}^2, s_{q2}^2, \ldots, s_{qP}^2]^T$, then

$$S_{qp}^2 = \frac{1}{U} \sum_{i,j} \left\{ F_{qp}(i, j) - \bar{F}_{qp} \right\}^2 \tag{5.64}$$

where $\bar{F}_{qp}$ is the average of $F_q(i, j)$.

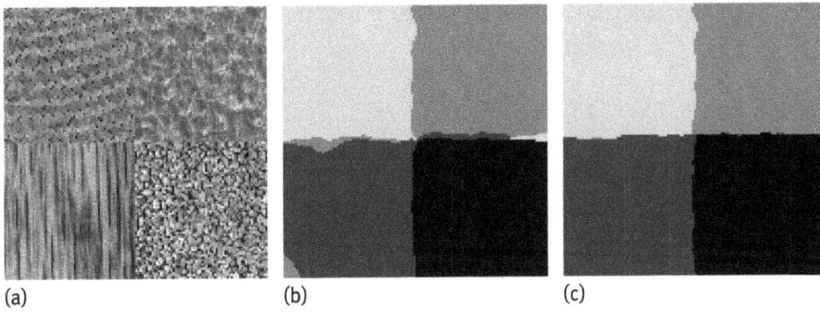

(a)　　　　　　　　　(b)　　　　　　　　　(c)

Figure 5.23: The segmentation of four types of Brodatz textures.

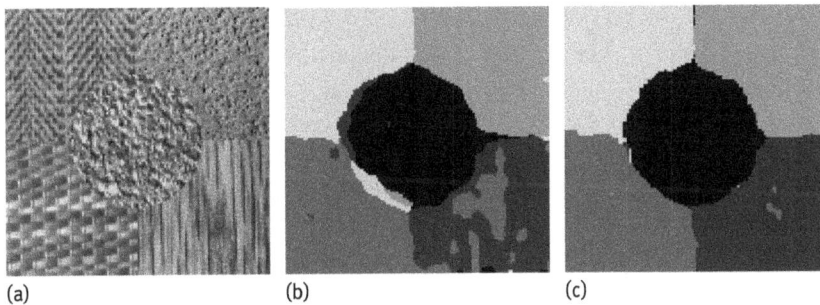

(a)　　　　　　　　　(b)　　　　　　　　　(c)

Figure 5.24: The segmentation of five types of Brodatz textures.

As in the classification, using the variance to weight the features and using the minimum classifier to classify the feature vector space can finally provide the segmented images. Two examples are shown in Figures 5.23 and 5.24 for four types and five types of Brodatz textures, respectively. In these two figures, both figures (a) are the original images, both figure (b) are the results of the presegmentation step, and both figures (c) are the results of the **postsegmentation** step.

5.5.2 Unsupervised Texture Segmentation

In real applications, the number of texture types often does not know *a priori*. In these cases, **unsupervised texture segmentation** is required (*e. g.*, see Khotanzad, 1989; Jain, 1991; Chen, 1995; Panda, 1997). One of the key problems is to determine the number of texture types. In the following, an unsupervised segmentation technique, which uses the information provided by wavelet packet transform to combine the determination of the number of texture types and the texture segmentation process, is given.

5.5.2.1 Feature Extraction and Rough Segmentation
The feature extraction process follows the previous procedure. In **rough segmentation**, the histogram of each feature image is computed, which has a Gaussian-like

distribution. Suppose that the threshold value for segmenting regions R_i and R_{i+1} is T_i, $i = 1, 2, \ldots, N - 1$, then T_i should satisfy

$$\frac{1}{\sqrt{2\pi}\sigma_i} \exp\left\{\frac{(T_i - \mu_i)^2}{2\sigma_i^2}\right\} = \frac{1}{\sqrt{2\pi}\sigma_{i+1}} \exp\left\{\frac{(T_i - \mu_{i+1})^2}{2\sigma_{i+1}^2}\right\} \tag{5.65}$$

where μ_i is the gray-level mean of the region and σ_i is the gray-level variance of the region. Solving eq. (5.65) yields

$$T_i = \frac{\mu_i\sigma_{i+1}^2 + \mu_{i+1}\sigma_i^2}{\sigma_{i+1}^2 - \sigma_i^2} \pm \frac{\sigma_i^2\sigma_{i+1}^2\sqrt{(\mu_i - \mu_{i+1})^2 + 2(\sigma_i^2 - \sigma_{i+1}^2)\ln(\sigma_{i+1}/\sigma_i)}}{\sigma_{i+1}^2 - \sigma_i^2} \tag{5.66}$$

5.5.2.2 Fusion of Segmentation Results

In the process of a rough segmentation, every channel image, obtained by wavelet packet decomposition, is segmented into several regions. Now, the fusion of these images (with a quad-tree structure) is required, which consists of three levels.

1. **Subchannel fusion**: This is for channels belonging to the same father node.
2. **Intralevel fusion**: This is for channels belonging to the same level of the quad-tree.
3. **Interlevel fusion**: This is for channels belonging to different levels of the quad-tree.

Among the above three levels of fusion, the latter fusion should be based on the former fusion. Consider the subchannel fusion. Suppose that the two roughly segmented images are $L_1(x, y)$ and $L_2(x, y)$, in which $L_1(x, y)$ has been classified into N_1 classes and $L_2(x, y)$ has been classified into N_2 classes, then the fused label image $L_{1,2}(x, y)$ is

$$L_{1,2}(x, y) = \max(N_1, N_2) \times L_1(x, y) + L_2(x, y) \tag{5.67}$$

in which the maximum number of texture types is $(N_1 \times N_2)$.

The processes for intralevel fusion and interlevel fusion are similar. For interlevel fusion, those fusing regions should be resized to the same size.

5.5.2.3 Fine Segmentation

Random noise and/or edge effects often cause uncertainty for some pixels, so a fine-tune process is necessary. Corresponding to the three levels in the fusion process of segmentation results, the **fine segmentation** process also includes three levels. For fine segmentation in the channel level, if the fusion result is $L(x, y)$, $L(x, y) = 1, 2, \ldots, N$, the feature values for four channels are $F_{LL}(x, y)$, $F_{LH}(x, y)$, $F_{HL}(x, y)$, and $F_{HH}(x, y)$, respectively, and the feature vector space is

$$\boldsymbol{F}(x, y) = [F_{LL}(x, y)\, F_{LH}(x, y)\, F_{HL}(x, y)\, F_{HH}(x, y)]^{\mathrm{T}} \tag{5.68}$$

For a given texture class i, $i = 1, 2, \ldots, N$, its center of cluster μ_i is given by

$$\mu_i = \frac{1}{\#(L(x, y) = i)} \sum_{L(x,y)=i} F(x, y) \tag{5.69}$$

where $\#(L(x, y) = i)$ represents the area of the ith texture region. For any uncertainty pixel (x_0, y_0), the Euclidean distance between its feature vector and each center of cluster will be computed, and the relabeling can be performed according to the following equation

$$L(x_0, y_0) = i \quad \text{if} \quad d_i = \min_{j=1}^{N}(d_j) \tag{5.70}$$

The processes of fine segmentation in intralevels and interlevels are similar as for channel levels. However, the corresponding feature spaces are different. For the fine segmentation in the intralevel, suppose that the wavelet packet decomposition is performed in the Jth level, then the feature space is composed of 4^{J-1} feature vectors similar to the form of eq. (5.68) and has a dimension of 4^J

$$F(x, y) = \left[F_{LL}^j(x, y), F_{LH}^j(x, y), F_{HL}^j(x, y), F_{HH}^j(x, y) \right]^T_{j=1, 2, \cdots, 4^{J-1}} \tag{5.71}$$

For fine segmentation in the interlevel, suppose that the segmentation is performed in the first K levels of wavelet packet decomposition, then the feature space is composed of K feature vectors similar to the form of eq. (5.71) and has a dimension of $\sum_{k=1}^{K} 4^k$

$$F(x, y) = \left\{ \left[F_{LL}^j(x, y), F_{LH}^j(x, y), F_{HL}^j(x, y), F_{HH}^j(x, y) \right]^T_{j=1,2,\cdots,4^{k-1}} \right\}_{k=1,2,\cdots,K} \tag{5.72}$$

5.5.2.4 Diagram and Results of Segmentation

The diagram of the whole segmentation procedure is shown in Figure 5.25. The process starts from the first-level wavelet packet decomposition to obtain the feature images for four channels, segments them roughly, performs channel fusion and fine segmentation, and obtains the primary segmentation results. Then, the process continues to decompose the four channels obtained from the first-level wavelet packet decomposition, which is equivalent to performing the second level wavelet packet decomposition to obtain the feature images for 16 channels. For four channels in the second level that are decomposed from the same channels in the first level, the above process for the first level is repeated four times. Once the four segmentation results are obtained, the channel fusion and fine segmentation are performed to get the segmentation results for the second level. Now, the process for inter-level fusion and fine segmentation is carried on for the results from two levels. The above process is iterated until the condition $N_{12..J} \leq N_{12..J-1}$ is satisfied. At this time, the number of texture types in the original image is obtained,

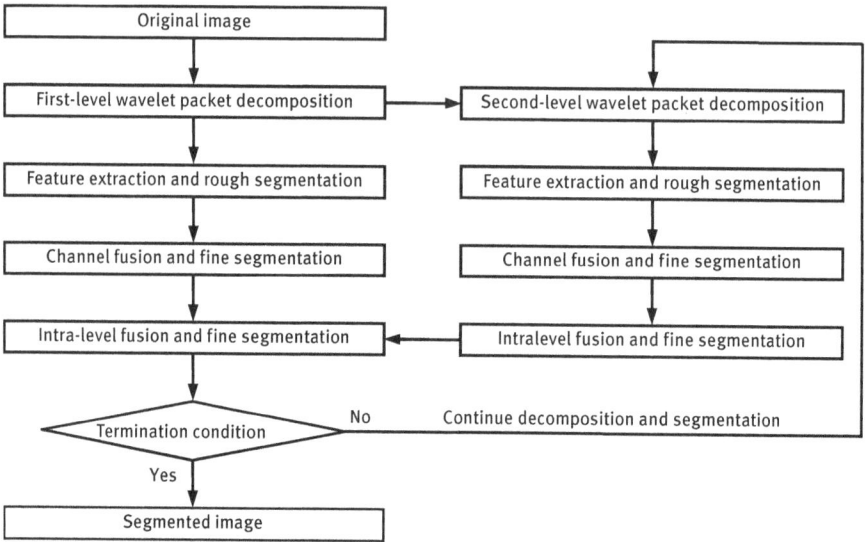

Figure 5.25: Diagram of unsupervised texture segmentation.

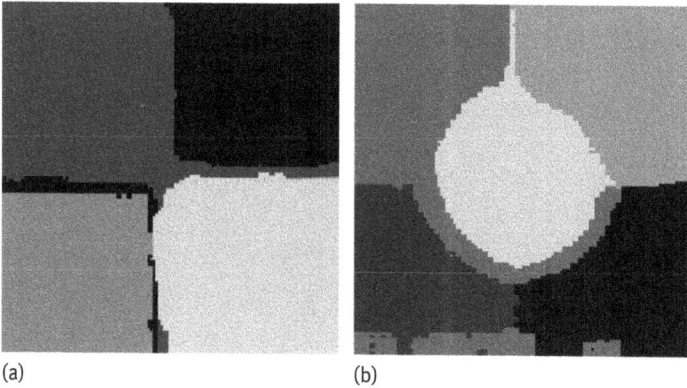

(a)   (b)

Figure 5.26: Results of unsupervised texture segmentation.

and the segmentation process is stopped. Otherwise, the above process will be continued.

The segmentation results obtained by using the above procedure for Figures 5.23(a) and 5.24(a) are shown in Figure 5.26(a, b), respectively. Compared to the results shown in Figures 5.23 and 5.24 that have used supervised segmentation techniques, the results shown in Figure 5.26 are worse, but the requirements for *a priori* knowledge are less.

5.5.3 Texture Classification Based on Wavelet Transformation

Image segmentation can be considered as a pixel classification problem. The following technique segments the texture image based on the wavelet features of image (Wu, 1999). It has several stages.

5.5.3.1 Feature Extraction and Weighting

The goal of feature extraction is to obtain a set of texture measurements that can be used to discriminate among different textures. The image has first undergone wavelet transform, which decomposes the image into a number of channel images. The l_1-norm for one channel image is given by

$$e = \frac{1}{M^2} \sum_{m,n=1}^{M} |w(m, n)| \tag{5.73}$$

where the channel is of dimension M by M, and m and n are the rows and columns of the channel and w is a wavelet coefficient with the channel. For a K-level wavelet composition, the dimension of the feature vector is $3K + 1$.

Generally, an extracted feature is a random variable with a certain degree of dispersion. The degree of dispersion of one feature can be used to measure the accuracy of describing the corresponding texture, and this degree is inversely proportional to the accuracy of the description. If the accuracy of the description is related to the effect on the texture classifications, the performance of the classification can be improved. Because the degree of the dispersion of a random variable can be represented by its variance, such an adjustment of the effect can be implemented by weighting the feature according to the variance.

Suppose that N is the sample number of one texture in the training set, $x_i = [x_{i,1}, x_{i,2}, \ldots, x_{i,P}]^T$, where $i = 1, 2, \ldots, N$ is its feature vector, $s_i^2 = [s_1^2, s_2^2, \ldots, s_P^2]^T$ is the variance vector, and P is the dimension size of the vector, then

$$s_p^2 = \frac{1}{N-1} \sum_{i=1}^{N} (x_{i,p} - \bar{x}_p)^2 \qquad p = 1, 2, \ldots, P \tag{5.74}$$

where $\bar{x}_p$ is the mean value of the feature $x_{i,p}$, $i = 1, 2, \ldots, N$, which is given by

$$\bar{x}_p = \frac{1}{N} \sum_{i=1}^{N} x_{i,p} \tag{5.75}$$

Weighting $x_{i,p}$ can get the weighted feature $x_{i,p}{}^W$, which is given by

$$x_{i,p}^W = \frac{x_{i,p}}{s_p} \qquad i = 1, 2, \ldots, N \tag{5.76}$$

Thus, the weighted feature vector x_i^W is obtained.

It can be seen from eq. (5.74) that the feature value with a relatively small degree of dispersion (s_P^2) is increased, which means that its effect on classification is enhanced. On the other hand, the feature value with a relatively large degree of dispersion is decreased, which means that its effect on classification is weakened.

5.5.3.2 Texture Classification

A simple minimum-distance classifier is used for **texture classification**. Assume that the number of textures is represented by Q.

Training Stage

(i) Given N samples obtained from the texture m, decompose each sample with the standard pyramid-structured wavelet transform using Daubechies four-tap wavelet filter coefficients, calculate the l_1-norm of each channel of wavelet decomposition, and then obtain the feature vector $x_{q,i}$, where $i = 1, 2, \ldots, N$.

(ii) Calculate the standard deviation vector s_q of $x_{q,i}$, weight $x_{q,i}$ by s_q, and then obtain the weighted feature vector $x_{q,i}^W$.
Calculate the clustering center vector X_q of the texture q as

$$X_q = \frac{1}{N} \sum_{i=1}^{N} x_{q,i}^W \tag{5.77}$$

(iii) Repeat the process for all Q textures.

Classification Stage

(i) Decompose an unknown texture with the standard pyramid-structured wavelet transform using Daubechies four-tap wavelet filter coefficients, and calculate the l_1-norm of each channel of wavelet decomposition to obtain the feature vector x.

(ii) For texture m in the training database, weight x by s_q, and obtain the weighted feature vector x^W.

(iii) Compute the Euclidean distance D_q between x^W and the clustering center X_q.

(iv) Repeat steps (ii) and (iii) for all Q textures.

(v) Assign the unknown texture to texture T if

$$T = \arg \left[\min_{q=1}^{Q} (D_q) \right] \tag{5.78}$$

5.6 Problems and Questions

5-1 Provide a list of the different properties used in texture description. Under what circumstances are they used?

5-2 For the image shown in Example 5.2, compute matrices $P_{(0,1)}$, $P_{(1,0)}$, and $P_{(-1,1)}$.

5-3* Suppose the space resolution of a map for an object is 512×512 and the gray-level resolution is 256. When the object is intact, the average value of the gray levels for the whole image is 100. When the object is broken, the average value of the gray levels for some blob regions will be different at 50 from the average value of the gray levels for the whole image. If the area of such a region is larger than 400, the object will be discarded. Solve this problem by texture analysis methods.

5-4 Suppose that the gray level value at the top-left corner of a 5×5 image is 0. Define the position operators W as one pixel to the right and two pixels to the right, respectively. Obtain the co-occurrence matrices for these two cases.

5-5* Consider a chessboard image formed by $m \times n$ black and white blocks. Obtain the position operators that can produce diagonal matrices as co-occurrence matrices.

5-6 Implement the algorithm for computing co-occurrence matrices. Find several images, and compute their matrices $P_{(0,1)}$, $P_{(1,0)}$, and $P_{(-1,1)}$.

5-7 Based on the results obtained in Problem 5-4, compute the values of 14 texture descriptors given in Section 5.2. Which descriptor has the biggest discrimination power?

5-8 Given an image with circular texture primitives as shown in Figure Problem 5.8, use structural approaches to describe the texture patterns. The model and re-writing rules can be self-defined. Draw its Voronoi grid.

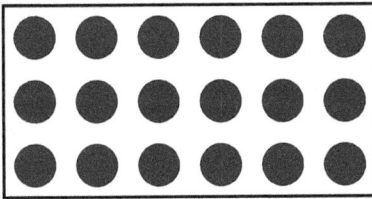

Figure Problem 5.8

5-9 Select some images, and compute their coarseness, contrast, roughness, and regularity. Do the results correspond to the intuitive feeling?

5-10 What is the particularity of texture segmentation versus normal image segmentation?

5-11 Implement a supervised texture segmentation algorithm, and test it with some texture images. What is the influence of the number of types on texture classes and on the segmentation results?

5-12 Implement an unsupervised texture segmentation algorithm, and test it with some texture images. Check the result to see if the number of segmented regions in the image is correct.

5.7 Further Reading

1. **Concepts and Classification**
   - Texture analysis has been discussed in many books about image processing and analysis (Jähne, 1997; Russ, 2002), computer vision (Forsyth, 2003), and content-based image retrieval (Zhang, 2003b).
   - Techniques for shapes from texture can be found in Forsyth (2012).
2. **Statistical Approaches**
   - Besides the gray-level co-occurrence matrix, a gray-level-gradient co-occurrence matrix can also be computed, based on which some other texture descriptors can be defined (Zhang, 2002b).
   - Texture images can be generated from the parameters obtained by using statistical methods for random processes (Russ, 2002).
3. **Structural Approaches**
   - Structural approaches are suitable for describing regular textures. The fractal concept is based on the regularity in different scales. For texture descriptions using a fractal model, please refer to Keller (1989).
4. **Spectral Approaches**
   - Two texture descriptors proposed by international standard MPEG-7, the homogeneous texture descriptor and the texture browsing descriptor, are based on spectrum properties (Zhang, 2003b). A comparison and evaluation of these two texture descriptors and another MPEG-7 texture descriptor, edge histogram descriptor, can be found in Xu (2006).
   - One example of using the spectral coefficients for texture properties in image retrieval can be found in Huang (2003).
   - Further discussions and image examples on the Gabor spectrum can be found in Forsyth (2003).
5. **Texture Segmentation**
   - Texture segmentation can be based on fractal models (Chaudhuri, 1995).
   - The classification of texture images in the Album "Texture" (Brodatz, 1966) has become a common test for checking the performance of texture classification.
   - One example of using wavelet transform for texture classifications can be found in Wu (1999).
   - One particular method for segmenting bi-texture regions using the Gabor filter can be found in Zhang (2001c).
   - More information on texture categorization and generation methods can be found in Rao (1990). Some other methods can be found in Forsyth (2003, 2012).

6 Shape Analysis

Shapes play an important role in describing the objects in images. Shape analysis is an essential branch of image analysis. The analysis of a shape should be based on shape properties and various available techniques. On the one hand, a property can be described by several techniques/descriptors. On the other hand, a technique can describe several properties of a shape.

The sections of this chapter are arranged as follows:

Section 6.1 overviews first the description and definition of shape and then summarizes the main research works for shape.

Section 6.2 describes a classification scheme for the shape of a plane object. The concept and property of each class are further explained to provide an overall understanding of the shape description.

Section 6.3 focuses on the description of the same shape features with different theoretical techniques. In particular, the methods for describing the compactness of shape with the ratio of appearance ratio, shape factor, spherical shape, circularity and eccentricity, and the methods for describing the shape complexity with simple descriptors, saturation and histogram of fuzzy graphs are presented.

Section 6.4 discusses the methods of describing different shape characteristics based on the same technique. The approaches with shape number comparison and region marker based on polygon representation, and the descriptor based on curvature calculation are introduced.

Section 6.5 focuses on the wavelet boundary descriptors that are formed by the coefficients of the wavelet transform of the object boundary. Its properties are provided and compared with Fourier boundary descriptors.

Section 6.6 describes the concept of fractal and the calculation method of fractal dimension (box counting method). Fractal is a mathematical tool suitable for texture and shape analysis, whose dimension is closely related to the scale of representation.

6.1 Definitions and Tasks

What is a shape, or what is the shape of an entity in the real world, or what is the shape of a region in an image all seem to be simple questions but are inherently difficult to answer. Many considerations have been taken into account, and various opinions exist. Some of them are listed in the following.

The precise meaning of the word **shape**, as well as its synonym **form**, cannot be easily formalized or translated into a mathematical concept (Costa, 2001). A shape is a word that is promptly understood by many, but which can be hardly defined by any.

DOI 10.1515/9783110524123-006

For example, some people say, "I do not know what a shape precisely is, but I know it once I see one."

A shape is not something that human languages are well equipped to deal with Russ (2002). Few adjectives could describe shapes, even in an approximate way (*e. g.,* rough vs. smooth, or fat vs. skinny). In most conversational discussion of shapes, it is common to use a prototypical object instead ("shaped like a ... "). Finding numerical descriptors of shapes is difficult because the correspondence between them and our everyday experience is weak, and the parameters all have a "made-up" character.

There is an important distinction between an **object shape** and an **image shape** (Gauch, 1992). An object shape traditionally involves all aspects of an object except its position, orientation, and size. For example, a triangle is still "triangle shaped" no matter how it is positioned, oriented, or scaled. Extending the notion of shape to images requires invariance to similar transformations in the intensity dimension. Thus, an image shape involves all aspects of an image except its position, orientation, size, mean gray-level value, and intensity-scale factor. For example, an intensity ramp (which is bright on one side and gradually becomes dark on the other side) is said to be "ramp shaped," which is independent of the maximum and minimum intensities in the grayscale image, as long as the intensity varies linearly. For more complex images, shape involves the spatial and intensity branching and bending of structures in images and their geometric relationships.

6.1.1 Definition of Shape

There are many **definitions of shape** in dictionaries. Four such definitions are given as follows (Beddow, 1997):
1. A shape is the external appearance as determined by outlines or contours.
2. A shape is that having a form or a figure.
3. A shape is a pattern to be followed.
4. A shape is the quality of a material object (or geometrical figure) that depends upon constant relations of position and proportionate distance among all the points composing its outline or its external surface.

From the application point of view, the definition of a shape should tell people how to measure the shape. The following points should be considered:
1. A shape is a property of the surface or the outline of the object.
2. An object's surface or outline consists of a set of points.
3. To measure a shape, one has to sample the surface points.
4. The relationship between the surface points is the pattern of the points.

The shape of an object can be defined as the pattern formed by the points on the boundary of this object. According to this **operable definition**, the analysis of a shape consists of four steps:

1. Determine the points on the surface or on the outline of the object.
2. Sample these points.
3. Determine the pattern formed by these sample points.
4. Analyze the above pattern.

6.1.2 Shape Analysis Tasks

Shape analysis is generally carried out by starting from an intermediate representation that typically involves the segmented image (where the object shape has been located) and/or special shape descriptors (Costa, 1999).

Computational shape analysis involves several important tasks, from image acquisition to shape classification, which can be broadly divided into three classes (Costa, 2001).

6.1.2.1 Shape Preprocessing

Shape preprocessing involves acquiring and storing an image of the shape and separating the object of interest from other nonimportant image structures. Furthermore, digital images are usually corrupted by noise and other undesirable effects (such as occlusion and distortions). Therefore, techniques for reducing and/or removing these effects are required.

6.1.2.2 Shape Representations and Descriptions

Once the shape of interest has been acquired and processed, a set of techniques can be applied in order to extract information from the shape, so that it can be analyzed. Such information is normally extracted by applying suitable representation and description techniques, which allow both the representation of the shape in a more appropriate manner (with respect to a specific task) and the extraction of measurements that are used by classification schemes.

6.1.2.3 Shape Classifications

There are two particularly important aspects related to shape classification. The first is the problem of, given an input shape, deciding whether it belongs to some specific predefined class. This can also be thought of as a shape recognition problem, which is usually known as **supervised classification**. The second equally important aspect of shape classification is how to define or identify the involved classes in a population of previously unclassified shapes. This represents a difficult task, and problems for acquiring expert knowledge are usually involved. The latter situation is known as **unsupervised classification** or **clustering**. Both supervised and unsupervised classifications involve comparing shapes, that is, deciding how similar two shapes are, which is performed, in many situations, by **matching** some important corresponding points in the shapes (typically landmarks).

6.2 Different Classes of 2-D Shapes

There are various shapes existing. Some of them are 2-D shapes, while others are 3-D shapes.

6.2.1 Classification of Shapes

Figure 6.1 summarizes a possible classification of planar shapes (Costa, 2001). The shapes are first divided as thin and thick (boundary and region, see below) ones. The former is further subdivided into shapes involving a **single-parametric curves** or **composed parametric curves**; the latter is characterized by those shapes containing several merged parametric curves (see below). The single curve shapes can also be classified as being open or closed, smooth or not (a smooth curve is such that all its derivatives exist), Jordan or not (a Jordan curve is such that it never intersects itself, see below), and regular or not (see below). Observe that thick shapes can be almost completely understood in terms of their thin versions, and composed parametric curves can be addressed in terms of single parametric curves.

6.2.2 Further Discussions

Some additional explanations are given in the following.

6.2.2.1 Thick and Thin Shapes

Generic shapes can be grouped into two main classes: those including or not including filled regions, which are henceforth called **thick shapes** and **thin shapes**, respectively. They correspond to region and boundary representations. Note the trace (trajectory if considering that a curve is formed by the moving of a particle) defined by

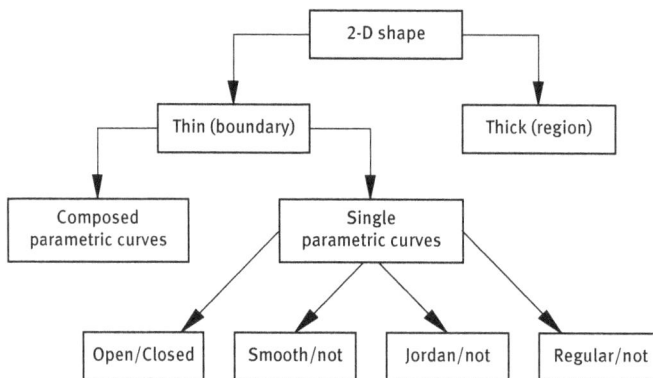

Figure 6.1: A possible classification of planar shapes.

any connected parametric curve can also be verified to be a thin shape. More formally, a thin shape can be defined as a connected set that is equivalent to its boundary, which is never verified for a thick shape. Consequently, any thin shape has an infinitesimal width (it is extremely thin).

It is worth noting that even the 2-D **object silhouette** often conveys enough information to allow recognition of the original object. This fact indicates that the 2-D shape analysis methods can often be applied for the analysis of 3-D objects. While there are many approaches for obtaining the full 3-D representation of objects in computer vision, such as by reconstruction (from stereo, motion, shading, etc.) or by using special devices (*e. g.*, 3-D scanners), dealing with 3-D models still is computationally expensive. In many cases, this is even prohibited. Hence, it is important to use 2-D approaches for such situations. Of course, there are several objects, such as characters, which are defined in terms of 2-D shapes and should therefore be represented, characterized, and processed as such.

In a more general situation, 2-D shapes are often the **archetypes** of the objects belonging to the same pattern class. Some examples are given in Figure 6.2. In spite of the lack of additional important pictorial information, such as color, texture, depth, and motion, the objects represented by each of the silhouettes in this image can be promptly recognized. Some of these 2-D shapes are abstractions of complex 3-D objects, which are represented by simple connected sets of black points on the plane.

6.2.2.2 Parametric Curve
A parametric curve can be understood in terms of the evolution of a point (or a very small particle) as it moves along a 2-D space. Mathematically, the position of the particle on the plane can be expressed in terms of the position vector $p(t)$ = $[x(t)\ y(t)]$, where t is real and is called the parameter of the curve. For example, $p(t)$ = $[cos(t)\ sin(t)]$ specifies a trajectory of the unitary radius circle for $0 < t < 2\pi$. In this interval, the curve starts at $t = 0$, defining a reference along its trajectory, and the parameter t corresponds to the angles with respect to the x-axis.

Consider now the situation depicted in Figure 6.3, which indicates the particle position with a given parameter value t, represented as $p(t)$ = $[x(t)\ y(t)]$, and the position at $t + dt$ (*i. e.*, $p(t + dt)$).

By relating these two positions as $p(t+dt) = p(t)+dp$, and assuming that the curve is differentiable, the speed of the particle with the parameter t is defined as

Figure 6.2: Some typical and easily recognizable 2-D shapes.

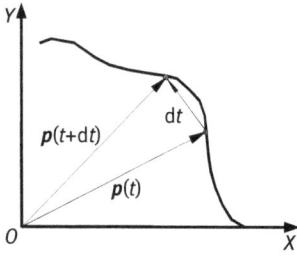

Figure 6.3: Two nearby points along the trajectory of a generic parametric curve.

$$\boldsymbol{p}'(t) = \frac{d\boldsymbol{p}}{dt} = \lim_{dt \to 0} \frac{\boldsymbol{p}(t + dt) - \boldsymbol{p}t}{dt} = \begin{bmatrix} \lim_{dt \to 0} \dfrac{x(t + dt) - x(t)}{dt} \\ \lim_{dt \to 0} \dfrac{y(t + dt) - y(t)}{dt} \end{bmatrix} = \begin{bmatrix} x'(t) \\ y'(t) \end{bmatrix} \tag{6.1}$$

For instance, the speed of the parametric curve given by $\boldsymbol{p}(t) = [\cos(t) \ \sin(t)]$ is $\boldsymbol{p}'(t) = [x'(t) \ y'(t)] = [-\sin(t) \ \cos(t)]$.

6.2.2.3 Regular Curve and Jordan Curve

When the speed of a parametric curve is never zero (the particle never stops), the curve is said to be a **regular curve**. When the curve never crosses itself, the curve is said to be a **Jordan curve**.

One of the most important properties of the speed of a regular parametric curve is the fact that it is a vector tangent to the curve at each of its points. The set of all possible tangent vectors to a curve is called a tangent field. Provided the curve is regular, it is possible to normalize its speed in order to obtain unit magnitudes for all the tangent vectors along the curve, which is given by

$$\boldsymbol{\alpha}(t) = \frac{\boldsymbol{p}'(t)}{\|\boldsymbol{p}'(t)\|} \Rightarrow \|\boldsymbol{\alpha}(t)\| = 1 \tag{6.2}$$

A same trace can be defined by many (actually an infinite number of) different parametric curves. A particularly important and useful parameterization is unitary magnitude speed. Such a curve is said to be parameterized by the arc length. It is easy to verify that, in this case,

$$s(t) = \int_0^t \|r'(t)\| dt = \int_0^t 1 dt = t \tag{6.3}$$

That is, the parameter t corresponds to the arc length. Consequently, a curve can be parameterized by the arc length if and only if its tangent field has unitary magnitudes. The process of transforming the analytical equation of a curve into an arc length parameterization is called arc length reparameterization, which consists of expressing the parameter t as a function of the arc length, and substituting t in the original curve for this function.

6.3 Description of Shape Property

One property of a shape region can be described by using different techniques/descriptors. This section discusses two important properties of shapes, for which people have designed/developed a number of different descriptors to depict them.

6.3.1 Compactness Descriptors

Compactness is an important property of shapes, which is closely related to the **elongation** of a shape. Many different descriptors have been proposed to give people an idea of the compactness of a region.

6.3.1.1 Form Factor (*F*)

The **form factor** of a region is computed from the perimeter B and area A of this region,

$$F = \frac{B^2}{4\pi A} \tag{6.4}$$

In continuous cases, the F value of a circular region is one and the F value of other forms of regions is greater than one. For digital images, it has been proven (see Haralick, 1992) that the F value of an octagon region achieves the minimum if the perimeter is computed according to 8-connectivity, while the F value of a diamond region achieves the minimum if the perimeter is computed according to 4-connectivity. In other words, which digital shape gives the minimum F value depends on whether the perimeter is computed as the number of its 4-neighboring border pixels or as the length of its border counting 1 for vertical or horizontal moves and $\sqrt{2}$ for diagonal moves.

The form factor could describe, in a certain sense, the compactness of a region. However, in a number of situations, using only the form factor could not distinguish different shapes. One example is given in Figure 6.4, in which all three shapes have the same area ($A = 5$) and same perimeter ($B = 12$). According to eq. (6.4), they have the same form factors. However, it is easy to see from Figure 6.4 that these shapes are quite different.

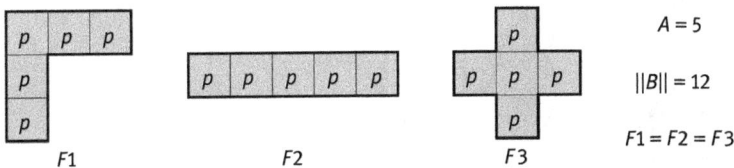

Figure 6.4: Different shapes with the same form factors.

6.3.1.2 Sphericity (*S*)

Sphericity originates from 3-D, which is the ratio of the surface to the volume. In 2-D, it captures the shape information by the radius of the inscribed circle of object r_i and the radius of the circumscribed circle r_c, which is given by

$$S = r_i/r_c \tag{6.5}$$

One example is shown in Figure 6.5 (the centers of these circles do not generally coincide with each other or with the centroid).

The value of this descriptor attains its maximum for a circular object, and this value will decrease as the object extends in several directions.

6.3.1.3 Circularity (*C*)

Circularity is another shape descriptor that can be defined by

$$C = \frac{\mu_R}{\sigma_R} \tag{6.6}$$

where μ_R is the mean distance from the centroid of the region to the boundary pixels (see Figure 6.6) and σ_R is the standard variance of the distance from the centroid to the boundary pixels:

$$\mu_R = \frac{1}{K} \sum_{k=0}^{K-1} \|(x_k, y_k) - (\bar{x}, \bar{y})\| \tag{6.7}$$

$$\sigma_R^2 = \frac{1}{K} \sum_{k=0}^{K-1} [\|(x_k, y_k) - (\bar{x}, \bar{y})\| - \mu_R]^2 \tag{6.8}$$

The value of this descriptor attains its maximum ∞ for a perfect circular object and will be less for more elongated objects.

Circularity has the following properties (Haralick, 1992):
1. As the digital shape becomes more circular, its value increases monotonically.
2. Its values for similar digital and continuous shapes are similar.
3. It is orientation and area independent.

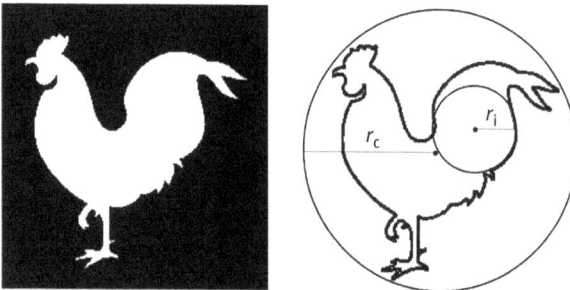

Figure 6.5: Illustrating the computation of sphericity.

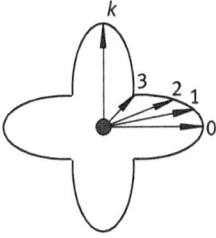

Figure 6.6: Illustrating the computation of circularity.

6.3.1.4 Eccentricity (E)

Eccentricity could also describe, in a certain sense, the compactness of a region. There are several formulas for computing eccentricity. For example, a simple method is to take the ratio of the diameter of the boundary to the length of the line (between two intersection points with a boundary), which is perpendicular to the diameter, as the eccentricity. However, such a boundary-based computation is quite sensitive to noise. Better methods for eccentricity take all the pixels that belong to the region into consideration.

In the following, an eccentricity measurement based on the **moment of inertia** is presented (Zhang, 1991a). Suppose a rigid body consists of N points, with masses $m_1, m_2, \ldots, m_N$ and coordinates $(x_1, y_1, z_1), (x_2, y_2, z_2), \ldots, (x_N, y_N, z_N)$. The moment of inertia, I, of this rigid body around an axis L is

$$I = \sum_{i=1}^{N} m_i d_i^2 \tag{6.9}$$

where d_i is the distance between the point m_i and the rotation axis L. Without loss of generality, let L pass the origin of the coordinate system and have the direction cosines α, β, γ, eq. (6.9) can be rewritten as

$$I = A\alpha^2 + B\beta^2 + C\gamma^2 - 2F\beta\gamma - 2G\gamma\alpha - 2H\alpha\beta \tag{6.10}$$

where

$$A = \sum m_i(y_i^2 + z_i^2) \tag{6.11}$$
$$B = \sum m_i(z_i^2 + x_i^2) \tag{6.12}$$
$$C = \sum m_i(x_i^2 + y_i^2) \tag{6.13}$$

are the moments of inertia of this rigid body around the x-axis, y-axis, and z-axis, respectively, and

$$F = \sum m_i y_i z_i \tag{6.14}$$
$$G = \sum m_i z_i x_i \tag{6.15}$$
$$H = \sum m_i x_i y_i \tag{6.16}$$

are called the inertia products.

Equation (6.10) can be interpreted in a simple geometric manner. First, look at the following equation

$$Ax^2 + By^2 + Cz^2 - 2Fyz - 2Gzx - 2Hxy = 1 \tag{6.17}$$

It represents a second-order surface centered on the origin of the coordinate system. If the vector from the origin to this surface is $\boldsymbol{r}$, which has the direction cosines α, β, γ, then taking eq. (6.10) into eq. (6.17) gives

$$r^2(A\alpha^2 + B\beta^2 + C\gamma^2 - 2F\beta\gamma - 2G\gamma\alpha - 2H\alpha\beta) = r^2 I = 1 \tag{6.18}$$

Now let us consider the right side of eq. (6.18). As I is always greater than zero, r should be limited (*i. e.,* the surface is closed). Since the surface under consideration is a second-order surface (conical surface), it must be an ellipsoid and can be called the **ellipsoid of inertia**. This ellipsoid has three orthogonal major axes. For a homogenous ellipsoid, any profile with two coplane axes is an ellipse, which can be called the **ellipse of inertia**. If taking a 2-D image as a planar rigid body, then, for each region in this image, a corresponding ellipse of inertia, which reveals the distribution of the points belonging to this region, can be obtained.

To determine the ellipse of inertia of a region, the directions and lengths of the two major axes of this region are required. The direction can be determined, for example, with the help of eigenvalues. Suppose the slopes of the two axes are k and l,

$$k = \frac{1}{2H}\left[(A - B) - \sqrt{(A - B)^2 + 4H^2}\right] \tag{6.19}$$

$$l = \frac{1}{2H}\left[(A - B) + \sqrt{(A - B)^2 + 4H^2}\right] \tag{6.20}$$

In addition, based on k and l, the lengths of the two axes of the ellipse of inertia, p and q, can be obtained by

$$p = \sqrt{2/[(A + B) - \sqrt{(A - B)^2 + 4H^2}]} \tag{6.21}$$

$$q = \sqrt{2/[(A + B) + \sqrt{(A - B)^2 + 4H^2}]} \tag{6.22}$$

The eccentricity of the region is just the ratio of p to q. Such defined eccentricity should be less sensitive to noise than boundary-based computations. It is easy to see that such a defined eccentricity would be invariant to the translation, rotation, and scaling of the region.

The values of the above-discussed descriptors F, S, C, and E for several common shapes are listed in Table 6.1.

The above descriptors are discussed in the continuous domain. Their computations for digital images are illustrated with the help of the following example sketches

Table 6.1: The values of F, S, C, and E for several common shapes.

Object	F	S	C	E
Square (side length = 1)	$4/\pi(\approx 1.273)$	$\sqrt{2}/2(\approx 0.707)$	9.102	1
Hexagon (side length = 1)	1.103	0.866	22.613	1.010
Octagon (side length = 1)	1.055	0.924	41.616	1
Rectangle (length × width = 2 × 1)	1.432	0.447	3.965	2
Ellipse (length × width = 2 × 1)	1.190	0.500	4.412	2

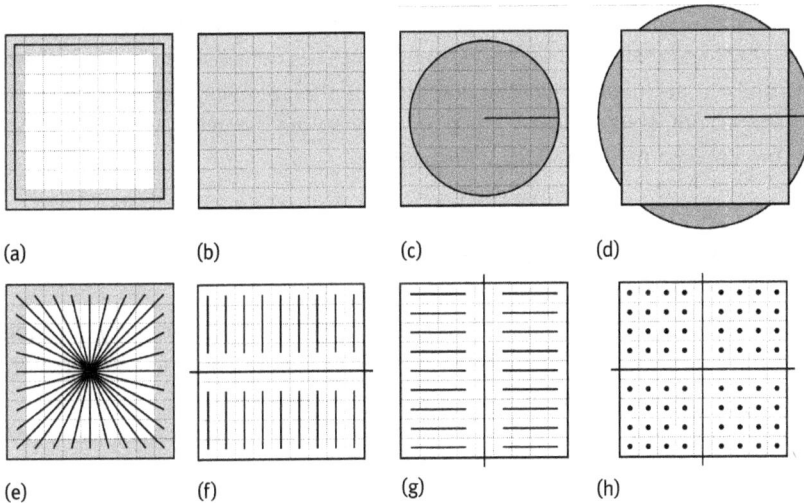

Figure 6.7: Illustrating the computation of F, S, C, and E for digital images.

for a square in Figure 6.7. Figure 6.7(a, b) are for computing B and A in form factors (F), respectively. Figure 6.7(c, d) are for computing r_i and r_c in sphericity (S), respectively. Figure 6.7(e) are for computing μ_R in circularities (C). Figure 6.7(f–h) are for computing A, B, and H in eccentricities (E), respectively.

6.3.2 Complexity Descriptors

Another important property of shapes is their complexity. Indeed, the classification of objects based on shape complexity arises in many different situations (Costa 2001). For instance, neurons have been organized into morphological classes by taking into account the complexity of their shapes (especially the dendritic tree). While complexity is a somewhat ambiguous concept, it is interesting to relate it to other geometric properties, such as **spatial coverage**. This concept, also known as the space-filling capability, indicates the capacity of a biological entity for sampling or interacting/filling the surrounding space. In other words, spatial coverage defines the shape interface with the external medium and determines important capabilities of the biological entity. For instance, a bacterium with a more complex shape (and hence a higher spatial coverage) will be more likely to find some food. At a larger spatial scale, the amount of water a tree root can drain is related to its respective spatial coverage of the surrounding soil. Yet another example of spatial coverage concerns the power of a neural cell to make synaptic contacts (see, e. g., Murray, 1995).

In brief, shape complexity is commonly related to spatial coverage in the sense that the more complex the shape, the larger its spatial covering capacity.

Unfortunately, though the term shape complexity has been widely used by scientists and other professionals, its precise definition does not exist yet. Instead, there are many different useful shape measurements that attempt to capture some aspects related to shape complexity. Some of them are listed in the following, in which B and A denote the shape perimeter and area, respectively.

6.3.2.1 Simple Descriptors for Shape Complexity

1. **Area to perimeter ratio,** which is defined as A/B.
2. **Thinness ratio,** which is inversely proportional to the circular property of shape and defined as $4\pi(A/B^2)$. The multiplying constant, 4π, is a normalization factor; hence, the thinness ratio is the inverse of the form factor.
3. $(B - \sqrt{B^2 - 4\pi A})/(B + \sqrt{B^2 - 4\pi A})$, which is related to the thinness ratio and to the circularity.
4. **Rectangularity,** which is defined by $A/area$ (MER), where MER stands for minimum enclosing rectangle.
5. **Mean distance to the boundary,** which is defined by A/μ_R^2.
6. **Temperature,** which is defined by $T = \log_2[(2B)/(B-H)]$, where H is the perimeter of the shape convex hull (see Section 10.2). The contour temperature is defined based on a thermodynamic formalism.

6.3.2.2 Histogram of Blurring Images

An interesting set of complexity measurements can be defined by the histogram analysis of blurred images. Recall that simple histogram measurements cannot be used as shape features because the histogram does not take into account the spatial distribution of the pixels, but only the gray-level frequencies.

For instance, consider the two images in Figure 6.8(a, b), where two different shapes are presented in these two images, respectively. The two shapes have the same area and, as a result, the same number of black pixels. Therefore, they have identical histograms, which are shown in Figure 6.8(c, d), respectively. It is now to consider blurring these two shape images by an averaging filter. The results obtained are shown in Figure 6.9(a, b), respectively. These results are influenced by the original shapes in images, and thus, their corresponding image histograms become different, which are easily verified from Figure 6.9(c, d). Therefore, shape features can be defined by

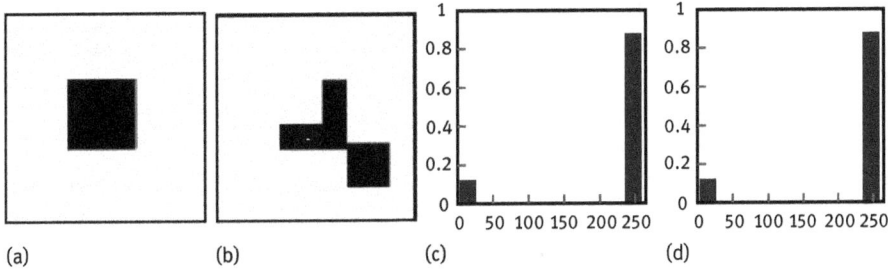

Figure 6.8: Two different shapes and their respective histograms.

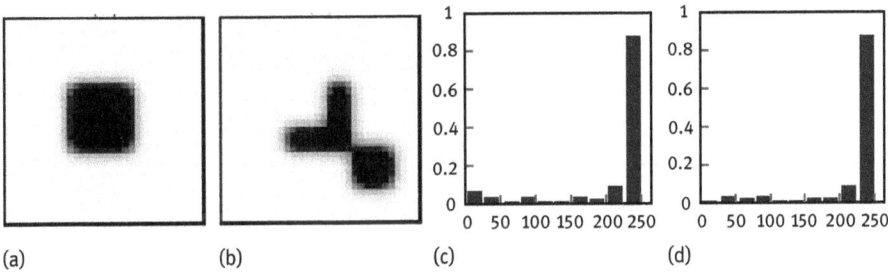

Figure 6.9: Two blurred versions of the shapes in Figure 6.8 and their histograms.

extracting information from the latter type of histograms (*i. e.*, those obtained after image blurring).

A straightforward generalization of this idea is the definition of a multi-scale family of features (Costa 2001). These features can be calculated based on the convolution of the input image with a set of multi-scale Gaussians $g(x, y, s) = \exp[-(x^2 + y^2)/(2s^2)]$. Let $f(x, y, s) = f(x, y) * g(x, y, s)$ be the family of the multi-scale (blurred) images indexed by the scale parameters and obtained by 2-D convolutions between f and g for each fixed value of the scale parameter. Examples of the multiscale histogram-based complexity features are listed in the following:

1. **Multiscale entropy:** It is defined as $E(s) = -\sum_i p_i(s) \ln p_i(s)$, where $p_i(s)$ is the relative frequency of the ith gray level in the blurred image $f(x, y, s)$ for each value of s.

2. **Multiscale standard deviation:** It is defined as the square root of the variance of $f(x, y, s)$ for each scale s. The variance itself can also be used as a feature.

6.3.2.3 Fullness

There are certain relationships between the compactness and complexity of objects. In many cases, objects with relatively compact distribution often have relatively simple shapes.

Fullness in some senses reflects the compactness of the object, which takes into account the full extent of an object in a box around the object. One method to compute the fullness of an object is to calculate the ratio of the number of pixels belonging to the object with the number of pixels in the entire circumference box contains the object.

Example 6.1 Fullness of objects.
In Figure 6.10, there are two objects and their corresponding circumference boxes, respectively. These two objects have the same out-boundary, but the object in Figure 6.10(b) has a hole inside. The fullnesses of these two objects are 81/140 = 57.8% and 63/140 = 45%, respectively. By comprising the values of fullness, it is seen that the distribution of pixels in the object of Figure 6.10(a) is more concentrated than the distribution of pixels in the object of Figure 6.10(b). It seems that the density of pixel distribution in Figure 6.10(a) is bigger than that in Figure 6.10(b), while the complexity of the object of Figure 6.10(b) is higher that of the object in Figure 6.10(a).

The above obtained statistics is similar to statistics provided by histogram, which does not reflect the spatial distribution of information. It does not provide a general sense of shape information. To solve this problem, the lateral histogram of objects

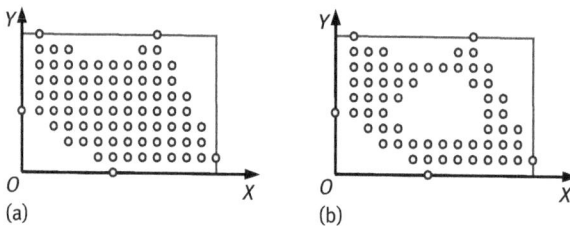

Figure 6.10: Two objects for fullness comparison.

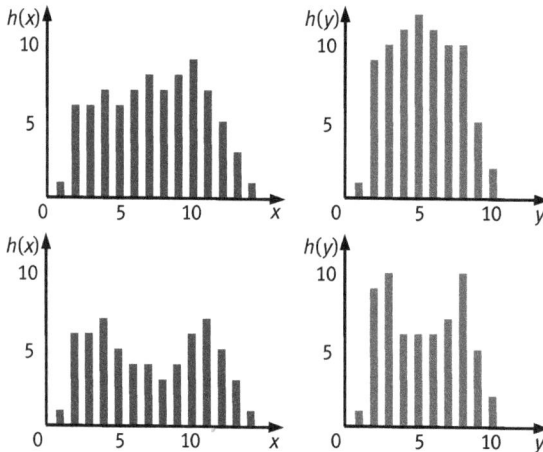

Figure 6.11: The lateral histogram of X-coordinate and Y-coordinate.

can be used. Here, the lateral histogram of X-coordinate is obtained by adding up the number of pixels along columns, while the lateral histogram of Y-coordinate is obtained by adding up the number of pixels along rows. The X-coordinate and Y-coordinate histograms obtained from Figure 6.10(a, b) are shown in Figure 6.11 up row and bottom row, respectively. The histograms of bottom row are both nonmonotonic histograms. There are obvious valleys in the middle, which are caused by the hole inside the object. ◻

6.4 Technique-Based Descriptors

Different properties of a shape region can be described by using a particular technique, or by using several descriptors based on a particular representation. This section shows some examples.

6.4.1 Polygonal Approximation-Based Shape Descriptors

A shape outline can be properly represented by a polygon (Section 3.2). Based on such a representation, various shape descriptors can be defined.

6.4.1.1 Directly Computable Features
The following shape features can be extracted directly from the results of polygonal approximation by straightforward procedures.
1. The number of corners or vertices.
2. The angle and side statistics, such as their mean, median, variance, and moments.
3. The major and minor side lengths, side ratio, and angle ratio.
4. The ratio between the major angle and the sum of all angles.
5. The ratio between the standard deviations of sides and angles.
6. The mean absolute difference of the adjacent angles.

6.4.1.2 Shape Numbers
Shape number is a descriptor for the boundary of an object based on chain codes. Although the first-order difference of a chain code is independent to rotation, in general, the coded boundary depends on the orientation of the image grid. One way to normalize the grid orientation is to align the chain-code grid with the sides of the basic rectangle. In practice, for a desired shape order, the rectangle of an order n, whose eccentricity best approximates that of the basic rectangle, is determined and this new rectangle is used to establish the grid size (see Section 3.5.2 for generating shape number).

Based on the shape number, different shape regions can be compared. Suppose the similarity between two closed regions A and B is the maximum common shape number of the two corresponding shape numbers. For example, if

$S_4(A) = S_4(B), S_6(A) = S_6(B), \ldots, S_k(A) = S_k(B), S_{k+2}(A) \neq S_{k+2}(B), \ldots$, then the similarity between A and B is k, where $S(\cdot)$ represents the shape number and the subscript corresponds to the order. The self-similarity of a region is ∞.

The distance between two closed regions A and B is defined as the inverse of their similarity

$$D(A, B) = 1/k \tag{6.23}$$

It satisfies

$$\begin{aligned} & D(A, B) \geq 0 \\ & D(A, B) = 0 \quad \text{iff} \quad A = B \\ & D(A, C) \leq \max\left[D(A, B), D(B, C)\right] \end{aligned} \tag{6.24}$$

One example of the shape number matching shape description is shown in Figure 6.12 (Ballard, 1982).

6.4.1.3 Shape Signatures

One-dimensional signatures can be obtained from both contour/boundary-based and region-based shape representations. Contour-based signatures have been discussed in Section 3.2. The following brief discussions are for region-based signatures.

Region-based signatures use the whole shape information, instead of using only the shape boundary as in the contour-based approach. The basic idea is still to take the projection along different directions. By definition, a projection is not an information-preserving transformation. However, enough projections allow reconstruction of the boundary or even the region to any desired degree of accuracy (this is the basis for computer-assisted tomography as introduced in Chapter 5 of Volume I).

One example is shown in Figure 6.13. Two letters, S and Z, are in the forms of 7×5 dot-matrix representations, which are to be read automatically by scanning. The vertical projections are the same for these two letters. Therefore, a horizontal projection in addition would be necessary to eliminate the ambiguity.

This projection-based signature actually integrates the image pixel values along lines perpendicular to a reference orientation (e. g., projection signatures that are

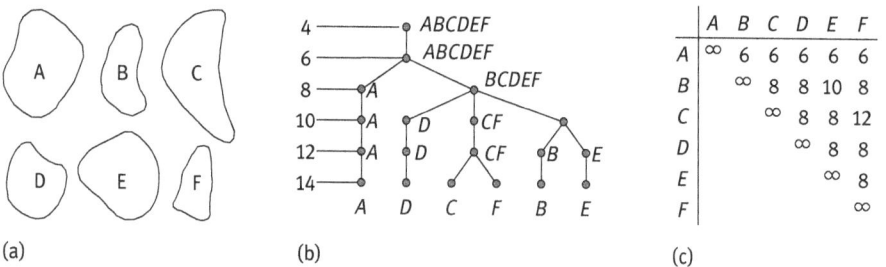

(a) (b) (c)

Figure 6.12: Illustrating matching based on shape numbers.

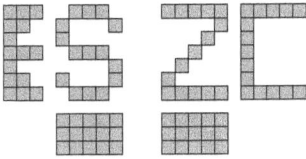

Figure 6.13: Illustration of shape signatures.

defined along the x and y axes). This concept is also closely related to the Hough transform.

There are many different ways to define signatures. It is important to emphasize that, as shape signatures describe shapes in terms of a 1-D signal, they also allow the application of 1-D signal processing techniques for shape analysis.

6.4.2 Curvature-Based Shape Descriptors

The **curvature** is one of the most important features that can be extracted from contours. Strong biological motivation has been identified for studying the curvature, which is apparently an important clue explored by the human vision system.

6.4.2.1 Curvature and Geometrical Aspects

As far as computational shape analysis is concerned, curvature plays an important role in the identification of many different geometric shape primitives. Table 6.2 summarizes some important geometrical aspects that can be characterized by considering curvatures (Levine, 1985).

6.4.2.2 Discrete Curvature

In the discrete space, curvature commonly refers to the changes of direction within the discrete object under study. Hence, a discrete curvature only makes sense when it is possible to define an order for a sequence of discrete points. Therefore, the study of **discrete curvature** will be restricted to digital arcs and curves. The changes in directions will be measured via the angles made between the specific continuous segments defined by points on the discrete object in question.

Table 6.2: Curvatures and their geometrical aspects.

Curvature	Geometrical Aspect
Curvature local absolute value maximum	Generic corner
Curvature local positive maximum	Convex corner
Curvature local negative minimum	Concave corner
Constant zero curvature	Straight line segment
Constant nonzero curvature	Circle segment
Zero crossing	Inflection point
Average high curvature in absolute or squared values	Shape complexity, related to the bending energy

A formal definition of a discrete curvature is as follows (Marchand 2000). Given a set of discrete points $P = \{p_i\}_{i=0,...,n}$ that defines a digital curve, $\rho_k(p_i)$, the k-order curvature of P at point $p_i \in P$ is given by $\rho_k(p_i) = |1 - \cos \theta_k^i|$, in which θ_k^i = angle (p_{i-k}, p_i, p_{i+k}) is the angle made between the continuous segments $[p_{i-k}, p_i]$ and $[p_i, p_{i+k}]$, where $k \in \{i, ..., n - i\}$.

The integer order k is introduced to make the curvature less sensitive to local variations in the directions followed by the digital arc or curve. A discrete curvature of a high order will reflect more accurately the global curvature of the continuous object approximated by the set of discrete points. Figure 6.14 illustrates the process of calculating $\rho_3(p_{10})$, the third-order discrete curvature of the digital arc $P_{pq} = \{p_i\}_{i=0,...,17}$ at point p_{10}.

Figure 6.15 displays the resulting curvature values obtained for successive instances of the order k ($k = 1, ..., 6$) when using the digital arc presented in Figure 6.14. Clearly, the first-order curvature is not an accurate representation for discrete curvature since it only highlights local variations. This example also shows that as the order k increases, the curvature tends to illustrate the global behavior of the digital arc. The location of the peak (at point p_8 or p_9) corresponds to the place where a dramatic change in the global direction occurs in the digital arc.

6.4.2.3 Curvature Computation

The curvature $k(t)$ of a parametric curve $c(t) = [x(t), y(t)]$ is defined as

$$k(t) = \frac{x'(t)y''(t) - x''(t)y'(t)}{(x'(t)^2 + y'(t)^2)^{3/2}} \tag{6.25}$$

It is clear from this equation that estimating the curvature involves the derivatives of $x(t)$ and $y(t)$, which is a problem where the contour is represented in a digital (spatially sampled) form. There are two basic approaches used to circumvent this problem.

1. Defining alternative curvature measurements based on the angles between vectors
   This approach consists of estimating the angle defined at $c(n_0)$ by vectors along the contour. Various estimations differ in the way that these vectors are fitted or by the method that the estimation is applied. Let $c(n) = [x(n), y(n)]$ be a discrete

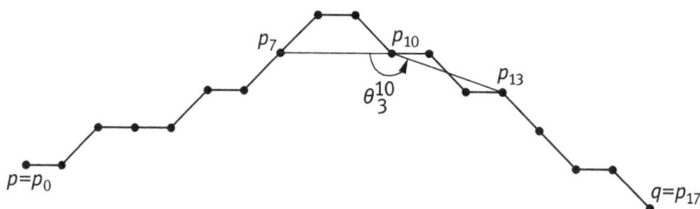

Figure 6.14: Discrete curvature calculation.

curve. The following vectors can be defined:

$$u_i(n) = [x(n) - x(n-i), y(n) - y(n-i)]$$
$$v_i(n) = [x(n) - x(n+i), y(n) - y(n+i)]$$

(6.26)

These vectors are defined between $c(n_0)$ (the current point) and the ith neighbors of $c(n)$ to the left and to the right, as shown in Figure 6.16, where the left and the right vectors represent $u_i(n)$ and $v_i(n)$, respectively (Costa 2001).

The high curvature points can be obtained by the following equation

$$r_i(n) = \frac{u_i(n)v_i(n)}{\|u_i(n)\| \|v_i(n)\|}$$

(6.27)

where $r_i(n)$ is the cosine of the angle between the two vectors $u_i(n)$ and $v_i(n)$. Therefore, $-1 \leq r_i(n) \leq 1$, with $r_i(n) = -1$ for straight lines and $r_i(n) = 1$ when the angle becomes $0°$ (the smallest possible angle). In this sense, $r_i(n)$ can be used as a measurement that is capable of locating high curvature points.

2. Interpolating $x(t)$ and $y(t)$ followed by differentiation
Suppose the curvature at $c(n_0)$ is to be estimated. The interpolation approach consists of interpolating the neighbor points around $c(n_0)$, as shown in Figure 6.17.

There are different approaches for approximating or interpolating the contour in order to analytically derive the curvature. The simplest approach is to approximate the derivatives of $x(n)$ and of $y(n)$ in terms of finite differences, that is,

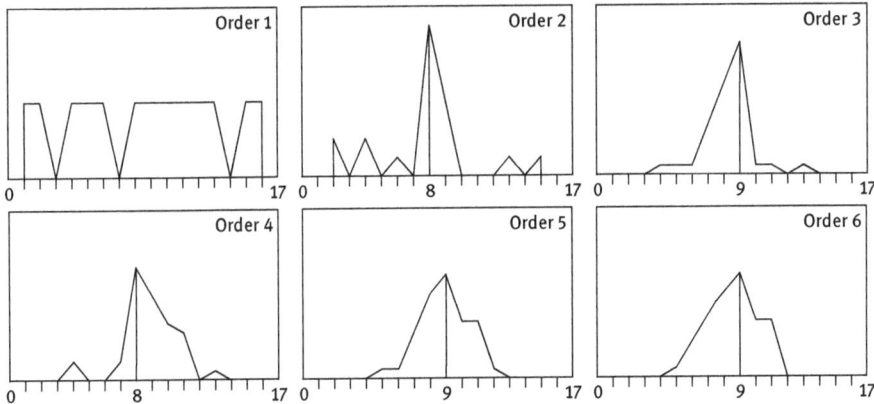

Figure 6.15: Curvature of different orders.

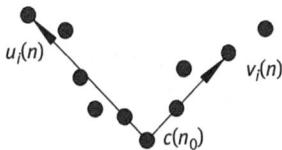

Figure 6.16: Angle-based curvature computation.

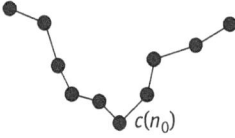

$c(n_0)$

Figure 6.17: Interpolation-based curvature computation.

$$
\begin{aligned}
x'(n) &= x(n) - x(n-1) \\
y'(n) &= y(n) - y(n-1) \\
x''(n) &= x'(n) - x'(n-1) \\
y''(n) &= y'(n) - y'(n-1)
\end{aligned}
\tag{6.28}
$$

The curvature can then be estimated by taking the above-calculated values into eq. (6.25). Although this simple approach can be efficiently implemented, it is sensitive to noise.

A more elaborate technique has been proposed (Medioni, 1987), which is based on approximating the contour in a piecewise fashion in terms of cubic B-splines. Suppose that it is required to approximate a contour segment by a cubic parametric polynomial in t, with $t \in [0, 1]$. The curve approximating the contour between two consecutive contour points $A(t = 0)$ and $B(t = 1)$ is defined as

$$
\begin{aligned}
x(t) &= a_1 t^3 + b_1 t^2 + c_1 t + d_1 \\
y(t) &= a_2 t^3 + b_2 t^2 + c_2 t + d_2,
\end{aligned}
\tag{6.29}
$$

where a, b, c, and d are taken as the parameters of the approximating polynomial. By taking the respective derivatives of the above parametric curve into eq. (6.25):

$$
k = 2\frac{c_1 b_2 - c_2 b_1}{(c_1^2 + c_2^2)^{3/2}}
\tag{6.30}
$$

Here, it is supposed that the curvature is to be estimated around the point $A(0)$ (*i. e.*, for $A(t = 0)$).

6.4.2.4 Descriptors Based on Curvatures

While the curvature itself can be used as a feature vector, this approach presents some serious drawbacks including the fact that the curvature signal can be too long (involving thousands of points, depending on the contour) and highly redundant. Once the curvature has been estimated, the following shape measurements can be calculated in order to circumvent these problems.

Curvature Statistics

The histogram of curvatures can provide a series of useful global measurements, such as the curvature's mean, median, variance, standard deviation, entropy, moments, etc.

Maxima, Minima, and Inflection Points
The fact that not all points along a contour are equally important (in the sense of conveying information about the shape) motivates the analysis of dominant points such as those points where the curvature is either a positive maximum, or a negative minimum, or an inflection point. The number of such points, their position along the contour, and their curvature values (in case of maxima and minima curvature points) can be used for shape measurements.

Symmetry Measurement Metric
A **symmetry measurement metric** for polygonal segments (*i. e.*, curve segments) is defined as

$$S = \int_0^L \left(\int_0^t k(l)dl - \frac{A}{2} \right) dt \tag{6.31}$$

where t is the parameter along the curve, the inner integration is a measurement metric of the angular change until t, A is the total angular change of the curve segment, L is the length of the curve segment, and $k(l)$ can be taken as the curvature along the contour.

Bending Energy
In the continuous case, the mean **bending energy**, also known as the **boundary energy**, is defined as

$$B = \frac{1}{P} \int [k(t)]^2 dt \tag{6.32}$$

where P is the contour perimeter. The above equation means that the bending energy is obtained by integrating the squared curvature values along the contour and dividing the result by the curve perimeter. Note that the bending energy is also one of the most important global measurements related to the shape complexity.

6.4.2.5 Curvatures on Surfaces
From differential geometry, it is known that the curvature K at a point p on a curve is given by

$$K = \frac{g''(t)}{[1 + g'(t)^2]^{3/2}} \tag{6.33}$$

Curvatures on surfaces are more difficult to define than on planes, as there are infinitely many curves through a point on the surface (Lohmann, 1998). Therefore, it cannot directly generalize the definition of the curvature for planar curves to the curvature of points on a surface. However, it can identify at least one direction in which the curvature is the maximal curvature and another direction in which the curvature is a minimal curvature. Figure 6.18 illustrates this point. These two

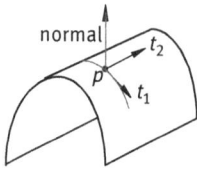

Figure 6.18: Principal directions of a curvature.

directions are generally called the principal directions and are denoted t_1 and t_2. Given the principal directions, it can identify a curve through point p along those directions and employ eq. (6.33) to compute the curvature along these curves. The magnitudes of the curvature along the t_1 and t_2 directions are denoted k_1 and k_2.

Note that sometimes – for instance, for entirely flat surfaces – more than two principal directions exist. In this case, any arbitrary curve of the maximal or minimal curvature can be used to compute t_1 and t_2, k_1 and k_2.

The two principal directions give rise to the following definitions of the curvature on a surface. The term

$$K = k_1 k_2 \qquad (6.34)$$

is called the **Gaussian curvature**, which is positive if the surface is locally elliptic and negative if the surface is locally hyperbolic. The term

$$H = (k_1 + k_2)/2 \qquad (6.35)$$

is called the **mean curvature**. It determines whether the surface is locally convex or concave.

By a combined analysis of the signs of Gaussian curvature K and mean curvature H, different classes of surfaces can be distinguished, as shown in Table 6.3 (Besl, 1988).

Example 6.2 Eight types of surfaces.
One respect example for each of the eight types of surfaces listed in Table 6.3 are shown in Figure 6.19. The relative arrangement of these examples is same with respect to the relative arrangement in Table 6.3. Figure 6.19(a) is for saddle ridge. Figure 6.19(b) is for minimal. Figure 6.19(c) is for saddle valley. Figure 6.19(d) is for

Table 6.3: Classification of surfaces according to the signs of curvatures.

	$H < 0$	$H = 0$	$H > 0$
$K < 0$	Saddle ridge	Minimal	Saddle valley
$K = 0$	Ridge	Flat/planar	Valley
$K > 0$	Peak		Pit

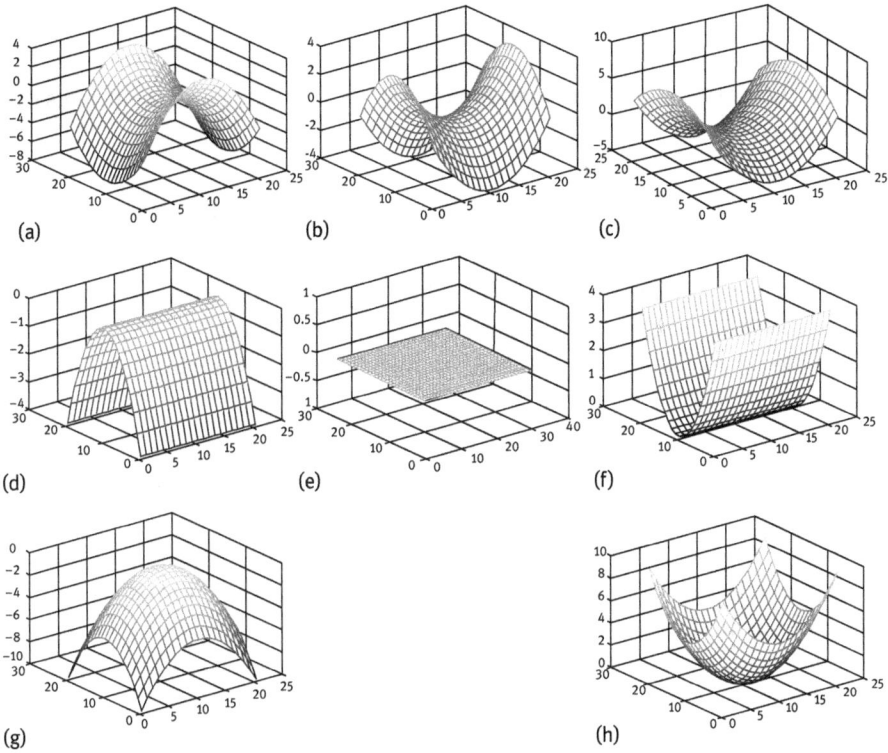

Figure 6.19: Eight types of surfaces.

ridge. Figure 6.19(e) is for flat/planar. Figure 6.19(f) is for valley. Figure 6.19(g) is for peak. Figure 6.19(h) is for pit.

6.5 Wavelet Boundary Descriptors

Wavelet boundary descriptors are descriptors formed by the coefficients of the wavelet transform of the object boundary.

6.5.1 Definition and Normalization

A wavelet family can be defined as

$$W_{m,n}(t) = \frac{1}{\sqrt{2^m}} W\left(\frac{t - n2^m}{2^m}\right) = 2^{-m/2} W(2^{-m}t - n) \qquad (6.36)$$

For a given boundary function $f(t)$, its wavelet transform coefficients are

$$D_{m,n} = \langle f, W_{m,n} \rangle \equiv \int f(t) \bar{W}_{m,n}(t) dt \qquad (6.37)$$

The formula for the reconstruction of $f(t)$ from its wavelet transform coefficients can be written as

$$f(t) = \sum_{m=m_0+1}^{\infty} \sum_{n=-\infty}^{\infty} \langle f, W_{m,n} \rangle W_{m,n}(t) + \sum_{m=-\infty}^{m_0} \sum_{n=-\infty}^{\infty} \langle f, W_{m,n} \rangle W_{m,n}(t) \qquad (6.38)$$

where m_0 depends on the precision required when the coefficient series is truncated. Suppose the scale function $S_{m,n}(t)$ is given by $S_{m,n}(t) = 2^{-m/2} S_{m,n}(2^{-m}t-n)$, and the first term on the right side of eq. (6.38) is represented by a linear combination. Equation (6.38) can be written as

$$f(t) = \sum_{n=-\infty}^{\infty} \langle f, S_{m,n} \rangle S_{m,n}(t) + \sum_{m=-\infty}^{m_0} \sum_{n=-\infty}^{\infty} \langle f, W_{m,n} \rangle W_{m,n}(t) \qquad (6.39)$$

According to the properties of the wavelet transform, the first term on the right side of eq. (6.39) can be considered a coarse map and the second term on the right side of eq. (6.39) corresponds to other details. Name $\langle f, S_{m,n} \rangle$ as a scale coefficient and $\langle f, W_{m,n} \rangle$ as a wavelet coefficient, both of them together form the wavelet boundary descriptors corresponding to the boundary $f(t)$.

In practical computation of the wavelet boundary transformation, the following points should be considered.

6.5.1.1 Normalization for Translation
Scale coefficients and wavelet coefficients have the following relations

$$\int_{-\infty}^{\infty} S_{m,n}(t) dt = \int_{-\infty}^{\infty} 2^{-m/2} S(2^{-m}t - n) dt = 2^{m/2} \int_{-\infty}^{\infty} S(t) = 2^{m/2} \qquad (6.40)$$

$$\int_{-\infty}^{\infty} W(t) dt = 0 \qquad (6.41)$$

Suppose the boundary is translated by δ. The translated boundary is $f_t(t) = f(t) + \delta$ and

$$\langle f_t(t), W(t) \rangle = \int_{-\infty}^{\infty} [f(t) + \delta] W(t) dt = \langle f(t), W(t) \rangle + 0 \qquad (6.42)$$

$$\langle f_t(t), S(t) \rangle = \langle f(t), S(t) \rangle + 2^{m/2} \delta = \int_{-\infty}^{\infty} [f(t) + \delta] S_{m,n}(t) dt \qquad (6.43)$$

It can be seen from eq. (6.42) that the translation of a boundary has no effect on the wavelet coefficients that describe the details. However, it can be seen from eq. (6.43) that the translation of a boundary has effect on the scale coefficients. In other words, though wavelet coefficients are invariant to translation, translation normalization for scale coefficients is required.

6.5.1.2 Normalization for Scale

Suppose the boundary is under scaling by a parameter A. The scaled boundary is $f_s(t) = Af(t)$, where the scaling center is also the boundary center. Both scale coefficients and wavelet coefficients are influenced by scaling as

$$\langle f_s(t), S(t) \rangle = \int_{-\infty}^{\infty} Af(t)S(t)dt = A \langle f(t), S(t) \rangle \tag{6.44}$$

$$\langle f_s(t), W(t) \rangle = \int_{-\infty}^{\infty} Af(t)W(t)dt = A \langle f(t), W(t) \rangle \tag{6.45}$$

So, a scale normalization with a parameter A for both scale coefficients and wavelet coefficients are required to ensure that the descriptor is invariant to scale change.

6.5.1.3 Truncation of Coefficients

If taking all coefficients as descriptors, no boundary information will be lost. However, this may be not needed in practical situations. It is possible to keep all scale coefficients and to truncate certain wavelet coefficients. In such a case, the precision for describing the boundary could be kept at a certain level, which is determined by m_0. A high value of m_0 causes more details to be lost, while a low value of m_0 yields less compression.

6.5.2 Properties of the Wavelet Boundary Descriptor

The wavelet boundary descriptor has two basic properties.
1. Uniqueness
   The wavelet transform is a one-to-one mapping, so a given boundary corresponds to a unique wavelet boundary descriptor. On the other hand, a wavelet boundary descriptor corresponds to a unique boundary.
2. Similarity measurable
   The similarity between two descriptors can be measured by the value of the inner product of the two vectors corresponding to the two descriptors.

Some experiments in the content-based image retrieval framework have been conducted to test the descriptive ability of the wavelet boundary descriptor with the distance between two descriptors.
1. Difference between two boundaries
   This can be seen in Figure 6.20, in which Figure 6.20(a) is the querying boundary, and Figure 6.20(b–d) is the first three returned results. Distance between two descriptors is computed by first calculating the ratio of the value of the inner product of the two vectors corresponding to the two descriptors and then normalizing the ratio with the maximal possible value of the inner product. The obtained measurement is a percentage of the distance measurement (DIS). The DISs for Figure 6.20(b–d) are 4.25%, 4.31%, and 5.44%, respectively. As the images

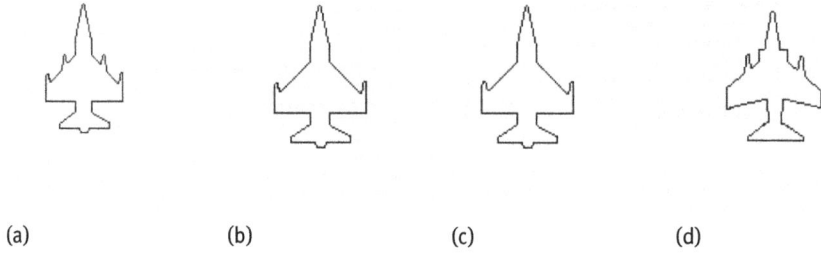

(a)   (b)   (c)   (d)

Figure 6.20: The relation between the distance and the difference of the boundary.

in Figure 6.20(b, c) are mainly different from Figure 6.20(a) by the position of the boundary, while Figure 6.20(d) is quite different from Figure 6.20(a) by the boundary form, it is evident that the distance between two descriptors is related to the difference between two boundaries.

2. Invariance to the displacement of the boundary

   After normlalization for translation, the wavelet boundary descriptor is invariant to the displacement of the boundary. This can be seen from Figure 6.21, in which Figure 6.21(a) is the querying boundary, and Figure 6.21(b–d) is the first three returned results. The DISs for Figure 6.21(b–d) are 0%, 1.56%, and 1.69%, respectively. From Figure 6.21(b), it can be seen that pure translation has no influence on the distance measurement. From Figure 6.21(c, d), it is evident that some small changes in the form of the boundary have more influence than that caused by translation.

3. Invariance to the scaling of boundary

   After normalization for the scaling, the wavelet boundary descriptor is invariant to the scaling of the boundary. This can be seen from Figure 6.22, in which Figure 6.22(a) is the querying boundary, and Figure 6.22(b–d) is the first three returned results. The DISs for Figure 6.22(b–d) are 1.94%, 2.32%, and 2.75%, respectively. From Figure 6.22(b), it can be seen that the scale change has little effect on the distance measurement. Compared with Figure 6.22(c) and 6.22(d),

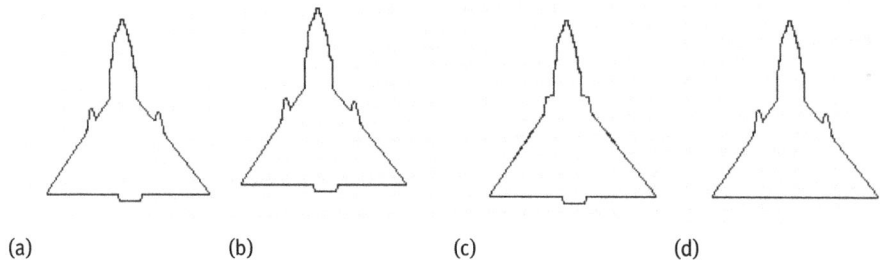

(a)   (b)   (c)   (d)

Figure 6.21: Illustrating the invariance of the boundary descriptor to translation.

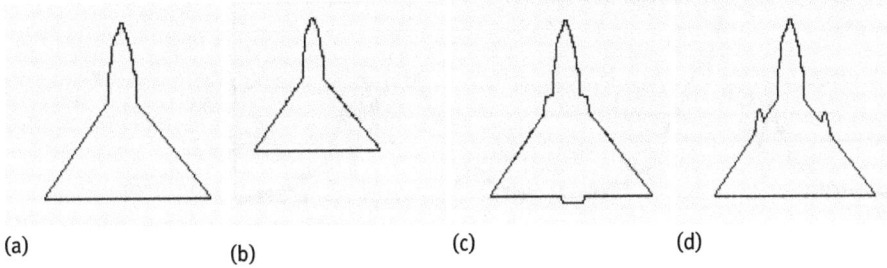

(a)
(b)
(c)
(d)

Figure 6.22: Illustrating the invariance of the boundary descriptor to scale changes.

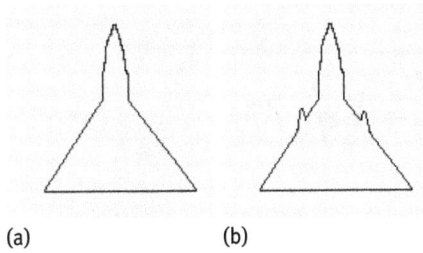

(a)
(b)

Figure 6.23: Two test boundaries with a tiny difference in two locations.

the changes in the form of the boundary have more influence than that caused by scale changes.

6.5.3 Comparisons with Fourier Descriptors

Compared to **Fourier boundary descriptors**, wavelet boundary descriptors have several advantages, which are detailed below.

6.5.3.1 Less Affected by Local Deformation of the Boundary
A wavelet boundary descriptor is composed of a sequence of coefficients and each coefficient is more locally related to certain parts of the boundary than the coefficients of the Fourier boundary descriptor. The local deformation merely has influence on the corresponding coefficients but no other coefficients.

The following experiment is conducted to show this advantage. Two test boundaries are made with a tiny difference in two locations, as shown in Figure 6.23(a, b).

Suppose the descriptor for boundary A is $\{x_a\}$ and $\{y_a\}$, the descriptor for boundary B is $\{x_b\}$ and $\{y_b\}$, and their difference is defined as $\delta^2 = (x_a - x_b)^2 + (y_a - y_b)^2$. Figures 6.24 and 6.25 show this difference for the wavelet boundary descriptor and the Fourier boundary descriptor, respectively. In both figures, the horizontal axis corresponds to the number of coefficients (started from the top of boundary), and the vertical axis is for the difference.

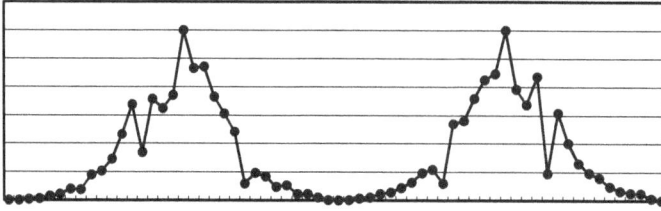

Figure 6.24: The difference as a function of the coefficients for the wavelet boundary descriptor.

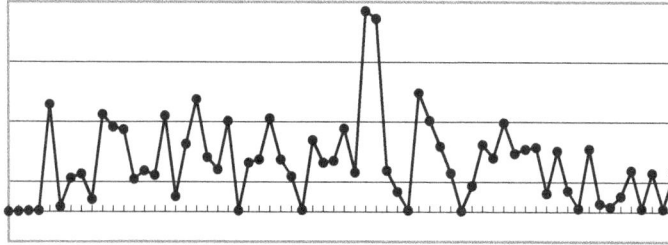

Figure 6.25: The difference as a function of the coefficient for the Fourier boundary descriptor.

It is easy to see that the local deformation of the boundary causes a quite regular and local change of the difference in the wavelet coefficients as shown in Figure 6.24, while this deformation of the boundary causes a quite irregular and global change of the difference in the Fourier coefficients as shown in Figure 6.25.

6.5.3.2 More Precisely Describe the Boundary with Fewer Coefficients

One example is given in Figure 6.26, where Figure 6.26(a) shows an original boundary and the reconstructed boundary with 64 wavelet coefficients and Figure 6.26(b) shows the same original boundary and the reconstructed boundary with 64 Fourier coefficients. The Fourier coefficients just provide the information on 64 discrete points. To get more information that is precise on the boundary, more coefficients are needed.

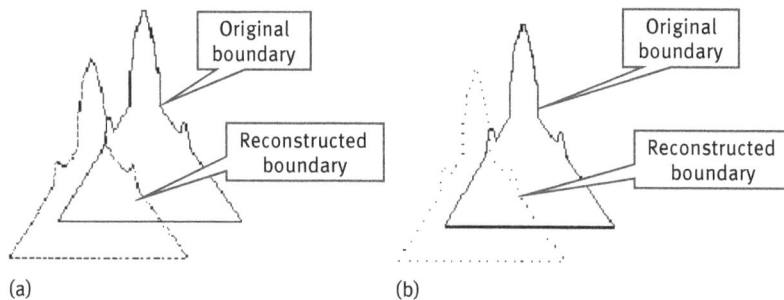

(a)   (b)

Figure 6.26: Reconstruction of the boundary using different descriptors.

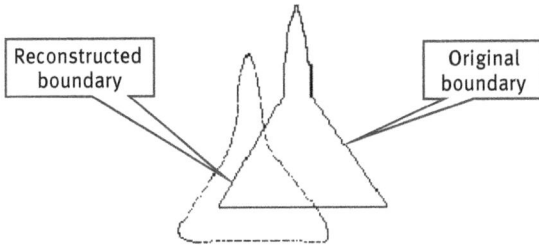

Figure 6.27: Reconstruction of the boundary with only 16 coefficients.

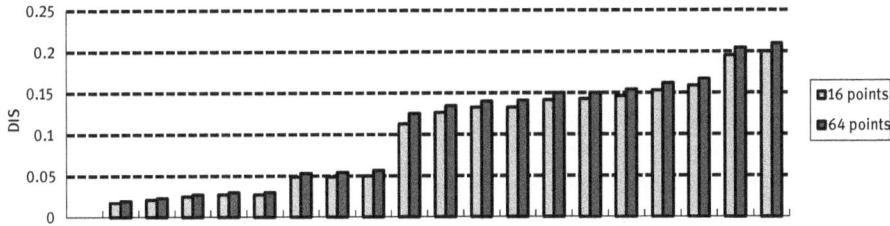

Figure 6.28: Comparing the DIS values obtained with 64 coefficients and 16 coefficients.

Figure 6.27 shows the results of using only 16 wavelet coefficients for representing the original boundary and for reconstructing the boundary. With fewer coefficients, the descriptive precision is decreased (*e. g.*, sharp angle becomes rounded), but the retrieval performance is less influenced.

Figure 6.28 gives the values of DIS for the first 20 returned results with 64 coefficients and 16 coefficients, respectively. It can be seen that the change in the DIS values with fewer coefficients is not very significant.

Figure 6.29 provides the retrieval results using 16 coefficients of a wavelet boundary descriptor. Figure 6.29(a) is the querying boundary, and Figure 6.29(b–d) is the first three returned results. The DISs for Figure 6.29(b–d) are 1.75%, 2.13%, and 2.50%, respectively. Though the DIS values here have some difference with respect to Figure 6.22, the retrieved boundary images are exactly the same. It seems that, with fewer coefficients, the boundary can still be described sufficiently.

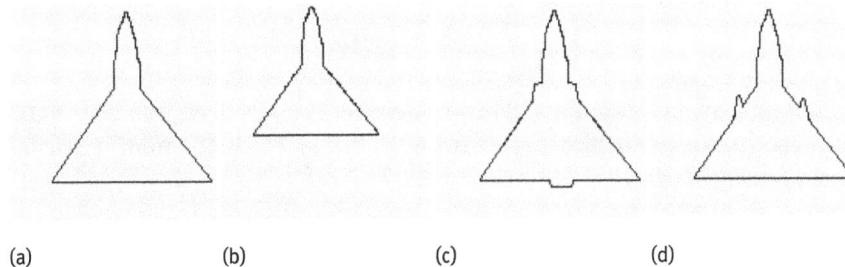

(a)   (b)   (c)   (d)

Figure 6.29: Retrieval results with 16 coefficients.

6.6 Fractal Geometry

Fractal geometry is a tool to study irregularities. It can also describe self-similarity. As far as shape analysis is concerned, fractal measurements are useful in problems that require complexity analysis of self-similar structures across different scales (Costa, 2001).

6.6.1 Set and Dimension

There are two possible distinct definitions of the dimension of a set in the Euclidean space R^N.

1. The **topological dimension** coincides with the number of degrees of freedom that characterize a point position in the set, denoted d_T. Therefore, the topological dimension is 0 for a point, 1 for a curve, 2 for a plane, and so on.
2. The **Hausdorff–Besicovitch dimension**, d, was formulated in 1919 by the German mathematician Felix Hausdorff.

For a set in R^N, it has $d \geq d_T$. Both of the two dimensions are within the interval $[0, N]$. The topological dimension always takes an integer value, which is not necessarily the case for the Hausdorff–Besicovitch dimension. In the case of planar curves (such as image contours), the Hausdorff–Besicovitch dimension is an important concept that can be applied to complexity analysis, where the closer d is to 2, the more the curve fills the plane to which it belongs. Therefore, this dimension can be used as a curve complexity measurement.

The notion of a **fractal set** is defined as a set whose Hausdorff-Besicovitch dimension is greater than its topological dimension Mandelbrot (1982). Since d can take a non-integer value, the name *fractal dimension* is used. It is nevertheless important to say that the formal definition of the Hausdorff-Besicovitch dimension is difficult to introduce and hard to calculate in practice. Therefore, alternative definitions for the fractal dimension are taken into account. Frequently, these alternative definitions lead to practical methods to estimate the fractal dimensions of experimental data.

6.6.2 The Box-Counting Approach

Let S be a set in R^2, and $M(r)$ be the number of open balls of a radius r that are necessary to cover S. An open ball of a radius r centered at (x_0, y_0), in R^2, can be defined as a set of points $\{(x, y) \in R^2 | [(x - x_0)^2 + (y - y_0)^2]^{1/2} < r\}$. The **box-counting fractal dimension** d is defined as

$$M(r) \sim r^{-d} \tag{6.46}$$

Therefore, the box-counting dimension is defined in terms of how the number of open balls, which are necessary to cover S, varies as a function of the size of the balls (the radius r, which defines the scale).

An example of how the box-counting dimension can be used in order to define the fractal dimension of Koch's triadic curve is illustrated in Figure 6.30. As shown in the figure, this curve can be constructed from a line segment by dividing it into three identical portions and substituting the middle portion by two line segments having the same length. The complete fractal curve is obtained by recursively applying *ad infinitum* the above construction process to each of the four resulting segments. Figure 6.30 shows the results of the four sequential steps of the Koch curve construction process. The self-similarity property of fractal structures stems from the fact that these structures are similar irrespective of the spatial scale used for their visualization.

Recall that the definition of the box-counting dimension involves covering the set with $M(r)$ balls for each radius r. While analyzing the Koch curve, 1-D balls that cover the line segments are taken. For the sake of simplicity, a unit length is also assumed for the initial line segment. Consequently, a single ball of the radius $r = 1/2$ is needed to cover the Koch curve segment (see Figure 6.31(a)). In the case where smaller balls are used, when $r = 1/6$, $M(r) = 4$ balls are needed to cover the Koch curve; whereas, when $r = 1/18$, a total number of $M(r) = 16$ balls is required, as illustrated in Figure 6.31(b, c). An interesting conceptual interpretation of this effect is as follows. Suppose the line segments of Figure 6.31(a–c) represent measuring devices with the smallest measuring units. This means that, if one chooses to measure the Koch curve in Figure 6.31 with the measuring device of a length 1, the conclusion is that the length of the curve is 1.

This is because that it cannot take into account the details that are less than 1. On the other hand, if a measuring device with a length 1/3 were used instead, four segments would be needed, yielding a total length of $4(1/3) \approx 1.33$. Table 6.4 summarizes this measuring process of $M(r)$ as a function of r.

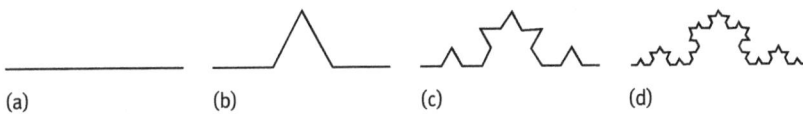

(a)   (b)   (c)   (d)

Figure 6.30: The steps in constructing Koch's triadic curve.

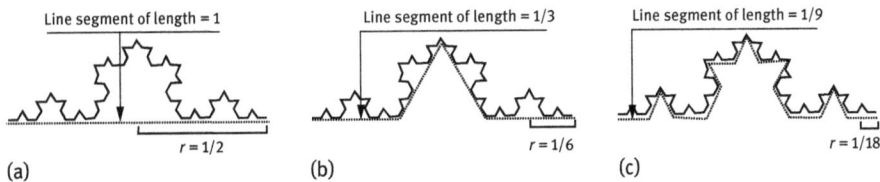

Line segment of length = 1   Line segment of length = 1/3   Line segment of length = 1/9

$r = 1/2$ $r = 1/6$ $r = 1/18$

(a)   (b)   (c)

Figure 6.31: The length for a fractal curve depends on the measuring device.

Table 6.4: The number $M(r)$ for balls of radius r necessary to cover the Koch curve in Figure 6.30.

r	$M(r)$	Measured curve length
$1/2 = (1/2)(1) = (1/2)(1/3)^0$	$1 = 4^0$	1
$1/6 = (1/2)(1/3) = (1/2)(1/3)^1$	$4 = 4^1$	1.33
$1/18 = (1/2)(1/9) = (1/2)(1/3)^2$	$16 = 4^2$	1.78
......	......	......

This analysis indicates that when r is decreased by a factor of 1/3, $M(r)$ is increased by a factor of 4. From eq. (6.46), this means that

$$4 \sim (1/3)^{-d} \tag{6.47}$$

Therefore, it has $d = \log(4)/\log(3) \approx 1.26$, which actually is the fractal dimension of the Koch curve. The Koch curve is **exactly self-similar**, as it can be constructed recursively by applying the generating rule, which substitutes each segment by the basic element in Figure 6.27(b), as explained above.

6.6.3 Implementing the Box-Counting Method

The basic algorithm for estimating the box-counting dimension is based on partitioning the image into square boxes of size $L \times L$ and counting the number $N(L)$ of boxes containing at least a portion (whatever small) of the shape. By varying L (the sequence of the box sizes, starting from the whole image, is usually reduced by 1/2 from one level to the next), it is possible to create the plot representing $\log[N(L)]$ versus $\log(L)$. The fractal dimension can then be calculated as the absolute value of the slope of the line interpolated to the $\log[N(L)]$ versus $\log(L)$ plot, as illustrated in Figure 6.32. The sequence of the box sizes, starting from the whole image, is usually reduced by 1/2 from one level to the next.

It is important to note that truly fractal structures are idealizations that neither exist in nature nor can be completely represented by computers. The main reasons for this are (1) *ad infinitum* self-similarity is never verified for shapes in nature (both at the microscopic scale, where the atomic dimension is always eventually achieved,

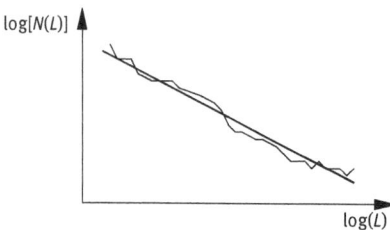

Figure 6.32: Ideal line and a real curve for the box-counting method.

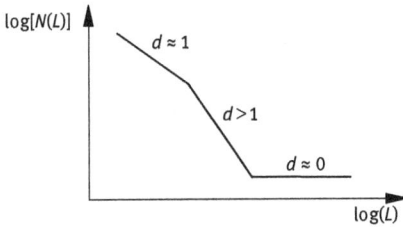

Figure 6.33: Diagrams of different Hausdorff dimensions for the box-counting method.

and the macroscopic scale, since the shape has a finite size) and (2) the limited resolution adopted in digital imaging tends to filter the smallest shape detail. Therefore, it is necessary to assume that "fractal" shapes present limited fractality (i. e., they are fractal with respect to a limited scale interval).

This is reflected in the $\log[N(L)]$ versus $\log(L)$ plot in Figure 6.33 by the presence of three distinct regions (i. e., a non-fractal region ($d \approx 1$), a fractal region ($d > 1$), and a region with the dimension of zero ($d \approx 0$)). This general organization of the curve can be easily understood by taking into account the relative sizes of the boxes and the shape. When the boxes are too small, they tend to "see" the shape portions as straight lines, yielding a dimension of 1. As the size of the boxes increases, the complex details of the shape start to become perceived, yielding a fractal dimension, which is possibly greater than 1 and less than or equal to 2. Finally, when the boxes become too large, the whole shape can be contained within a single box, implying a dimension of zero. Consequently, the line should be fitted to the fractal portion of the $\log(N(L))$ versus $\log(L)$ plot in order to obtain a more accurate estimation of the limited fractal behavior of the shape under analysis.

6.6.4 Discussions

The fractal dimension attempts to condense all of the details of the boundary shape into a single number that describes the roughness in one particular way. There can be an unlimited number of visually different boundary shapes with the same fractal dimension or **local roughness** (Russ, 2002).

The fractal dimension of a surface is a real number greater than 2 (the topological dimension of the surface) and less than 3 (the topological dimension of the space in which the surface exists). A perfectly smooth surface (dimension 2.0) corresponds to the Euclidean geometry, and a plot of the actual area of the measured area as a function of the measurement resolution would not change. However, for real surfaces, an increase in the magnification or a decrease in the resolution with which it is examined will reveal more nooks and crannies and the surface area will increase. For a surprising variety of natural and man-made surfaces, a plot of the area as a function of the resolution is linear on a log-log graph, and the slope of this curve gives the dimension d. As the measurement resolution becomes smaller, the total measured area increases.

6.7 Problems and Questions

?

6-1 In Section 6.1, several definitions for a shape were given. What are their common points? What are their differences? Provide some examples to show the application areas for each of them.

6-2 Given a square and a circle with the same area. Can they be distinguished by using the form factor, sphericity, circularity, and eccentricity?

6-3* In Figure Problem 6-3, a circle is approximated by an octant formed by discrete points. Compute the form factor, sphericity, circularity, and eccentricity of the octant.

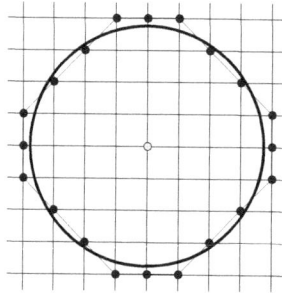

Figure Problem 6-3

6-4 What are the meanings of points, lines, and circles in Figure 6.7?

6-5 Prove the following two conclusions for a digital image.

  (1) If the boundary is 4-connected, and then the form factor takes the minimum value for an octant region.

  (2) If the boundary is 8-connected, and then, the form factor takes the minimum value for a diamond region.

6-6 If the boundary of a region in a digital image is 16-connected, what region form gives the minimum value for the form factor?

6-7 Provide some examples to show that the six simple descriptors listed in Section 6.3 reflect the complexity of the region form.

6-8 Compute the shape number for three regions in Figure 6.4 and compare them. Which two are more alike?

6-9 Following the examples in Figure 6.11, draw the horizontal projection and vertical projection of 26 capital English letters that are represented by 7×5 dot-matrix:

  (1) Compare the horizontal projection and vertical projection. What are their particularities?

  (2) Using only these projections, can the 26 capital English letters be distinguished?

6-10 Back to the octant in Figure Problem 6-3:

  (1) Compute its curvatures (take $k = 1, 2, 3$);

  (2) Compute the corresponding bending energy.

6-11 Can the fractal dimension be used to distinguish a square from a circle?

6-12* Consider the pattern sequence in Figure Problem 6-12. The left pattern is a square, and dividing it into nine equal small squares and removing the central small square yields the middle pattern in Figure Problem 6-12. Further dividing the rest of the eight squares into nine equal small squares and removing the central small square yields the right pattern in Figure Problem 6-12. Continuing this way, a square with many holes (from big ones to small ones) can be obtained. What is the fractal dimension of this square?

Figure Problem 6-12

6.8 Further Reading

1. **Definitions and Tasks**
    - One deep study on shape analysis based on contours can be found in Otterloo (1991).
    - An application of the wavelet-based shape description for image retrieval can be found in Yao (1999).
2. **Different Classes of 2-D Shape**
    - Shape classification is still studied widely, and some discussions can be found in Zhang (2003b).
3. **Shape Property Description**
    - Detailed discussion on the form factor of an octagon can be found in Rosenfeld (1974).
    - The computation of data listed in Table 6.1 can be found in Zhang (2002b).
4. **Technique-Based Descriptors**
    - The computation of curvature for 3-D images can be found in Lohmann (1998).
    - The structure information of a region has close relation with the shape information and the relationship among objects (Zhang, 2003b).
5. **Wavelet Boundary Descriptors**
    - More details on the wavelet boundary descriptors can be found in Zhang (2003b).
6. **Fractal Geometry**
    - Besides box-counting methods, another method for computing fractal dimensions is based on the area of influence or the area of spatial coverage (Costa, 1999; 2001).
    - According to fractal dimensions, 2-D signatures can be described (Taylor, 1994).

7 Motion Analysis

In order to analyze the change of the image information or moving objects, an image sequence (also known as motion pictures) should be used. Image sequence is a composition of a series of consecutive time variant 2-D (space) images or is often considered as a kind of 3-D images that can be expressed as $f(x, y, t)$. Compared to the still image $f(x, y)$, the time variable t has been added. When t takes a certain value, an image (a frame) in the sequence is obtained. Different from a single image, the image sequence (video is one kind of commonly used and will be used in the following) is continuously collected to reflect the change of moving objects and change of scenery. On the other hand, objective things are constantly moving and changing. The movement is absolute, while the stationary is relative. The scene changes and object motion in the image sequence was quite obvious and clear.

The analysis of motion in image sequence could be based on the analysis of a still image. However, the expansion in technology, change in the means, and the broadened of purpose are required.

The sections of this chapter are arranged as follows:

Section 7.1 overviews the subjects of the motion studies, including motion detection, locating and tracking moving objects, moving object segmentation and analysis as well as 3-D scene reconstruction and motion/scene understanding.

Section 7.2 discusses the motion information detection problem. The principles of motion detection using the image difference and using the model-based motion detection are introduced.

Section 7.3 focuses on motion detection especially moving object (target) detection. It introduces the techniques and effects of background modeling, and the computing of optical flow field and its application in the motion analysis.

Section 7.4 discusses the segmentation of moving objects in image sequences. First, the relation between moving object segmentation and motion information extraction is analyzed, and then, the dense optical flow algorithm based on brightness gradient is introduced. The ideas of segmentation and practical methods based on parameters and models are described.

Section 7.5 describes the typical motion tracking or moving object tracking technology, including the Kalman filter, particle filter, mean shift, and kernel tracking techniques as well as the tracking strategy using subsequences decision.

7.1 The Purpose and Subject of Motion Analysis

Compared to texture and shape, the concept of motion is relatively straightforward. The objective and tasks of **motion analysis** may include the following.

DOI 10.1515/9783110524123-007

7.1.1 Motion Detection

Motion detection means to detect whether there is movement information in the scene (including **global motion** and **local motion**). In this case, only a single fixed camera is usually required. A typical example is the security surveillance, then any factors leading to changes in the image are required to be taken into account. Of course, due to the changes in lighting is relatively slow while the changes caused by the movement of objects would be relatively quick, these changes should be further distinguished.

7.1.2 Locating and Tracking Moving Objects

The concern here is mainly to discover whether there are moving objects in the scene; if an object is existing, what is the current position; further, it may need to find its **trajectory** and to predict its next moving direction and tendency as well as future moving route. For such a situation, only a single fixed camera could be generally sufficient. In practice, the actual camera may be still while the object is moving, or the camera moving while the object is stationary. If both of them are moving, this would be the most complex case.

According to different research purposes, different techniques may be employed. If it is required only to determine the position of moving objects, then some motion segmentation methods can be used. In this case, the initial segmentation of moving objects can be made by means of the motion information. If further determination of the direction of movement and the trajectory of the moving object are required, or even the prediction of motion tendency is wanted, then some matching techniques would be used to establish the relationship among the object (target) image data, the object characteristics, or the graph representing a moving object.

Moving object localization is often considered as the synonymous with moving object detection, but it is more concerned about the object location instead of object's characteristics. Often some assumptions are made:

1. Maximum speed: If the position of a moving object is known in the previous frame, then its position in current frame would be inside a circle whose center is at the position in last frame and whose radius is the maximum speed of object motion;
2. Small acceleration: the change of the object velocity is limited, it is often predictable as it has more or less some regularity;
3. Mutual correspondence: rigid object will maintain a stable pattern in the image sequence, and the object in scene has correspondence with point in image;
4. Joint motion: If there are multiple object points, their modes of motion are related (similarity).

7.1.3 Moving Object Segmentation and Analysis

The **moving object segmentation** goes further than just moving object location. More than the position determination of the object, this process needs to accurately extract the object so as to obtain its pose. The moving object analysis will provide the identity and motion parameters of the object, so as to get the movement law, and determine the motion type, etc. In this situation, it is often needed to use a video camera to obtain an image sequence, to make the distinction between global motion and local motion. Also it is often needed to obtain 3-D characteristics of objects, or to further identify the category of the moving object.

7.1.4 Three-Dimensional Scene Reconstruction and Motion Understanding

On the basis of moving information and object motion information, the depth/distance of 3-D objects from camera, the surface orientation of objects, and the occlusion of several objects can be computed. On the other hand, by combining moving information and other information extracted from images, the causal relationship of motion could be determined. Further aided by scene knowledge, the interpretation of scene and motion, the understanding of behavior of moving objects could also be learned. This will be discussed in Volume III of this book set.

7.2 Motion Detection

To understand the change of a scenario, it is first required to detect motion information, which determines whether there is movement, what are moving and where are changed. Secondly, it is required to estimate the motion, that is, to determine the parameters of movement (magnitude and direction, etc.). The second step is also known as motion estimation, but in many cases, these two steps are still collectively known as motion detection. Continuously, motion detection is very unique for video image processing and is the basis for many tasks using video as input.

Detection of motion often means to detect the motion information of the whole image. As indicated in the previous section, there are both foreground and background motions in video, so motion detection is necessary to detect changes caused by the movement of the whole scene, and also to detect changes caused by the movement of the specific objects.

7.2.1 Motion Detection Using Image Differences

In a video, the difference between the two consecutive (before and after) frame images can be found by a pixel-by-pixel comparison. Suppose lighting conditions do not change substantially in multiple frames, then the place where the difference image has nonzero values indicates that the pixel at this place has been moved/changed (to be noted that at the place where the difference image has zero value the pixel may also be moved). In other words, taking the difference between two adjacent frame images can find the location and shape change of moving objects in the image.

7.2.1.1 Calculating a Difference Image

Referring to Figure 7.1(a), suppose the object is brighter than the background. With the help of **difference image**, one positive region in the front of movement and one negative region in the rear of movement can be obtained. The moving information of the object, or the shape of some parts of an object can be further obtained. If taking the differences for all adjacent two images in a series, and making the logic AND of positive regions and negative regions, respectively, the shape of an entire object can finally be found. Figure 7.1(b) illustrates this procedure, a rectangular region is gradually moved down and sequentially travel across the different parts of the oval object, combining the results of all times, a complete ellipse object is produced.

If a series of images have been collected under the relative motion between the image acquisition device and the scene being shot, then the pixels that have been changed can be determined with the help of the motion information presented in the image. Suppose the two images $f(x, y, t_i)$ and $f(x, y, t_j)$ were collected at the time t_i and t_j, respectively, then the difference image obtained is

$$d_{ij}(x, y) = \begin{cases} 1 & |f(x, y, t_i) - f(x, y, t_j)| > T_g \\ 0 & \text{Otherwise} \end{cases} \tag{7.1}$$

where T_g is the gray-level threshold. The pixel with value 0 in difference image corresponds to the place where no change occurs between the time before and after the current time (for changing arising from motion). The pixel with value 1 in difference image corresponds to the place where change occurs between the time before and after the current time. This change is often caused by the motion of object. However, the pixel with value 1 may also arising from other different circumstances, such as $f(x, y, t_i)$ is a pixel belonging to the moving object while $f(x, y, t_j)$ is a pixel belonging to background, or vice versa. Other examples include that $f(x, y, t_i)$ is a pixel belonging to a moving object, while $f(x, y, t_j)$ is a pixel belonging to another moving

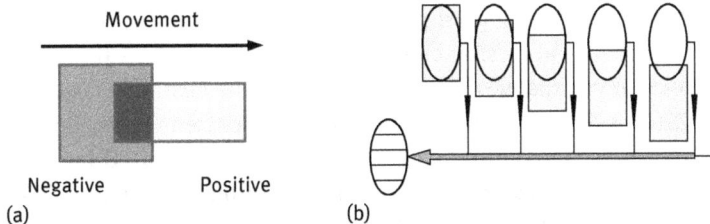

Figure 7.1: Using difference image to extract object.

object, or $f(x, y, t_j)$ is a pixel belonging to the same moving object but at another location (so the gray-level may be different). Example 3.4 in Volume I of this book set has shown an instance that the object moving information can be detected by using image difference.

The threshold T_g in eq. (7.1) is used to determine whether the gray-levels of two consecutive images exists obvious difference. Another method is to use the likelihood ratio to identify if there is significant difference:

$$\frac{\left[\frac{\sigma_i + \sigma_j}{2} + \left(\frac{\mu_i - \mu_j}{2} \right)^2 \right]^2}{\sigma_i \cdot \sigma_j} > T_s \tag{7.2}$$

where μ_i, μ_j and σ_i, σ_j are the mean and variance of two collected images at time t_i and t_j, respectively; T_s is the significance threshold.

In real applications, due to the influence of random noise, the places where no pixels to move could also have some nonzero difference values between two images. To distinguish the movement of pixels with the influence of noise, it is possible to use a larger threshold value for the difference image, that is, when the difference is greater than a pre-determined threshold, then the pixels would be considered as had moved. In addition, as the pixels with value 1 that caused by noise in image are generally more isolated, so these pixels can also be removed by connectivity analysis. However, such an approach could sometimes remove also those pixels belonging to smaller objects and/or belonging to slow motion objects.

7.2.1.2 Calculating a Cumulative Difference Image

To overcome the above-mentioned problem with random noise, using multiple images can be considered. If the change at one location only appears occasionally, it can be judged as noise. Let a series of image be $f(x, y, t_1)$, $f(x, y, t_2)$, ..., $f(x, y, t_n)$, and the first image $f(x, y, t_1)$ be the reference image. By comparing the first image with each of subsequent image, a **cumulative difference image** (ADI) can be obtained. In this image, the value at each location is the sum of the number of changes in each comparison.

One example for cumulative difference image ADI is given in Figure 7.2. Figure 7.2(a) shows an image captured at time t_1, there is a square object inside, which is moved horizontally to the right one pixel per unit time. Figure 7.2(b, c) represents the images captured at time t_2 and t_3 (after one and two time units), respectively. Figure 7.2(d, e) corresponds to cumulative difference images for the next time t_2 and t_3, respectively. Figure 7.2(d) is the common difference image discussed earlier, the left square marked with 1 corresponds to the gray-level difference (as a unit) between the trailing edge of the object in Figure 7.2(a) and the background in Figure 7.2(b), and the right square marked with 1 corresponds to the gray-level difference (also as a unit) between the background in Figure 7.2(a) and the leading edge of the object in

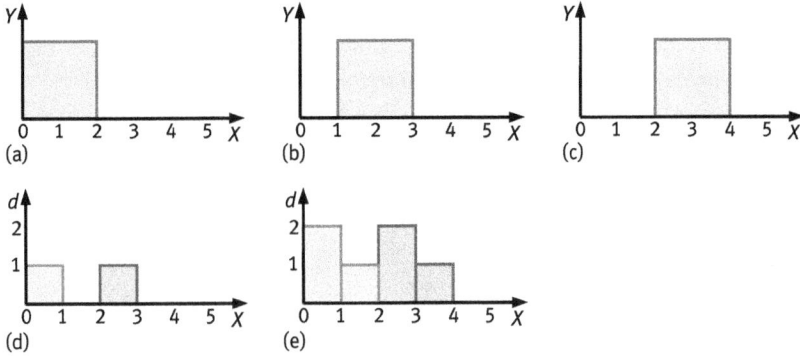

Figure 7.2: Using accumulated difference image to extract object.

Figure 7.2(b). Figure 7.2(e) can be obtained by adding Figure 7.2(d) to the gray-level difference of Figure 7.2(a, c), in which the gray difference between 0 and 1 is two units, and the gray difference between 2 and 3 is also two units.

Referring to the example above, it is shown that the cumulative difference image ADI has three functions:

1. In ADI, the gradient relationship between the values of adjacent pixels can be used to estimate the velocity vector of object movement, where the gradient direction is the direction of the velocity, and the gradient magnitude is proportional to the magnitude of velocity.
2. The pixel values in ADI can help to determine the size and the moving distance of moving object.
3. ADI includes all the historical data of the object motion, and is helpful for detecting slow motion and motion of smaller objects.

In practical applications, three types of ADI can be distinguished Gonzalez (2008): absolute ADI ($A_k(x, y)$), positive ADI ($P_k(x, y)$) and negative ADI ($N_k(x, y)$). Assuming the gray-level of moving object is larger than that of background, then for $k > 1$, the three types of definitions of ADI (taking $f(x, y, t_1)$ as reference, and T_g is as above) are:

$$A_k(x, y) = \begin{cases} A_{k-1}(x, y) + 1 & |f(x, y, t_1) - f(x, y, t_k)| > T_g \\ A_{k-1}(x, y) & \text{Otherwise} \end{cases} \tag{7.3}$$

$$P_k(x, y) = \begin{cases} P_{k-1}(x, y) + 1 & [f(x, y, t_1) - f(x, y, t_k)] > T_g \\ P_{k-1}(x, y) & \text{Otherwise} \end{cases} \tag{7.4}$$

$$N_k(x, y) = \begin{cases} N_{k-1}(x, y) + 1 & [f(x, y, t_1) - f(x, y, t_k)] < -T_g \\ N_{k-1}(x, y) & \text{Otherwise} \end{cases} \tag{7.5}$$

The above three types of ADI values are all the results of the pixel counting, they are initially zero. The following information can be obtained from them:

1. The nonzero area of positive ADI is equal to that of moving object.
2. The position corresponding to the moving object in ADI is that of moving object in the reference image.
3. When the moving object in positive ADI is moved to the place that is not coincide with the moving object in reference image, the counting of positive ADI stops.
4. The absolutely ADI includes all object regions in both positive ADI and negative ADI.
5. The movement direction and speed of moving objects can be determined on the basis of the absolute ADI and the negative ADI.

7.2.2 Model-Based Motion Detection

Motion detection can also be carried out by means of the motion model. In the following, the camera model is considered for global motion detection.

Assuming that the global motion vector (MV) $[u, v]^T$ at a point (x, y) in image can be calculated by its spatial coordinates and a set of model parameters $(k_0, k_1, k_2, \ldots)$, the common model can be expressed as

$$\begin{cases} u = f_u(x, y, k_0, k_1, k_2, \cdots) \\ v = f_v(x, y, k_0, k_1, k_2, \cdots) \end{cases} \tag{7.6}$$

When the model parameters were estimated, a sufficient number of observation points should first be selected from adjacent frames. It is followed by using some matching algorithm to derive the MVs of these points, and the parameter fitting method is finally applied to estimate model parameters. Many methods have been proposed to estimate the global motion model, each of them have their own characteristics in selecting the observation points, in aspects of matching algorithms, motion models and motion estimation methods.

Equation (7.6) represents a general model. In practice, often more simplified models are used. Commonly considered motion types of camera have six kinds:

1. **Panning**: the camera (axis) rotates horizontally;
2. **Tilting**: the camera (axis) rotates vertically;
3. **Zooming**: the camera changes its focal length (long or short);
4. **Tracking**: the camera moves horizontally (laterally);
5. **Booming**: the camera moves vertically (transversely);
6. **Dolling**: the camera moves forth and back (horizontally).

These six types of camera movement can also combined to form new operations that constitute three categories (Jeannin 2000):

1. Shifting operation;
2. Rotating operation;
3. Scaling operation.

For general applications, the linear affine model with 6 parameters is:

$$\begin{cases} u = k_0 x + k_1 y + k_2 \\ v = k_3 x + k_4 y + k_5 \end{cases}$$
(7.7)

Affine model belongs to the linear polynomial parameter model that is mathematic-ally easier to handle. In order to improve the description ability of the global motion model, some extensions can be made on the basis of affine model. For example, by adding the quadratic term xy into the polynomial model, a bilinear parameter model can be obtained:

$$\begin{cases} u = k_0 xy + k_1 x + k_2 y + k_3 \\ v = k_4 xy + k_5 x + k_6 y + k_7 \end{cases}$$
(7.8)

A global MV detection method, based on bilinear model, is as follows Yu (2001b). To make an estimate of the eight parameters in bilinear models, it is required to find a group (more than four) MVs with observation values (eight equations can be provided). When obtaining the MV of observations, taking into account the relatively large global motion values of global motion, the whole frame image is often divided into a number of square pieces (such as 16×16), and then the observed MVs are com-puted by using a block matching method. By selecting a larger matching block size, the offset between the global MV and the matching MV caused by local motion can be reduced, in order to obtain a more accurate global motion observations.

Example 7.1 Global motion detection based on the bilinear model.
A real example of global motion detection based on bilinear model is presented in Figure 7.3, in which the MVs obtained by using the block matching algorithm are super-imposed on the original image (as a short line segment starting from the center of block) to express the movement of each block.

It is seen that in this image, the right part has some motions with higher velocity. This is because that the center of camera's zooming is located at (or the optical axis is pointed to) the left part of goalkeeper. On the other side, there are still some motions caused by local objects (football players), so at these locations the MVs computed by block matching algorithm would be different from the global MVs (such as shown in the vicinity of the location of each football player). In addition, the block matching method may generate some random error data in the low texture regions of the image, as in the background (near the grandstand). The presence of these errors could result the abnormal data points in the image. ◻

Figure 7.3: Motion vector values obtained by the direct block matching algorithm.

7.3 Moving Object Detection

In the above section, some basic motion detection methods have been introduced. Here two further technical categories are presented, which are more suitable for the detection of moving objects (with local motion).

7.3.1 Background Modeling

Background modeling is a wide-ranging idea for motion and object detection, and can be realized with different techniques, so it is also seen as a general term for a class of motion detection method.

7.3.1.1 Basic Principle

Motion detection is to find the motion information in the scene. An intuitive approach is to compare the current frame to be checked with the original background without motion information, the difference in the comparison indicates the result of movement. First consider a simple case. There is a moving object in a stationary scene (background), then the difference caused by the movement of this object in adjacent two video frames would appear at the corresponding place. In this case, by computing a difference image (as above) could detect moving objects and locate their positions.

Calculating a difference image is a simple and fast method for motion detection, but the result would not be good enough in a number of cases. This is because the calculation of difference images will detect all light variations, environment ups and downs (background noise), camera shake, etc. together with the object motion. This problem is more serious especially when the first frame is taken as the reference frame. Therefore, the true motion of objects can only be detected in very tightly controlled situations (such as in the unchanged environment and background).

A reasonable idea for motion detection is not to consider that the background is entirely stationary, but to calculate and maintain a dynamic (satisfy some models) background frame. This is the basic idea of background modeling.

There is a simple background modeling method, which uses the mean or median of N frames prior to the detection of the current frame, to determine and update the value of each pixel in the N cycle frame. One particular algorithm includes the following steps:

1. Acquire first pre-N frame images, determine the values of the N frames at each pixel location, and take their average value as the current background value;
2. Acquire then the $N + 1$ frame image, compute the difference between the current frame and the current background at each pixel (the difference can be thresholded to eliminate or reduce noise);
3. Use smooth or a combination of morphological operations to remove those very small difference and to fill holes in large regions. Preserved regions should represent the moving objects in the scene;
4. Update the average value at each pixel location by combining the information of the $N + 1$ frame image;
5. Return to step (2), and consider the next frame image.

This approach based on average values for maintaining the value of the background is relatively simple, with small amount of calculation, but the result would be not very good when there are multiple objects or slow motion objects in the scene.

7.3.1.2 Typical Methods in Applications

Here, some typical and basic background modeling methods used in real situations are introduced. They divide the foreground extraction process into two steps: model training and actual detection. A mathematical model for background is established through the training, and the model is used in detection to eliminate background and to obtain the foreground.

Approach Based on Single Gaussian Model Single Gaussian model-based approach considers that the values of the pixels follow a Gaussian distribution in the video sequence. Concretely, for each fixed pixel location, the mean μ and variance σ of the position of the pixel values in N frames of training sequence are calculated, from which a unique single Gaussian background model is identified. During the motion detection, the method of background subtraction is carried out to calculate the difference between the pixel values in current frame image and the pixel values of the background of the model, then the difference is compared to the threshold value T

(often take 3 times of variance), *i. e.* according to $|\mu_T - \mu| \le 3\sigma$, the pixel can be determined as belonging to foreground or background.

This model is relatively simple but requires more stringent application conditions. For example, it requires no significant changes in light intensity in a long time, while the moving foreground has small shadow in the background during the period of detection. Another disadvantage is the high sensitivity for the change of light (light intensity), which can cause the model does not hold (both mean and variance are changed). When moving foreground existing in the scene, it may cause large false alarm rate. This is because there is only one model, the moving foreground cannot be separated from stationary background.

Approach Based on Video Initialization   In the case that the background of the training sequence is stationary, but moving foreground exists, if it is able to first extract out the background values of each pixel, and to separate moving foreground from background, then the background modeling can be carried out, the foregoing problem would be overcame. This process can also be seen as initializing the training video before the background modeling, so that the influence of moving foreground on the background modeling could be filtered out.

In practice, a minimum length threshold T_1 is first set for N frames training images containing moving foreground, and then, this threshold is used to select the sequence of length N for each pixel location in the sequence, to obtain several sub-sequence $\{L_k\}$, $k = 1, 2, \ldots$, from which the sequence with longer length and smaller variance will be selected as the background sequence.

By this initialization, the case that background is stationary while moving foreground exists in the training sequence is transformed to the case that background is stationary and no moving foreground in the training sequence. In this situation, it is still possible to use the approach based on single Gaussian model (as discussed above) for background modeling.

Approach Based on Gaussian Mixture Model   When the movement exists also in the background of training sequence, the result based on the method of single Gaussian model is not very suitable. In this case a more robust and effective approach is modeling each pixel by mixed Gaussian distribution, namely **Gaussian mixture model** (GMM). In other words, the modeling is made for each state in background. The model parameters of states are updated according to the data belonging to the state, in order to solve the problem of moving background problem under the background modeling. According to local nature, some Gaussian distributions represent foreground while others represent background. The following algorithm can distinguish these Gaussian distributions.

The basic method based on GMM is sequentially reading N frames of training images, each time the iterative modeling is carried out for every pixel. Suppose that a pixel having a gray level $f(t)$, $t = 1, 2, \ldots$, at the time t, $f(t)$ can be modeled by using K

(K is the maximum number of pixels allowed for each model) Gaussian distributions $N(\mu_k, \sigma_k^2)$, where $k = 1, \ldots, K$. Gaussian distribution changes over time with the change of scene, so it is a function of time, and can be written as

$$N_k(t) = N[\mu_k(t), \sigma_k^2(t)], \quad k = 1, \ldots, K \tag{7.9}$$

The main concern for the choice of K is computational efficiency, and K is often taking values of 3–7.

At the beginning of the training, an initial standard deviation is set. When a new image is read, the pixel value of this image is used to update the original pixel values of the background model. For each Gaussian distribution a weight $w_k(t)$ is added (the sum of all weights is 1), so the probability of observing $f(t)$ is

$$P[f(t)] = \sum_{k=1}^{K} w_k(t) \frac{1}{\sqrt{2\pi}} \exp\left[\frac{-[f(t) - \mu_k(t)]^2}{\sigma_k^2(t)} \right] \tag{7.10}$$

EM algorithm can be used to update the parameters of the Gaussian distribution, but it is computationally expensive. An easy way is to compare each pixel with the Gaussian function, if it falls within the average range of 2.5 times the variance, then it is considered to be a match, that is, it fits to the model, so it can be used to update the values of the mean and variance of the model. If the number of models for current pixel is less than K, a new model is to be established for this pixel. If there are more than one matches appearing, the best can be chosen.

If a match is found, then for a Gaussian distribution l:

$$w_k(t) = \begin{cases} (1 - a)w_k(t - 1) & k \neq l \\ w_k(t - 1) & k = l \end{cases} \tag{7.11}$$

Then w is renormalized. In eq. (7.11), a is a learning constant, $1/a$ determines the rate of parameter change. The parameters to match the Gaussian function can be updated as follows:

$$\mu_k(t) = (1 - b)\mu_l(t - 1) + bf(t) \tag{7.12}$$
$$\sigma_k^2(t) = (1 - b)\sigma_l^2(t - 1) + b[f(t) - \mu_k(t)]^2 \tag{7.13}$$

where

$$b = aP[f(t)|\mu_l, \sigma_l^2] \tag{7.14}$$

If no match is found, the Gaussian distribution corresponding to the minimum weight can be replaced by a new Gaussian distribution with mean $f(t)$. Compared to other $K-1$ Gaussian distributions, it has a higher variance and lower weight, so it may become part of the local background. If K models have been checked and they do not meet the conditions, then the model corresponding to the smallest weight is replaced with a new model, the mean value of the new model is the pixel value, and an initial standard deviation is set. Continuous in this way until all the training images are trained.

At this moment, the Gaussian distribution the most likely to be assigned to the gray value of the current pixel can be determined. Then, it is required to make sure whether it belongs to the foreground or the background. This may be determined by means of a constant B corresponding to the whole process of observation. Suppose the ratios of background pixels in all frames are greater than B, then all Gaussian distributions can be ranked according to $w_k(t)/\sigma_k(t)$, a higher value indicates a large weight, or a small variance, or both. These cases correspond to the situations that the given pixel is likely to belong to the background.

Approach Based on Codebook   In the codebook based method, each pixel is represented by a codebook, a **codebook** may comprise one or more code words, and each code word represents a state (Kim, 2004). The initial code is learnt by means of a group of training frame images. There are no restrictions on the content of the training frame images that can contain moving foreground or moving background. Next, a time-domain filter is used for filtering out the code words representing a motion of the foreground and keeping the code words representing background; and a spatial filter is used to recover those codewords wrongly filtered out, so as to reduce the false alarm caused by the occurrence of sporadic foreground regions in the background. Such codebook represents a compressed form of a background model in video sequence.

7.3.1.3 Some Experiment Results
Background modeling is a training–testing process. It uses some earlier frame images in the beginning of a sequence to train a background model. This model is then applied to the rest frame images, and the motion is tested according to the difference between the current frame and the background model. In the simplest case, the background is stationary in the training sequence, and there is no moving foreground. Some complicated situations include: the background is stationary, but there is moving foreground in training sequence; the background is not static, but there is no moving foreground in the training sequence. The most complicated situation is that the background is not static, and there is also moving foreground in the training sequence. In the following, some experiment results obtained by using background modeling for the first three cases are illustrated (Li, 2006).

Experimental data are from three sequences of an open access, universal video library (Toyama, 1999). There are a total of 150 color images in a sequence, each of them has a resolution of 160×120. During the experiment, image editing software has been used to provide the binary reference results, then each of the background modeling method is used to detect the moving object in the sequence, and a binary test result is obtained. For each sequence, 10 images are selected for testing. The test results are compared with the reference results, the average detection rate (the ratio of the number of foreground pixels detected over the real number of foreground pixels) and the average false alarm rate (the ratio of the number of no-foreground pixels detected over the number of pixels detected as foreground) are collected.

Results for Stationary Background with No Moving Foreground   A set of experiment results are shown in Figure 7.4. In the sequence used, the initial scene consists of only stationary background, the goal is to detect the person subsequently entered in the scene. Figure 7.4(a) is a scene after the entering of a person. Figure 7.4(b) shows the reference result. Figure 7.4(c) gives the test results obtained with a method based on a single Gaussian model. The detection rate of this method is only 0.473, while the false alarm rate is 0.0569. It is seen from Figure 7.4(c) that a lot of pixels have not been detected (in regions of low gray-level pixels), some error detected pixels on the background have be found, too.

Results for Stationary Background with Moving Foreground   A set of experiment results are shown in Figure 7.5. In the sequence used, the initial scene has a person inside but was leaved later, the goal is to detect the leaved person. Figure 7.5(a) is a scene when the person has not left. Figure 7.5(b) shows the reference result. Figure 7.5(c) gives the test result obtained with a method based on video initialization. Figure 7.5(d) gives the test result obtained with a method based on codebook.

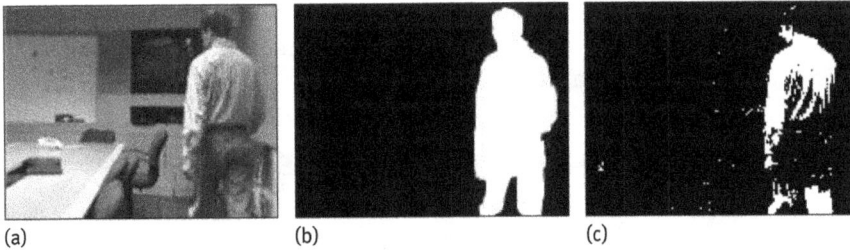

(a)   (b)   (c)

Figure 7.4: Results for stationary background with no moving foreground.

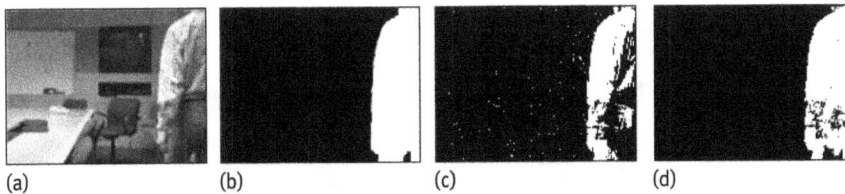

(a)   (b)   (c)   (d)

Figure 7.5: Results for stationary background with moving foreground.

Table 7.1: Statistical results for stationary background with moving foreground.

Method	Detection rate	False alarm rate
Based on video initialization	0.676	0.051
Based on codebook	0.880	0.025

The comparison of these two methods shows that the detection ratio of the method based on codebook is higher than that of the method based on video initialization, and the false alarm rate of the method based on codebook is lower than that of the method based on video initialization. This is because the codebook method has construct many codewords for each pixel so the detection rate is improved, while the spatial filter used in detection process reduces the false alarm rate. Some statistical results are shown in Table 7.1.

Results for Moving Background Without Moving Foreground   A set of experiment results are shown in Figure 7.6. In the sequence used, the initial scene has a shaking tree in the background, the goal is to detect the person getting in after then. Figure 7.6(a) is a scene after the person has entered. Figure 7.6(b) shows the reference result. Figure 7.6(c) gives the test result obtained with a method based on Gaussian mixing model. Figure 7.6(d) gives the test result obtained with a method based on codebook.

The comparison of these two methods shows that both methods have specifically designed models for moving background so that higher detection rates are achieved (The former has a little high rate than the latter). Because the former one does not have the processing steps corresponding to the spatial filter in the latter method, the false alarm rate of the former method is a little high than that of the latter. Specific statistical data are shown in Table 7.2.

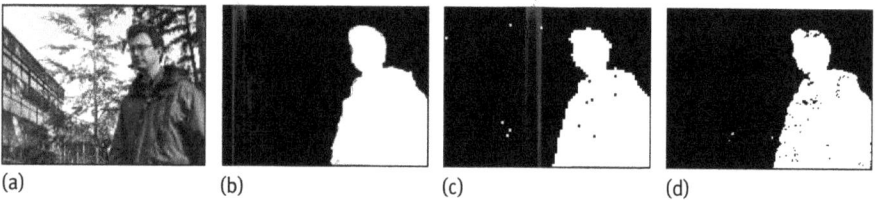

(a)   (b)   (c)   (d)

Figure 7.6: Results for moving background with no moving foreground.

Table 7.2: Statistical results for moving background without moving foreground.

Method	Detection rate	False alarm rate
Based on Gaussian mixing model	0.951	0.017
Based on codebook	0.939	0.006

Finally, it should be noted that the method based on single Gaussian model is relatively simpler, but the situations it could be used are less popular, as it can be only used for the cases of stationary background without moving foreground. Other methods have tried to overcome the limitations of single-Gaussian model-based method, but their common problem is that if the background needs to update, then the entire background model should be recalculated, rather than just updating parameters with a simple iteration.

7.3.2 Optical Flow

The movement of objects in scene can make the objects to appear in different relative positions in images captured during object movements. This difference in position may be called parallax, which corresponds to displacement vector (including magnitude and direction) reflected in the image. If the parallax is divided by the time difference, then the velocity vector (also called instantaneous displacement vector) will be obtained. All the velocity vectors together (may vary among them) constitute a vector field and in many cases may also be referred to as optical flow field (more distinctions will be discussed in Volume III of this book set).

7.3.2.1 Optical Flow Equation

Let a particular image point is at (x, y) at time t, this point will move to the point $(x + dx, y + dy)$ at time $t + dt$. If the time interval dt is small, it may be desirable (or assumed) that the gray level of this image point remains unchanged. In other words, there is:

$$f(x, y, t) = f(x + dx, y + dy, t + dt) \qquad (7.15)$$

The right-hand side may be expanded with a Taylor series, let $dt \to 0$, taking the limit and omitting the higher order terms available will produce

$$-\frac{\partial f}{\partial t} = \frac{\partial f}{\partial x}\frac{dx}{dt} + \frac{\partial f}{\partial y}\frac{dy}{dt} = \frac{\partial f}{\partial x}u + \frac{\partial f}{\partial y}v = 0 \qquad (7.16)$$

where u and v are the moving speed of the image point in the X and Y directions, respectively, which constitutes a velocity vector. Let

$$f_x = \partial f/\partial x \qquad f_y = \partial f/\partial y \qquad f_t = \partial f/\partial t \qquad (7.17)$$

Then, the **optical flow equation** is obtained:

$$[f_x, f_y] \cdot [u, v]^T = -f_t \qquad (7.18)$$

Optical flow equation shows that the gray level changing rate in time at a point is the product of the gray level changing rate in space at this point and the moving velocity of this point in space.

In practice, the gray-level changing rate in time can be estimated with the first-order differential average along the time axis:

$$f_t \approx \frac{1}{4} \left[f(x, y, t + 1) + f(x + 1, y, t + 1) + f(x, y + 1, t + 1) + f(x + 1, y + 1, t + 1) \right]$$
$$- \frac{1}{4} \left[f(x, y, t) + f(x + 1, y, t) + f(x, y + 1, t) + f(x + 1, y + 1, t) \right] \tag{7.19}$$

The gray-level changing rate in space can be estimated with the first-order differential average along X and Y directions:

$$f_x \approx \frac{1}{4} \left[f(x + 1, y, t) + f(x + 1, y + 1, t) + f(x + 1, y, t + 1) + f(x + 1, y + 1, t + 1) \right]$$
$$- \frac{1}{4} \left[f(x, y, t) + f(x, y + 1, t) + f(x, y, t + 1) + f(x, y + 1, t + 1) \right] \tag{7.20}$$

$$f_y \approx \frac{1}{4} \left[f(x, y + 1, t) + f(x + 1, y + 1, t) + f(x, y + 1, t + 1) + f(x + 1, y + 1, t + 1) \right]$$
$$- \frac{1}{4} \left[f(x, y, t) + f(x + 1, y, t) + f(x, y, t + 1) + f(x + 1, y, t + 1) \right] \tag{7.21}$$

7.3.2.2 Optical Flow Estimation Using Least Squares

Formulas (7.19)–(7.21), after substituting into eq. (7.18), can be used to estimate the optical flow components u and v with the help of least squares. In two adjacent images $f(x, y, t)$ and $f(x, y, t + 1)$, pixels with the same u and v are selected at N different positions. Let $\hat{f}_t^{(k)}$, $\hat{f}_x^{(k)}$, and $\hat{f}_y^{(k)}$ represent the estimations of f_t, f_x, and f_y, at kth position $(k = 1, 2, \ldots, N)$, respectively:

$$\boldsymbol{f}_t = \begin{bmatrix} -\hat{f}_t^{(1)} \\ -\hat{f}_t^{(2)} \\ \vdots \\ -\hat{f}_t^{(N)} \end{bmatrix} \qquad \boldsymbol{F}_{xy} = \begin{bmatrix} \hat{f}_x^{(1)} & \hat{f}_y^{(1)} \\ \hat{f}_x^{(2)} & \hat{f}_y^{(2)} \\ \vdots & \vdots \\ \hat{f}_x^{(N)} & \hat{f}_y^{(N)} \end{bmatrix} \tag{7.22}$$

Then the least squares estimations for u and v are:

$$[u \quad v]^{\mathrm{T}} = \left(\boldsymbol{F}_{xy}^{\mathrm{T}} \boldsymbol{F}_{xy} \right)^{-1} \boldsymbol{F}_{xy}^{\mathrm{T}} \boldsymbol{f}_t \tag{7.23}$$

Example 7.2 Optical flow detection.

An example of optical flows detected is given in Figure 7.7. Figure 7.7(a) is an image of a sphere ball with a pattern on surface. Figure 7.7(b) is the image obtained after the ball

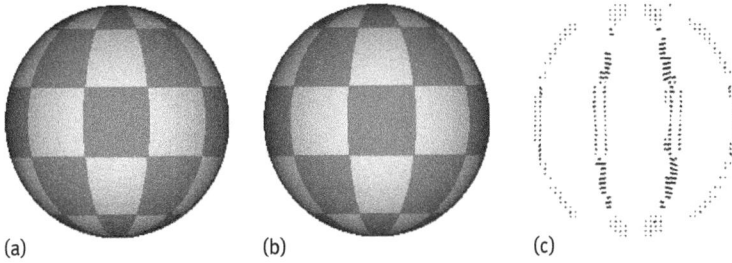

Figure 7.7: Optical flow detection.

rotating (around the vertical axis) to the right by a small angle. The ball movement in 3-D space has been reflected in the 2-D image almost as a horizontal translation, so the detected optical flow as shown in Figure 7.7(c) has a large part of the optical flow distribution along the meridian, reflecting the result of the horizontal movement of pattern edges.

7.3.2.3 Optical Flow in Kinematic Analysis

The motion trajectory of objects can be obtained by using image difference. Using optical flow cannot obtain the motion trajectory of objects but can obtain useful information for image interpretation. The analysis of optical flow can be used to solve a variety of motion issues: stationary camera with object motion, moving camera with stationary object, and moving camera with moving object.

The movement in image sequence is a combination of the following four basic movements, using optical flow for their detection and identification can be achieved with some simple operators on the basis of their characteristics (see Figure 7.8).

1. Translational movement (of different directions) in which the distance from the camera is constant: it composes a set of parallel MV (see Figure 7.8(a)).
2. Translational (isotropic symmetry) movement in which the camera moves in the depth direction along the sight line: it forms a set of vectors (see Figure 7.8(b)) with the same focus of expansion (FOE, see below).
3. Equidistant rotation around the sight line: it gives a set of concentric MV (see Figure 7.8(c)).

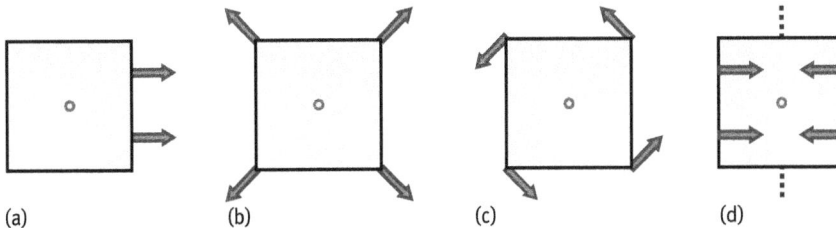

Figure 7.8: Identification of four basic movements.

4. Rotation of plane object that is perpendicular to the sight line: it gives one or more vectors starting from the line segments (see Figure 7.8(d)).

Example 7.3 Interpretation of optical flow.

Optical flow field reflects the scene movement. Figure 7.9 provides some examples of optical flow field as well as their interpretation (arrow length corresponding to velocity). Figure 7.9(a) corresponds to the rightward movement of only one object. Figure 7.9(b) corresponds to the camera forward (direction into the paper) movement; at this time, the fixed object in scene appears to depart from FOE and diverge outwardly, there is another object with horizontal movement has its own FOE. Figure 7.9(c) corresponds to the case that an object moves in the direction toward a fixed camera, its FOE is inside the boundary (if the object move from the camera, it looks as leaving the focus of contraction (FOC)). Figure 7.9(d) corresponds to the case where an object rotates around the sight line. Figure 7.9(e) corresponds to the case where an object rotates around a horizontal axis that is perpendicular to the sight line, the feature points on the object looks having up and down movement (the boundary may have concussion).

◻

7.3.3 Detecting Specific Movement Pattern

In many applications, certain movement patterns need to identify. The information based on images and based on motion can be combined. Motion information can

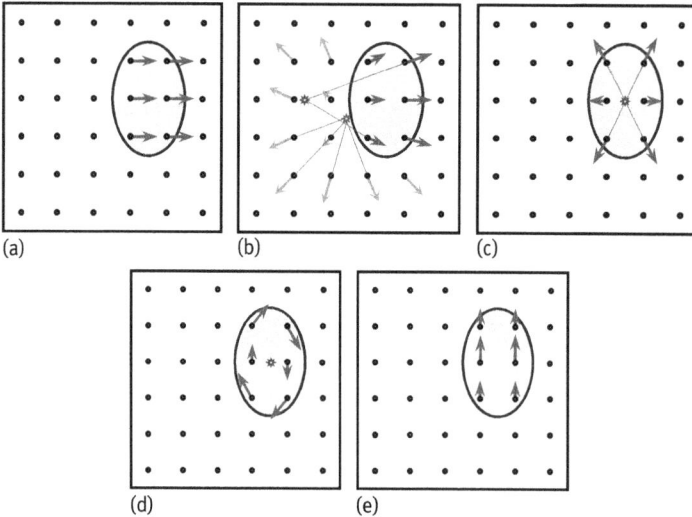

Figure 7.9: Interpretation of optical flow.

be obtained by a particular difference between images obtained consecutively. To improve the accuracy and to use spatial distribution information, the image is divided into a number of blocks, and then, two moving image blocks (one is collected at time t, another is collected at time $t + dt$) are considered. The direction of movement can be computed by the following four kinds of difference images:

$$U = |f_t - f_{t+dt\uparrow}|$$
$$D = |f_t - f_{t+dt\downarrow}|$$
$$L = |f_t - f_{t+dt\leftarrow}|$$
$$R = |f_t - f_{t+dt\rightarrow}|$$

(7.24)

where the arrow represents the direction of movement of the image, for example $\downarrow$ represents image frame $I_{t+\delta t}$ moving down with respect to the former one I_t.

The amplitude of movement can be obtained by summing block areas of image. This sum can be quickly calculated using the integral image described below.

Integral image is a matrix representation method that can maintain the global information of image (Viola 2001). In the integral image, the value $I(x, y)$ at position (x, y) represents the sum values of all pixels that are belong to the upper left part of this position in the original image $f(x, y)$.

$$f(x, y) = \sum_{p \le x, q \le y} f(p, q)$$

(7.25)

The construction of integral image can be achieved by using only one scan to image with iterations:

1. Let $s(x, y)$ represent the accumulation of one row of pixels, $s(x, y - 1) = 0$;
2. Let $I(x, y)$ represent an integral image, $I(-1, y) = 0$;
3. Make a progressive scan to the whole image, for each pixel (x, y) calculate iteratively the line accumulation $s(x, y)$ and the integral image $I(x, y)$:

$$s(x, y) = s(x, y - 1) + f(x, y)$$ (7.26)
$$I(x, y) = I(x - 1, y) + s(x, y)$$ (7.27)

4. When the pass of progressive scan of the whole image reaches the pixel at the bottom right corner, the integral image $I(x, y)$ has been constructed.

As shown in Figure 7.10, any sum of rectangles can be calculated by means of four reference arrays.

For example, the sum in rectangle D is:

$$D_{sum} = I(\delta) + I(\alpha) - [I(\beta) + I(\gamma)]$$

(7.28)

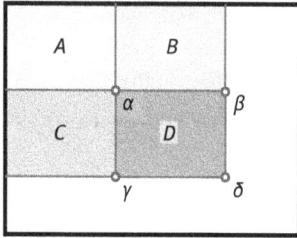

Figure 7.10: Computation in integral image.

where $I(\alpha)$ is the value of the integral image at point α, which is the sum value for pixels in rectangle A; $I(\beta)$ is the sum value of rectangle A and rectangle B; $I(\gamma)$ is the sum value of rectangle A and rectangle C; and, $I(\delta)$ is the sum value of all rectangles A, B, C, and D. Therefore, to compute the difference between two rectangles requires eight reference arrays. In practice, a look-up table can be created, and the computation will be completed with the help of the look-up table.

In object detection and tracking, Haar rectangle features, as shown in Figure 7.11, are commonly used. They can be quickly computed with the help of integral image, by means of subtracting the shaded rectangles from no shadow rectangles. For example, Figure 7.11(a) and Figure 7.11(b) need 6 times of referring look-up table; Figure 7.11(c) needs 8 times of referring look-up table; and Figure 7.11(d) needs 9 times of referring look-up table.

7.4 Moving Object Segmentation

Detecting and extracting moving objects from image sequences can be considered as a problem of spatial segmentation. For example, a segmentation of video image is mainly to detect and extract all independent motion regions (objects) from each frame. To solve this problem, either the time-domain information (gray-level variation between frames) in the sequence of images, or the spatial-domain information (gray-level variation inside frames) can be used. In addition, these information can be combined.

7.4.1 Object Segmentation and Motion Information Extraction

Segmentation of moving object and extraction of motion information are closely linked. Common strategy has the following three categories.

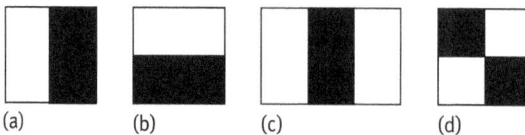

(a) (b) (c) (d)

Figure 7.11: Using integral image to compute Haar features.

7.4.1.1 Segmentation Then Calculating Motion Information

First performing object segmentation and then calculating the motion information can be seen as a direct segmentation approach. It mainly use the gray-level and gradient information of spatial-temporal image to directly segment images. One method is first to use grayscale or color information (or other characteristics) of motion regions to divide video frame image into different blocks, then to estimate the parameters of affine motion model by using MV field for each motion region. The advantage of this approach is that the object boundary could be well reserved, while the disadvantage is that for complex scenes, it often results in over-segmentation, as the same moving object may be composed by several regions.

Another method fits the entire change region first into a parameter model according to the least mean square criterion, and then detecting object while progressively dividing the big region into small regions.

7.4.1.2 Calculating Motion Information Then Segmentation

First calculating the motion information then performing segmentation can be seen as an indirect object segmentation approach. Another commonly used method is to first estimate optical flow between two or more frame images, then to perform segmentation based on the optical flow field. In fact, if the MV field of the whole image can be first calculated, then it can be used in segmentation. According to the assumption that the whole scene consists of a series of planes that can be expressed by 2-D or 3-D motion models, after obtaining dense optical flow field, it is possible to use Hough transform and split-merge procedure to divide MV field into different regions. On the other side, the MV field can also be seen as a Markov random field, the maximum *a posteriori* probability and global optimization with simulated annealing algorithm can be used to get the segmentation result. Finally, K-means clustering can be used to partition the MV field. Segmentation based on MV field can ensure that the boundary of segmented region is at the position with large differences of MV, the so-called movement border. For pixels with different color or texture properties, as long as they have similar MVs, they will still be classified into the same area. This method can reduce the possibility of oversegmentation and is more in line with people's understanding of the moving object.

7.4.1.3 Segmentation and Calculating Motion Information Simultaneously

Calculating MV field and segmenting motion region at the same time is often associated with Markov random field and the framework of the maximum *a posteriori* estimation (MAP). Generally, it require a considerable amount of calculation.

7.4.2 Dense Optical Flow Algorithm

In order to accurately calculate the local MV field, the **dense optical flow** algorithm (also known as the Horn–Schunck algorithm), which is based on the brightness gradient can be used (Ohm, 2000). It gradually approaches the MV of each pixel between adjacent frame images through an iterative approach.

7.4.2.1 Solving Optical Flow Equation

Dense optical flow algorithm is based on optical flow equation. It is seen from eq. (7.18) that for each pixel there is only one equation but with two unknowns (u, v), so solving optical flow equation directly is an ill-posed problem, and additional constraints are required to convert it into a solvable problem (Zhang, 2000b). Here, the introduction of optical flow error and velocity field gradient error could convert the problem solving for optical flow equation into an optimization problem. First, define the optical flow error e_{of} as the parts that do not meet the optical flow equation in MV field, namely:

$$e_{of} = \frac{\partial f}{\partial x} u + \frac{\partial f}{\partial y} v + \frac{\partial f}{\partial t} \tag{7.29}$$

Computing MV field is to make e_{of} to attend to square minimum in the entire frame image, that is, minimizing e_{of} means to calculate the MVs in line with the optical flow constraint equations. On the other side, the definition of the velocity field gradient error e_s^2 is

$$e_s^2 = \left(\frac{\partial u}{\partial x}\right)^2 + \left(\frac{\partial u}{\partial y}\right)^2 + \left(\frac{\partial v}{\partial x}\right)^2 + \left(\frac{\partial v}{\partial y}\right)^2 \tag{7.30}$$

Error e_s^2 describes the smoothness of optical flow field, the smaller e_s^2 is, the more even the optical flow field is, so minimizing e_s^2 means to make the entire MV field as smooth as possible. The dense optical flow algorithm takes into account the two restrictions, wants to seek an optical flow field (u, v) with minimum weighting sum of two types of errors in the entire frame image:

$$\min_{u,v(x,y)} \int_A [e_{of}^2(u, v) + \alpha^2 e_s^2(u, v)] \, dxdy \tag{7.31}$$

where A represents an image region, α represents the relative weighting between optical flow error and velocity field gradient error and can be used to adjust the influence of smoothness constrain.

7.4.2.2 Problems of Basic Algorithm

Basic dense optical flow algorithm in practical applications may encounter some problems.

Aperture Problem

It (see example below) leads to the large increase in solving optical flow equation. Having the aperture problem, the MV tends to direct to the direction of normal direction of edges at strong edge positions, and thus, it requires to form a correct MV through multiple iterations to propagate the information in local neighborhood to the current place. Such multiple iterations make the amount of calculations often very large for dense optical flow algorithm, and the motion information extraction based on this algorithm can only do off-line nonreal-time processing.

Example 7.4 Aperture problem.
Using local operators can calculate local gray-scale variation in space and time in the image. The functional scope of a local operators is in accordance with its mask size. However, as the size of the mask is limited, so the aperture problem will appear.

One example is given in Figure 7.12. It is seen that as the size of moving object is much bigger than the size of mask, so in Figure 7.12(a), only a vertical solid line of edge in the range of mask could be seen. This solid line moves to the position of a dotted line in Figure 7.12(b). The movement of edge can be described with MV. In Figure 7.12(a), there are different possibilities of movements, MV can direct from a point on the solid line to any point on the dotted line. If the MV is decomposed into two directions, the direction parallel to edge and the direction perpendicular to edge, only the normal component (perpendicular to the edge) of MV can be determined, while the other component cannot be determined. This problem is called aperture problem.

The cause for aperture problem comes from the inability to determine the corresponding points of the front and rear edges in the image sequence, or there is no reason to distinguish different points on the same edge. From this perspective, the aperture problem can be regarded as a special case of the more general problem of correspondence (correspondence will be discussed in detail in Volume III).

If the corresponding points on the images before and after motion can be determined, then it is possible to unambiguously determine the MV. A typical example is that a corner is located in the range of local operator mask, as shown in Figure 7.12(b). In this case, the MV at corner point can be completely determined, and MVs of other points may also be determined according to a comparable relationship. However, this case also shows that, with the local operator the motion information may only be obtained at a limited number of sparse points. ◻

Approximation Error Problem
When the scene has more violent motion, the MV will have a relatively large magnitude (the global MV is often more significant), and the optical flow error will be

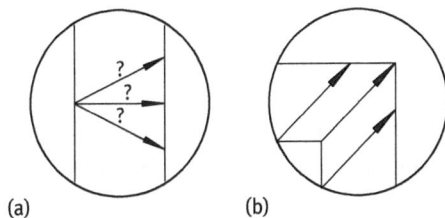

(a)   (b)

Figure 7.12: Aperture problem in motion analysis.

relatively large, too. This will lead to some large error for the result obtained by optimizing eq. (7.31).

An improved algorithm for dense optical flow uses displaced frame difference item $f(x + \bar{u}_n, y + \bar{v}_n, t + 1) - f(x, y, t)$, instead of the optical flow error term e_{of}. It also uses the average gradient terms:

$$f_x = \frac{1}{2}\left[\frac{\partial f}{\partial x}(x + \bar{u}_n, y + \bar{v}_n, t + 1) + \frac{\partial f}{\partial x}(x, y, t)\right] \tag{7.32}$$

$$f_y = \frac{1}{2}\left[\frac{\partial f}{\partial y}(x + \bar{u}_n, y + \bar{v}_n, t + 1) + \frac{\partial f}{\partial y}(x, y, t)\right] \tag{7.33}$$

to replace partial differentials $\partial f/\partial x$ and $\partial f/\partial y$, respectively. Thus, it can better approximate relatively large MV. In the specific implementation, if the position of pixel does not coincide with the pixel coordinates, then the displacement frame difference and the average gradient values requires both the use of interpolation.

Using items of displaced frame difference and average gradient obtained above, the calculation of incremental MVs $[\Delta u(x, y, t)_{n+1}, \Delta v(x, y, t)_{n+1}]$ in the $n + 1$ iteration can be represented by the following two equations:

$$\Delta u(x, y, t)_{n+1} = -f_x \frac{[f(x + \bar{u}_n, y + \bar{v}_n, t + 1) - f(x, y, t)]}{\alpha^2 + f_x^2 + f_y^2} \tag{7.34}$$

$$\Delta v(x, y, t)_{n+1} = -f_y \frac{[f(x + \bar{u}_n, y + \bar{v}_n, t + 1) - f(x, y, t)]}{\alpha^2 + f_x^2 + f_y^2} \tag{7.35}$$

Finally, as global smoothness constraints is used by the dense optical flow algorithm, so the MV at the moving object boundary will be smoothed into gradual transition, and the blurring of motion boundaries occur. Such a problem can be solved by using some segmentation techniques, as shown in the next section.

7.4.2.3 Global Motion Compensation

Global motion and local motion often mixed together. On the basis of obtaining the global motion parameters caused by the movement of the camera, it is possible to recover the global MV with the help of estimated parameters. In dense optical flow algorithm, by compensating the global MV, the MVs caused by local objects can be approached with progressive iteration Yu (2002).

In the actual calculation process, starting with the estimated global motion model, the global MV for each pixel can be calculated. Then, it is merged with the current local MVs, as the initial value of the input for the next iteration. Concretely, the process has the following steps:

1. Set the initial local MV (u_l, v_l) to be $(0, 0)$ for all points in image;
2. According to the global motion model, the global MV (u_g, v_g) at each point is calculated;

3. Calculating the actual MV for each pixel:

$$(\bar{u}_n, \bar{v}_n) = (\bar{u}_g, \bar{v}_g) + (\bar{u}_l, \bar{v}_l)_n \qquad (7.36)$$

where $(\bar{u}_l, \bar{v}_l)_n$ is the average value of the local MV in the neighborhood of this pixel after n times of iteration;

4. According to eqs. (7.34) and (7.35), calculating the correcting values of MV $(\Delta u, \Delta v)_{n+1}$ at this point;

5. If the magnitude of $(\Delta u, \Delta v)_{n+1}$ is greater than a threshold value T, then let

$$(u_l, v_l)_{n+1} = (u_l, v_l)_n + (\Delta u, \Delta v)_{n+1} \qquad (7.37)$$

and go to step (3); otherwise, ending the calculation.

One example of comparison between the direct block matching method (Section 7.2.2) and the improved algorithm for dense optical flow is given in Figure 7.13. For the same original image, the MV field obtained by using a block matching method is superposed on it, as shown in Figure 7.13(a); the MV field obtained by global motion estimation is superposed on it (the global smoothness constraint results in no-obvious motion boundary and lesser extent MV), as shown in Figure 7.13(b); the local MV field obtained by using a dense optical flow algorithm with **global motion compensation** method is superposed on it, as shown in Figure 7.13(c).

As can be seen from these figures, the impact of global motion in the block matching results has been successfully compensated, and the MV error in low texture background has been eliminated. The final result of the MVs are concentrated in the upward moving parts of player and ball, which is more in line with the local movement content in scene.

7.4.3 Parameter and Model-Based Segmentation

The basic principle of parameter and model-based segmentation method is as follows. Suppose there are K independent moving objects in image, each optical flow vector

(a) (b) (c)

Figure 7.13: Comparing the results of the block matching method and the improved iterative algorithm for dense optical flow.

thus calculated corresponds to the projection of an opaque object with rigid body motion in 3-D space. In this way, each moving object with independent movement can be accurately described by a set of mapping parameters. On the other hand, suppose K groups of vectors exist, each group defines an optical flow vector for every pixel. The optical flow vector defined by the mapping parameter is called model-based or synthetic optical flow vector; thus, there are K optical flow vectors at each pixel. According to this analysis, the optical flow based image segmentation for different moving regions can be seen as assign a synthetic optical flow vector label to the estimation of the optical flow vector. The problem here is that the number of parameters K and the mapping parameters in each group are not known in advance. If a special value is provided for K, then in cases that the estimation of optical flow vectors corresponding to all groups are known, the mapping parameters in the sense of minimum mean square error can be calculated. In other words, it is required to know mapping parameters to determine the segmentation label, it is also required to know the segmentation label to determine mapping parameters. This indicates that an iterative approach would be needed to perform the segmentation.

7.4.3.1 Using Hough Transform

If considering the optical flow as a region feature, the segmentation can be made by using Hough transform. **Hough transform** (see Section 2.2.1) can also be seen as a clustering technique. Here, a "vote" according to the selected feature is carried out in feature space, to find the most representative cluster. Because a plane moving as a rigid motion will generate affine flow field in orthogonal projection, so a method using directly Hough transform to segment optical flow employs an affine flow model with six parameters:

$$
\begin{aligned}
v_1 &= a_1 + a_2 x_1 + a_3 x_2 \\
v_2 &= a_4 + a_5 x_1 + a_6 x_2
\end{aligned}
\tag{7.38}
$$

where $a_1 = V_1 + z_0 R_2$; $a_2 = z_1 R_2$; $a_3 = z_2 R_2 - R_3$; $a_4 = V_2 - z_0 R_1$; $a_5 = R_3 - z_1 R_1$; $a_6 = -z_2 R_1$. Note that V_1 and V_2 are components of the translational velocity vector; R_1, R_2 and R_3 are components of the angular velocity vector; z_0, z_1 and z_2 are plane coefficients. If the optical flow is known in three or more points, parameters a_1, a_2, a_3, a_4, a_5, a_6 can be solved out. After determining the maximum and minimum values for each parameter, the 6-D feature space can be quantized to a certain parameter state. Thus, with each optical flow vector $v(x) = [v_1(x)\ v_2(x)]^T$, a set of quantitative parameters can be voted to minimize:

$$
\eta^2(x) = \eta_1^2(x) + \eta_2^2(x)
\tag{7.39}
$$

where $\eta_1(x) = v_1(x) - a_1 - a_2 x_1 - a_3 x_2$; $\eta_2(x) = v_2(x) - a_4 - a_5 x_1 - a_6 x_2$. The parameter set with more than a predetermined amount of votes is likely to represent a candidate movement, so the number of classes K for marking each optical flow vector and the corresponding parameter set can be determined.

The disadvantage of the above direct method is the amount of calculations is too large (high dimensional parameter space). To solve this problem, a two-step algorithm based on improved Hough transform can be used. In the first step, collecting the optical flow vectors nearby to form a single parameter set consistent with the elements. There are a variety of ways to simplify the calculations in the second step, including:

1. Dividing the parameter space into two nonoverlapping set $\{a_1, a_2, a_3\} \times \{a_4, a_5, a_6\}$ for two Hough transformations;
2. Using multiresolution Hough transform, performing space quantization at each resolution level based on the estimated parameters obtained in the preceding stage;
3. Using multiple Hough transforms, combining first the optical flow vectors the most consistent with the candidate parameters together. The most consistent elements obtained in the first step will be merged at least mean square sense to form the relevant fragment (segment) in a second step. Finally, if there are optical flow vectors not been merged, they will be attributed to their adjacent segments.

In summary, the improved Hough transform first aggregates optical flow vectors into groups, each of them agree with a small moving plane; these groups are then reintegrated into segments according to specific criteria to constitute the ultimate objects.

7.4.3.2 Using Region Growing

If considering the optical flow as a region feature, the segmentation can also be made by using **region growing** (Yu 2002). It is assumed that the moving object in the scene is composed of a series of planar regions having different motion characteristics, different regions may be represented by region motion models with different affine parameters (Ekin, 2000; Ohm, 2000). The 6-parameter affine motion model in eq. (7.38) is a first-order approximation model for complex region; in addition to express translational motion, it can also approximately describe region rotation and deformation.

The two key points of a region growing algorithm are the criteria for selecting seed points and for determining the growing of region, respectively. These two criteria can greatly influence the results of segmentation. Taking into account that all points within the segmented region should meet an affine motion model, the criterion for determining the growing of region is to judge whether the MVs near the growth point can be represented by the affine motion model in the region. In other words, the consistency criterion of region is whether the MV satisfies the affine motion model. This judgment can be made according to the difference between the MV calculated at this point and the MV estimated by the regional affine motion model, if difference is greater than a certain threshold, no longer continue to grow. The selection of seed points can be made by following to certain criteria, such as order selection, random selection, etc. (Yu, 2001a). Since the selections with both methods are likely to choose the seed points at moving edges, so it may not obtain a consistent region if it is grown in these places. To solve this problem, the seed points can be ranked according to the difference between their MVs with respect to that of points in its neighborhood, the difference $D_m(x, y)$ is defined as

$$D_m(x, y) = \sum_{(x',y') \in A} \sqrt{[u(x, y) - u(x', y')]^2 + [v(x, y) - v(x', y')]^2} \qquad (7.40)$$

where A is the neighborhood (e. g., 4- or 8-neighborhood) of point (x, y). The smaller the value of $D_m(x, y)$ is, the greater the likelihood of point (x, y) locating in a smooth region. At each time to select a seed point, the point with smallest $D_m(x, y)$ but have not yet been included in any region is selected so that the seeds would be in the central part of a smooth region, thus it can guarantee to obtain a more complete growth results. In practice, in order to reduce the amount of calculation, the local MV can be first set to zero, that is, the region without local motion will be classified as nonmovement region, so as to reduce the operating regions of segmentation.

Results obtained by the region growing algorithm is a series of moving region positions and their corresponding affine motion model parameters. In Figure 7.14, the segmentation result of the above algorithm for MV field in Figure 7.13 is given, in which all are the moving foreground regions (except black region). Different colors are used to represent regions with different parameters of motion model.

7.5 Moving Object Tracking

Tracking moving objects in video means to detect and locate the same object in each frame of the video image. Practice often meets several difficulties:
1. The object and background have some similarities, so it is not easy to capture the difference between the two;
2. The appearance of object would change over time. On one side, some objects are nonrigid ones, their appearances are inevitable to change over time; on the other

Figure 7.14: Segmentation results of motion vector field based on affine model.

side, the changes of lighting and other external conditions over time will lead to
the changes of object appearance, whether it is rigid or nonrigid;
3. During the tracking process, the change of relative position between the background and objects leads to situation where the object to be tracked becomes
occluded and complete object information would not be obtained. In addition, the
tracking needs to balance between the accuracy of object location and real-time
applications.

Moving object tracking often combines the positioning and representation of objects
with the trajectory filtering and data fusion. The former is mainly a bottom-up process
that needs to overcome the influence of object appearance, orientation, scale, and
lighting changes. The latter is primarily a top-down process that requires the consideration of the motion characteristics of the object, the utilization of a variety of prior
knowledge and motion models as well as the promotion of motion assumptions and
evaluation.

7.5.1 Typical Technology

Moving object tracking can use several different methods, including boundary-based
tracking, region-based tracking, mask-based tracking, feature-based tracking, motion
information-based tracking and the like. Motion information-based tracking can further be classified into two groups: tracking using the continuity of motion information
and tracking using the prediction of the next frame to reduce the search range. In
the following, several common techniques are introduced, in which Kalman filter and
particle filter are all methods related to the reducing of the search scope.

7.5.1.1 Kalman Filter
When tracking the objects in current frame, it is often desirable to be able to predict
their positions in the subsequent frames, so the prior information can be maximally used and the searches in subsequent frames can be minimized. In addition, the
prediction is also helpful for solving the problems caused by short-term occlusion
(Davies 2005). To do this, the position x and moving speed v of objects to be tracked
need to be continually updated:

$$x_i = x_{i-1} + v_{i-1} \qquad (7.41)$$

$$v_i = x_i - x_{i-1} \qquad (7.42)$$

Three quantities are required: the original position, the optimal estimated value (with
superscript −) for corresponding variables before observation (model parameters),
and the optimal estimated value (with superscript +) for corresponding variables after
observation. In addition, the noise should be considered. If m represents the noise of

position measurement, n represents the noise of moving speed estimation, then the two equations above become

$$x_i^- = x_{i-1}^+ + v_{i-1} + m_{i-1} \tag{7.43}$$

$$v_i^- = v_{i-1}^+ + n_{i-1} \tag{7.44}$$

When the moving speed is a constant and the noise is Gaussian noise, the optimal solution is

$$x_i^- = x_{i-1}^+ \tag{7.45}$$

$$\sigma_i^- = \sigma_{i-1}^+ \tag{7.46}$$

They are referred to as predictive equations, and there are

$$x_i^+ = \frac{x_i/\sigma_i^2 + (x_i^-)/(\sigma_i^-)^2}{1/\sigma_i^2 + 1/(\sigma_i^-)^2} \tag{7.47}$$

$$\sigma_i^+ = \left[\frac{1}{1/\sigma_i^2 + 1/(\sigma_i^-)^2}\right]^{1/2} \tag{7.48}$$

They are called calibration equations, in which $\sigma^\pm$ is the standard deviation obtained by using corresponding model to estimate $x^\pm$, σ is the standard deviation of the original measure x.

It is seen from the above equations that the estimation of position parameters can be improved and the errors based on these parameters can be reduced by repeating the measurements in each iteration. Because the noise is modeled as for the location, so the location earlier than $i-1$ can be ignored. In fact, a lot of position values can be averaged to improve the accuracy of the final estimation, which will be reflected in the values of x_i^-, σ_i^-, x_i^+, and σ_i^+.

The above algorithm is called **Kalman filter**, it is the best estimate for a linear system with zero mean Gaussian noise. However, since the Kalman filter is based on the average operation, it will produce a greater error when there are outliers. In most motion applications this problem will appear, so it is necessary to test each estimate to determine whether it is actually too far from real cases.

7.5.1.2 Particle Filter

Kalman filter requires the state equation being linear, and the state distribution being Gaussian, in practice, these requirements are not always met. **Particle filter** is an effective algorithm to solve nonlinear problems. The basic idea is to use random samples propagated in the state space (these samples are referred to as "particles") to approximate the posterior probability distribution function (PDF) of the system state, to obtain the estimated values of system status. Particle filter itself represents a sampling method, by means of which the specific distribution can be approximated with time structure. In the research of image techniques, it is also known as conditional density diffusion (CONDENSATION).

Suppose a system has a series of state $X_t = \{x_1, x_2, \ldots, x_t\}$, where the subscript represents the time. At time t, there is a probability density function that represents the possible case of x_t. This can be represented by a set of particles (a set of sample states), and the occurrence of particles is controlled by a probability density function. In addition, there are a series of observation $Z_t = \{z_1, z_2, \ldots, z_t\}$ associated with the probabilities of state X_t, and a Markov assumption that the probability of x_t depends on the previous state x_{t-1}, which can be expressed as the $P(x_t|x_{t-1})$.

Conditional density diffusion is an iterative process. It maintains a set of N samples s_i having weights w_i, namely

$$S_t = \{(s_{ti}, w_{ti})\}, \quad i = 1, 2, \ldots, N \quad \text{and} \quad \sum_i w_i = 1 \tag{7.49}$$

The sampling and the weight together can represent the probability density function of the state X_t under the given observation Z_t. Different from Kalman filter, it is not necessary here to satisfy the constrains of single mode distribution or Gaussian distribution, and multimode may be allowed. Now, it is required to deduce S_t from S_{t-1}.

Specific steps in particle filter are as follows (Sonka, 2008):
1. Suppose that a set of weighted samples $S_{t-1} = \{s_{(t-1)i}, w_{(t-1)i}\}$ are known at time $t-1$. Let the cumulative probability of weight is

$$\begin{aligned} c_0 &= 0 \\ c_i &= c_{i-1} + w_{(t-1)i}, \ i = 1, 2, \ldots, N \end{aligned} \tag{7.50}$$

2. Randomly selecting a number r from the uniform distribution [0 1], and determining $j = \arg[\min_i(c_i > r)]$ to calculate the first n samples of S_t. Diffusing the jth sample of S_{t-1}, this is called importance sampling, that is, giving the most weight to most probable sample.
3. Using Markov properties of x_t to derive s_{tn}.
4. Using observation Z_t to obtain $w_{tn} = p(z_t|x_t = s_{tn})$.
5. Returning to the second step, and perform N times iterations.
6. Normalizing $\{w_{ti}\}$, so that $\sum_i w_i = 1$.
7. Outputting the optimal estimation of x_t

$$x_t = \sum_{i=1}^{N} w_{ti} s_{ti} \tag{7.51}$$

Example 7.5 Iteration of particle filter.
Consider the 1-D case, where x_t and s_t are just scalar real numbers. Suppose at time t, x_t has a displacement v_t, and it is subject to a zero-mean Gaussian noise e, that is, $x_{t+1} = x_t + v_t + e_t$, $e_t \sim N(0, \sigma_1^2)$. Suppose further the observation z_t has Gaussian

distribution centered at x and with variance σ_2^2. The particle filter will make N times "guess" to obtain $S_1 = \{s_{11}, s_{12}, \ldots, s_{1N}\}$.

Now to generate S_2. Selecting a s_j from S_1 (irrespective of the value of w_{1i}), so that $s_{21} = s_j + v_1 + e$, where $e \sim N(0, \sigma_1^2)$. The above-described process is repeated N times to generate the particles at time $t = 2$. At this time, $w_{2i} = \exp[(s_{2i} - z_2)^2/\sigma_2^2]$. Renormalizing w_{2i}, and the iteration ends. The obtained estimation for x_2 is $\sum_i^N w_{2i} s_{2i}$.

7.5.1.3 Mean Shift and Kernel Tracking

Mean shift refers to the mean vector shifting. It is a nonparametric technique that can be used to analyze complex multimodal feature spaces and determine feature clustering. Mean shift technique can also be used to track moving objects; in this case, the region of interest corresponds to the tracking window, while the feature model should be built for tracked object. The basic idea in using mean shift technique for object tracking is to move and search the object model within the tracking window, and to calculate the correlation place with largest value. This is equivalent to determine the cluster center and move the window to the position coincides with the center of gravity (convergence).

To continuously track object from the previous frame to the current frame, the object model determined in the previous frame can be put on the center position x_c of local coordinate system in the tracking window, and let the candidate object in current frame be at position y. Description of the property of the candidate object can be characterized by taking advantage of the probability density function $p(y)$ estimated from the data out of current frame. The object model Q and the probability density function of candidate object $P(y)$ are defined as follows:

$$Q = \{q_v\} \quad \sum_{v=1}^{m} q_v = 1 \tag{7.52}$$

$$P(y) = \{p_v(y)\} \quad \sum_{v=1}^{m} p_v = 1 \tag{7.53}$$

where $v = 1, \ldots, m$, m is the number of features. Let $S(y)$ be the similarity function between $P(y)$ and Q:

$$S(y) = S\{P(y), Q\} \tag{7.54}$$

For an object-tracking task, the similarity function $S(y)$ is the likelihood that an object to be tracked in the previous frame is at position y in the current frame. Therefore, the local extreme value of $S(y)$ corresponds to the object position in the current frame.

To define the similarity function, the isotropic kernel can be used (Comaniciu 2000), in which the description of feature space is represented with kernel weights, then $S(y)$ is a smooth function of y. If n is the total number of pixels in the tracking window, wherein x_i is the ith pixel location, then the probability estimation for the candidate feature vector Q_v in candidate object window is

$$\hat{Q}_v = C_q \sum_i^n K(x_i - x_c)\delta\left[b(x_i) - q_v\right] \qquad (7.55)$$

where $b(x_i)$ is the value of the object characteristic function at pixel x_i; the role of δ function is to determine whether the value of x_i is the quantization results of feature vector Q_v; $K(x)$ is a convex and monotonically decreasing kernel function; C_q is the normalization constant:

$$C_q = 1\left/\sum_{i=1}^n K(x_i - x_c)\right. \qquad (7.56)$$

Similarly, the probability estimation of the feature model vectors P_v for candidate object $P(y)$

$$\hat{P}_v = C_p \sum_i^n K(x_i - y)\delta\left[b(x_i) - p_v\right] \qquad (7.57)$$

where C_p is a normalization constant (for a given kernel function, it can be calculated in advance):

$$C_p = 1\left/\sum_{i=1}^n K(x_i - y)\right. \qquad (7.58)$$

Bhattacharyya coefficient is commonly used to estimate the degree of similarity between the density of object mask and the density of candidate region. The more similar the distributions between the two densities is, the greater the degree of similarity. The center of the object position is

$$y = \frac{\sum_{i=1}^n x_i w_i K(y - x_i)}{\sum_{i=1}^n w_i K(y - x_i)} \qquad (7.59)$$

where w_i is a weighting coefficient. Note that from eq. (7.59), the analytic solution of y cannot be obtained, so an iterative approach is needed. This iterative process corresponds to a process of looking for the maximum value in neighborhood. The characteristics of **kernel tracking** method are: operating efficiency is high, easy modular, especially for objects with regular movement and low speed. It is possible to successively acquire new center of the object, and achieve the object tracking.

Example 7.6 Feature selection during tracking.
In object tracking, in addition to tracking strategies and methods, the choice of which kinds of object features is also very important (Liu, 2007). One example of using color histogram and edge direction histogram (EOH) under the mean shift framework is shown in Figure 7.15. Figure 7.15(a) is a frame image of a video sequence, in which the object to be tracked has similar color with background, the result obtained with color histogram is unsuccessful as shown in Figure 7.15(b), while the result obtained with edge direction histogram can catch the object as shown in Figure 7.15(c). Figure 7.15(d)

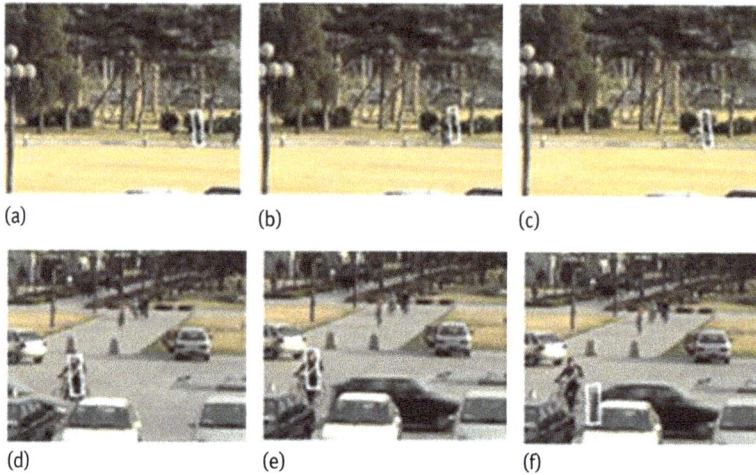

(a)   (b)   (c)

(d)   (e)   (f)

Figure 7.15: Examples of using a single feature alone.

is a frame image of another video sequence, in which the edge direction of object to be tracked is not obvious, the result obtained with color histogram can catch object as shown in Figure 7.15(e), while the result obtained with edge direction histogram is not reliable as shown in Figure 7.15(f). It is seen that using a feature alone under certain circumstances can lead to failure of the tracking result.

The color histogram mainly reveals information inside object, while the edge direction histogram mostly reflects information of object boundary. Combining the two, it is possible to obtain a more general effect. One example is shown in Figure 7.16, where the goal is to track the moving car. Since there are changes of object size, changes of viewing angle and object partial occlusion, etc., in the video sequence, so the car's color or outline will change over time. By combining color histogram and edge direction histogram, good result has been obtained.

7.5.2 Subsequences Decision Strategy

The method described in the previous subsection is conducted frame by frame during object tracking, the possible problem is only few information used to make decisions, and the small errors may spread and cannot be controlled. An improved strategy is to

Figure 7.16: Tacking with two features combined.

divide the entire tracking sequence into several sub-sequences (each with a number of frames), then an optimal decision for each frame is made on the basis of the information provided by each subsequence, and this policy is called subsequences decision strategy (Shen 2009b).

Subsequences decision includes the following steps:
1. The input video is divided into several subsequences;
2. The tracking is conducted in each subsequence;
3. If there is an overlap for adjacent subsequences, their results will be integrated together.

Subsequence decision can also be seen as an extension of traditional decision by frame. If each subsequence obtained by division is just one frame, the subsequence decision becomes a frame by frame decision.

Let S_i be the ith subsequence, wherein the jth frame is represented as $f_{i,j}$, the whole S_i comprises a total J_i frames. If the input video contains N frames and is divided into M subsequence, then

$$S_i = \{f_{i,1}, f_{i,2}, \ldots, f_{i,j}, \ldots, f_{i,J_i}\} \tag{7.60}$$

To ensure that any subsequence is not a subset of another sub-sequence, the following constraints are defined:

$$\forall m, n \quad S_m \subseteq S_n \Leftrightarrow m = n \tag{7.61}$$

Let $P_j = \{P_{j,k}\}$, $k = 1, 2, \ldots, K_j$ represent the K_j states of possible positions in the jth frame, then the frame by frame decision can be expressed as follows:

$$\forall P_{j,k} \in P_j \quad T(P_{j,k}) = \begin{cases} 1 & \text{Best } P_{j,k} \\ -1 & \text{Otherwise} \end{cases} \Rightarrow \text{Output}: \underset{P_{j,k}}{\arg}\left[T(P_{j,k}) = 1\right] \tag{7.62}$$

The subsequence decision can be expressed as follows:

$$\forall P_{i,j,k} \in P_{i,1} \times P_{i,2} \times \cdots \times P_{i,j} \times \cdots \times P_{i,J_i}$$
$$T_{\text{sub}}(P_{i,j,k}) = \begin{cases} 1 & \text{Best } P_{i,j,k} \\ -1 & \text{Otherwise} \end{cases} \Rightarrow \text{Output}: \underset{P_{i,j,k}}{\arg}\left[T_{\text{sub}}(P_{i,j,k}) = 1\right] \tag{7.63}$$

For subsequence S_i, it comprises a total of J_i frames, and there are K_j states of possible positions in each frame. The optimal search problem can be represented by graph structure, and then, this problem can be solved by means of dynamic programming (search for an optimal path).

Example 7.7 An example of subsequences decision.

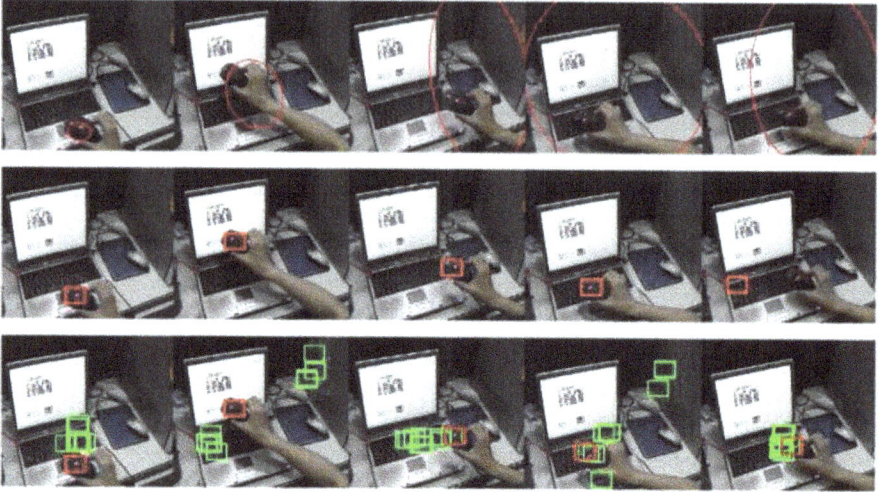

Figure 7.17: Tracking results of three different methods and strategies.

One example of subsequences decision is illustrated in Figure 7.17, in which the results obtained by three methods for a same video sequence are presented. The object to be tracked is a hand-held mobile mouse, which moves pretty fast and has similar colors with background. The dark box or an oval marks the final tracking results, while the light boxes mark the candidate positions.

The images in the first line are obtained by using mean shift method. The images in the second line are obtained by using particle filter-based approach. The images in the third line are obtained by using subsequence decision process (which only uses a simple color histogram to help detect the position of the candidate). It is seen from these images that both mean shift method and particle filter method have failed to maintain a continuous track, while only the subsequence decision method completes the entire track.

7.6 Problems and Questions

7-1 Summarize what factors make the gray value of the corresponding position 时 in the temporally adjacent image change.

7-2 Why does the method based on the median maintenance context described in Section 7.3.1 could not achieve very good results when there are a number of objects at the same time in the scene or the motion of object is very slow?

7-3 Use induction to list the main steps and tasks for calculating and maintaining a dynamic background frame using the Gaussian mixture model.

7-4 What assumptions are made in deriving the optical flow equation? What would happen when these assumptions are not satisfied?

7-5 Refer to Figure 7.10 to create a lookup table whose entries are needed to calculate the rectangle features in Figure 7.11.

7-6 Use the following two figures in Figure Problem 7-6 to explain the aperture problem.

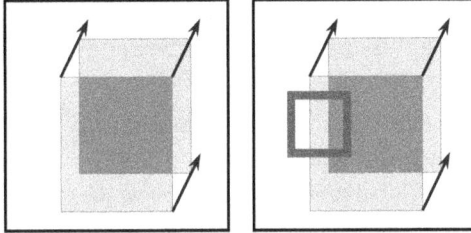

Figure Problem 7-6

7-7 What are the consequences of aperture problems and how can they be overcome?

7-8 If the observer and the observed object are both moving, how can the optical flow calculation be used to obtain the relative velocity between them?

7-9* Suppose that the motion vector of each point in an image is [2, 4], find the coefficient values in the six-parameter motion model.

7-10 Why is it possible to use the average gradient, instead of the partial derivative, for better approximating the larger magnitude of the motion vector? What could be the problem, and how to solve?

7-11* In most statistics-related applications, the total variance is the sum of the variances of the components. But why in eq. (7.48), the variance before the observation and the original measurement are so combined into the observed variance (equivalent to the parallel connection of resistance)?

7-12 Make a list to compare the suitable applications, the application requirements, the calculation speed, the robustness (anti-interference ability) and so on of Kalman filter and particle filter.

7.7 Further Reading

1. **The Purpose and Subject of Motion Analysis**
    - More information on motion analysis can be found in Gonzalez (2008), Sonka (2008), and Davies (2012).
    - A survey of face recognition based on motion information in video can be found in Yan (2009).
2. **Motion Detection**
    - More discussion on global motion detection and application can see Yu (2001b).

    – A method for fast prediction and localization of moving objects by combining the curve fitting of motion trajectories with the correlation motion detection tracking can be seen in Qin (2003).

3. Moving Object Detection

    – An adaptive subsampling method for moving object detection can be found in Paulus (2009).

    – Detection of pedestrians (a type of particular moving objects) can be seen in Jia (2008) and Li (2010).

    – One example of moving shadow detection is in Li (2011).

4. Moving Object Segmentation

    – An approach to global motion estimation can be found in Chen (2010).

    – The temporal segmentation of video image by means of the spatial domain information can be found in Zhang (2001c, 2003b, 2006b).

5. Moving Object Tracking

    – Examples of tracking the balls and players of table tennis (particular types of moving objects) can be found in Chen (2006).

    – A tracking method based on edge-color histogram and Kalman filter can be found in Liu (2008).

    – A multiple kernels tracking method based on gradient direction histogram features (Dalai 2005) can be found in Jia (2009).

    A 4-D scale space approach to solve the moving object tracking problem of adaptive scale can see Wang (2010).

    – A method of combining both mean shift and particle filter for the object tracking can be found in Tang (2011).

8 Mathematical Morphology

Mathematical morphology is a form-based mathematical tools suitable for image analysis (Serra, 1982). Its basic idea is to apply some forms of structure elements to measure and to extract the corresponding shape in image, in order to achieve image analysis and identification purposes. The applications of mathematical morphology can simplify image data, maintain their basic shape characteristics, and remove extraneous structures. Mathematical morphology algorithm has also a natural structure for parallel implementation.

The operands of mathematical morphology operation may be a binary image or a gray-scale image. In this chapter, they will be described separately.

The sections of this chapter will be arranged as follows:

Section 8.1 introduces the basic operations of binary image mathematical morphology. First of all, the most fundamental dilation and erosion operations are discussed in detail. On this basis, the opening and closing operations are also discussed. Finally, some basic properties of these basic operations are summarized.

Section 8.2 describes first another basic morphological operation, hit-or-miss transform. Then, some mathematical morphological operations composed of various basic operations for functions such as the construction of region convex hull, the thinning and thickening of region are introduced.

Section 8.3 discusses some practical binary morphological algorithms for specific image applications, such as noise filtering, corner detection, boundary extraction, target location, area filling, connected component extraction, and skeleton computation.

Section 8.4 focuses on the basic operations of mathematical morphology of grayscale image. A generalization from binary analogy is used to bring the basic operations (dilation and erosion, opening and closing) to grayscale image.

Section 8.5 presents some combined morphological operations, including morphological gradients, morphological smoothing, top-hat and bottom-hat transforms, morphological filters and soft morphological filters for grayscale images.

Section 8.6 discusses some practical grayscale morphology algorithms for specific image applications, which can be used for background estimation and removal, morphological detection of edges, fast segmentation of image clustering, watershed segmentation and texture segmentation.

8.1 Basic Operations of Binary Morphology

The operands of binary morphology are two sets, generally referred to as image set A and structure element set B (itself is still an image set and simply called **structure element**), the mathematical morphology operation uses B to operate A. In the following,

DOI 10.1515/9783110524123-008

the shaded region represents region with value 1, and the white region represents region with value 0, the operation is performed on image regions with value 1.

8.1.1 Binary Dilation and Erosion

A basic pair of binary morphology operations are dilation and erosion.

8.1.1.1 Binary Dilation
The operator of **dilation** is $\oplus$, dilating A by B is written $A \oplus B$ and is defined as

$$A \oplus B = \{x \,|\, [(\hat{B})_x \cap A] \neq \varnothing\} \tag{8.1}$$

The above equation shows that the process of dilating A by B is first to reflect B around the origin, then to shift this reflection by x, in requiring that the intersection of A and the reflection of B is not an empty set. In other words, the set obtained through dilating A by B is the set of origin positions of B when the displacement of B has at least intersected with a nonzero element in A. According to this explanation, eq. (8.1) can also be written as follows:

$$A \oplus B = \{x \,|\, [(\hat{B})_x \cap A] \subseteq A\} \tag{8.2}$$

This equation can help to understand the concept of dilation by means of a convolution operation. If the B is seen as a convolution mask, dilation is first to reflect B around the origin, then to move continuously its reflection on the image A.

Example 8.1 Illustration of dilation operation.
An example of the dilation operation is shown in Figure 8.1. Figure 8.1(a) is the shaded set A, Figure 8.1(b) is the shaded structure element B (the origin is marked with a "+"), the reflection of B is shown in Figure 8.1(c), and in Figure 8.1(d), the two shaded portions (the darker part represents the enlarged part) together form the set $A \oplus B$. It is seen that the original region has been expanded by dilation. ▫

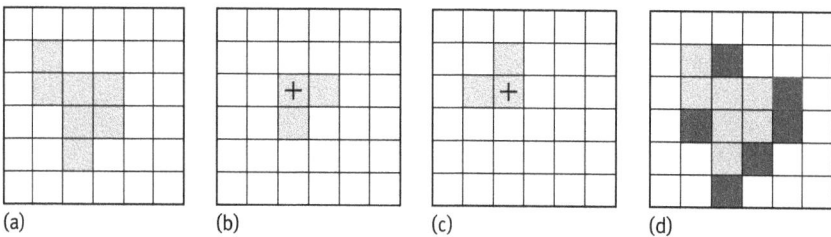

(a)   (b)   (c)   (d)

Figure 8.1: Illustration of dilation operation.

8.1.1.2 Binary Erosion

The operator of **erosion** is $\ominus$, eroding A by B is written $A \ominus B$ and is defined as follows:

$$A \ominus B = \{x \,|\, (B)_x \subseteq A\} \tag{8.3}$$

The above equation shows that the result of eroding A by B is the set of all x, where B after shifting x is still in A. In other words, the result of eroding A by B is a set consists of the origin of B when B is completely included in A. The above equation may also help to understand the erosion operation by means of related concepts.

Example 8.2 Illustration of erosion operation.
A simple example is given in Figure 8.2 for erosion operation, wherein the set A in Figure 8.2(a) and the structure element B in Figure 8.2(b) are the same as in Figure 8.1. The dark shaded region (only) in Figure 8.2(c) provides $A \ominus B$, while the light shaded region is the part that was belong to A but now eroded. It is seen that the original region has been shrunk by erosion. ▫

8.1.1.3 Dilation and Erosion with Vector Operations

In addition to the aforementioned intuitive definitions for dilation and erosion, respectively, there are a number of equivalent definitions. These definitions have their own characteristics, such as dilation and erosion operations can be achieved either by vector operation or shift operations, and even more convenient in using a computer to complete dilation and erosion operations in practice.

Look at first the vector operation, considering A and B as vectors, then the dilation and erosion can be expressed as

$$A \oplus B = \{x | x = a + b \quad \text{for some} \quad a \in A \quad \text{and} \quad b \in B\} \tag{8.4}$$
$$A \ominus B = \{x | (x + b) \in A \quad \text{for each} \quad b \in B\} \tag{8.5}$$

Example 8.3 Using vector operations to achieve dilation and erosion.
Referring to Figure 8.1, taking the upper left corner of the image as $\{0, 0\}$, then A and B can be expressed as: $A = \{(1,1),(1,2),(2,2),(3,2),(2,3),(3,3),(2,4)\}$ and

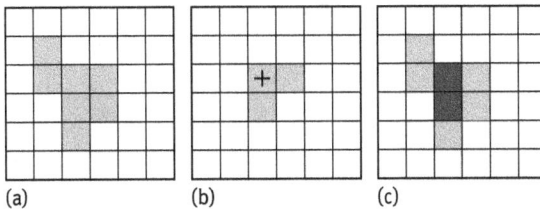

(a)   (b)   (c)

Figure 8.2: Illustration of erosion operation.

$B = \{(0, 0), (1, 0), (0, 1)\}$, respectively. Perform dilation using vector operation: $A \oplus B =$ $\{(1, 1),\ (1, 2),\ (2, 2),\ (3, 2),\ (2, 3),\ (3, 3),\ (2, 4),\ (2, 1),\ (2, 2),\ (3, 2),\ (4, 2),\ (3, 3),\ (4, 3),$ $(3, 4),\ (1, 2),\ (1, 3),\ (2, 3),\ (3, 3),\ (2, 4),\ (3, 4),\ (2, 5)\} = \{(1, 1),\ (2, 1),\ (1, 2),\ (2, 2),$ $(3, 2),\ (4, 2),\ (1, 3),\ (2, 3),\ (3, 3),\ (4, 3),\ (2, 4),\ (3, 4),\ (2, 5)\}$. The result is the same as in Figure 8.1(d). Similarly, if the erosion is obtained with the vector operation: $A \ominus B = \{(2, 2), (2, 3)\}$. Comparing to Figure 8.2(c) can verify the results here.　◙

8.1.1.4 Dilation and Erosion with Shift Operations

Shift operations and vector operations are closely linked, and the sum of vectors is a kind of shift operations. According to eq. (8.2), the shift computing formula for dilation can be obtained via eq. (8.4):

$$A \oplus B = \bigcup_{b \in B} (A)_b \tag{8.6}$$

This equation shows that the result of $A \oplus B$ is the union after shifting A for each $b \in B$. It is also interpreted as dilating A by B is to first shift A by every b and then OR up the results together.

Example 8.4 Dilation by means of shift.
An example of using shift operation to achieve dilation result is shown in Figure 8.3, wherein the set A in Figure 8.3(a) is the same as in Figure 8.1. Figure 8.3(b) is the structure element B. For simplicity, let the pixel at the origin not belong to B, that is, B contains only two pixels. Figure 8.3(c, d) provides the results of shift using the structure element points at the right and below of the origin, respectively. Figure 8.3(e) gives the result obtained by unifying Figure 8.3(c, d), that is, the union of two shaded parts. Comparing it with Figure 8.1(d), except point (1, 1), they are the same. The no-appearance of point (1, 1) in the dilation result is caused by the fact that B does not contain the origin.　◙

According to eq. (8.2), the shift computing formula for erosion can be obtained via eq. (8.5):

$$A \ominus B = \bigcap_{b \in B} (A)_{-b} \tag{8.7}$$

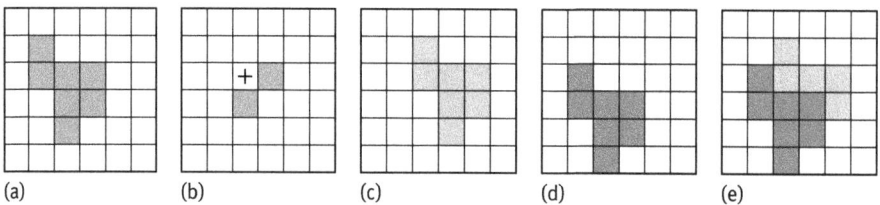

(a)　　　　　(b)　　　　　(c)　　　　　(d)　　　　　(e)

Figure 8.3: Dilation by means of shift.

This equation shows that the result of $A \ominus B$ is the intersection after reversely shifting A for each $b \in B$. It is also interpreted as: eroding A by B is to first reversely shifting A by every b and then AND up the results together.

Example 8.5 Erosion by means of shift.
An example of using shift operation to achieve erosion is shown in Figure 8.4. Wherein Figure 8.4(a, b) is the same as in Fig 8.3. Figure 8.4(c, d) provides the results of reverse shift using the structure element points at the right and below of the origin, respectively. The black region (two pixels) in Figure 8.4(e) gives the result obtained by AND the results in Figure 8.4(c, d). This result is same as that in Figure 8.2(c), so eq. (8.7) is verified. ◻

The AND set of shifting A by b is also equal to the AND set of shifting B by a, so the eq. (8.6) can be written as:

$$A \oplus B = \bigcup_{a \in A} (B)_a \tag{8.8}$$

8.1.1.5 Duality of Binary Dilation and Erosion

Dilation and erosion are two operations closely linked, one operation on the image object corresponds to another operation on the image background. According to the relation between the complement (represented by a superscript "c") and the reflection (represented by the top cap), the **duality of dilation and erosion** operations can be expressed as:

$$(A \oplus B)^c = A^c \ominus \hat{B} \tag{8.9}$$

$$(A \ominus B)^c = A^c \oplus \hat{B} \tag{8.10}$$

Example 8.6 Illustration of duality between dilation and erosion.
The duality of dilation and erosion can be explained with the help of Figure 8.5. Wherein Figure 8.5(a, b) gives A and B, respectively; Figure 8.5(c, d) gives $A \oplus B$ and $A \ominus B$, respectively; Figure 8.5(e, f) gives A^c and $\hat{B}$, respectively; Figure 8.5(g, h) gives $A^c \ominus \hat{B}$ and $A^c \oplus \hat{B}$ (a dark point in the dilation result represents the expanded

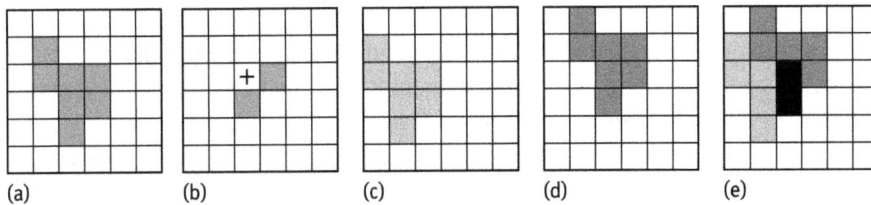

(a)   (b)   (c)   (d)   (e)

Figure 8.4: Erosion by means of shift.

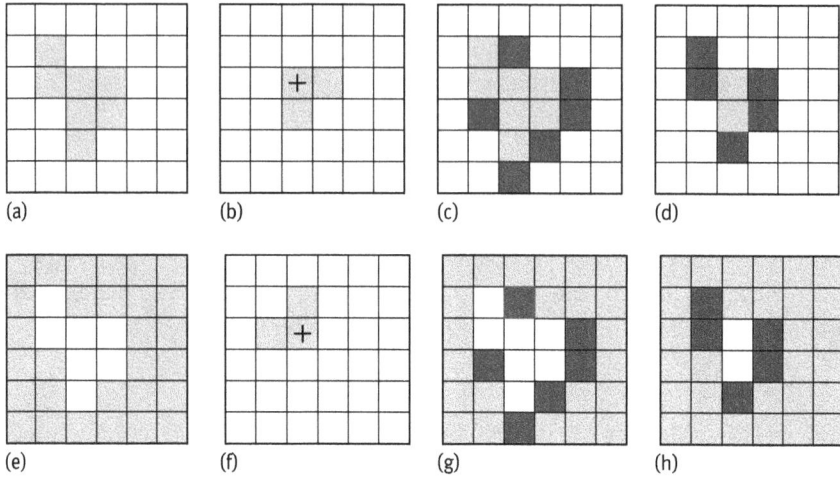

Figure 8.5: Duality of dilation and erosion.

point, and in the erosion result represents the etched point), respectively. Comparing Figure 8.5(c, g) can verify eq. (8.9), and comparing Figure 8.5(d, h) can verify eq. (8.10).

◻

Example 8.7 Verifying duality of dilation and erosion.
A set of real images corresponding to Figure 8.5 are shown in Figure 8.6, in which the duality of dilation and erosion is further verified. Figure 8.6(a) is a binary image obtained by thresholding Figure 1.1(b).

◻

8.1.2 Binary Opening and Closing

Opening and closing can also be considered as one basic pair of binary morphology operations.

8.1.2.1 Definitions

Dilation and erosion are not the inverse of each other, so they can be cascaded in combination. For example, the image may be first eroded and then the result dilated, or the image may be first dilated and then the result eroded (using the same structure elements). The former operation is called opening and the latter operation is called closing. They are also important mathematical morphology operations.

The operator for **opening** is ○, opening A by B is written $A \circ B$ and is defined as:

$$A \circ B = (A \ominus B) \oplus B \tag{8.11}$$

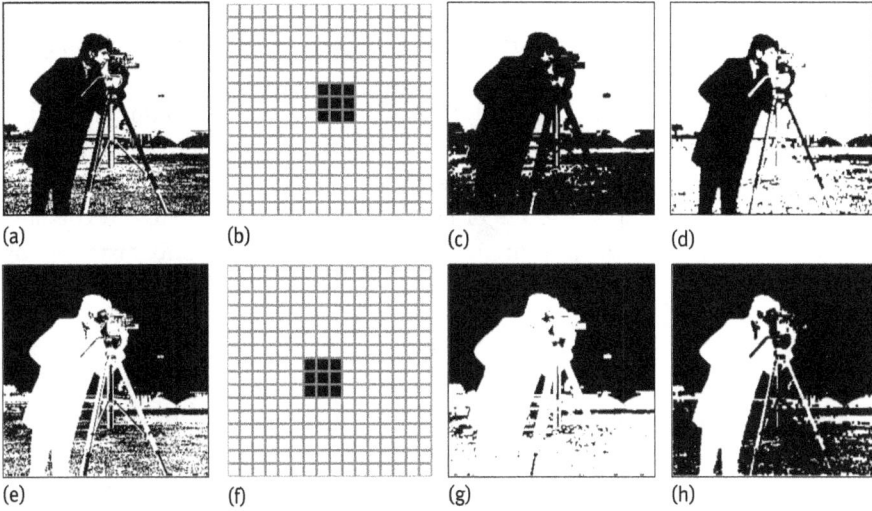

Figure 8.6: Verifying duality of dilation and erosion.

The operator for **closing** is •, closing A by B is written $A \bullet B$, and is defined as:

$$A \bullet B = (A \oplus B) \ominus B \qquad (8.12)$$

Both opening and closing are operations that can remove specific image details smaller than structure elements, while ensuring no global geometric distortion. Opening operation can filter off spines smaller than structure elements, cut out the elongated lap and play the role of separation. Closing operation can fill notches or holes smaller than structure elements and play the role of connection.

The ability of opening and closing for extracting shape matched to the structure elements in image can be obtained by the following characteristic theorems for opening and closing, respectively:

$$A \circ B = \{x \in A| \text{ for some } t \in A \ominus B, x \in (B)_t \text{ and } (B)_t \subseteq A\} \qquad (8.13)$$

$$A \bullet B = \left\{x | x \in \left(\hat{B}\right)_t \Rightarrow \left(\hat{B}\right)_t \bigcap A \neq \emptyset\right\} \qquad (8.14)$$

Equation (8.13) shows that opening A by B is to select some points in A and matched to B, these points can be completely obtained by the structure element B contained in A. Equation (8.14) shows that the result of closing A by B includes all points met the following conditions, that is, when the point is covered by the mapped and shifted structure elements, the intersection of A and the mapped and shifted structure elements is not zero.

8.1.2.2 Geometric Interpretation

Opening and closing can be given simple geometric interpretation by combining the method of the set theory. For opening, the structure element can be considered as a ball (on plane), the result of opening is outer edge of the ball rolling inside the set to be opened. Based on the filling property of opening operation, an implementation method based on set theory can be obtained. That is, opening A by B can be achieved by combining all results of filling B in A and then shift and compute their union. In other words, opening can be described using the following filling process:

$$A \circ B = \bigcup \{(B)_x \,|\, (B)_x \subset A\} \tag{8.15}$$

One example is given in Figure 8.7, in which Figure 8.7(a) gives A, Figure 8.7(b) gives B, Figure 8.7(c) shows B in several locations of A, Figure 8.7(d) gives the result of opening A by B.

There is a similar geometric interpretation for closing, the difference is now the structure elements are considered in the background. One example is given in Figure 8.8, in which Figure 8.8(a) gives A, Figure 8.8(b) gives B, Figure 8.8(c) shows B in several positions in A, Figure 8.8(d) gives the result of closing A by B.

8.1.2.3 Duality of Binary Opening and Closing

Similar to dilation and erosion, opening and closing also have duality. The **duality of opening and closing** can be expressed as

$$(A \circ B)^c = A^c \bullet \hat{B} \tag{8.16}$$

$$(A \bullet B)^c = A^c \circ \hat{B} \tag{8.17}$$

This duality can be derived from the duality of dilation and erosion represented by eqs. (8.9) and (8.10).

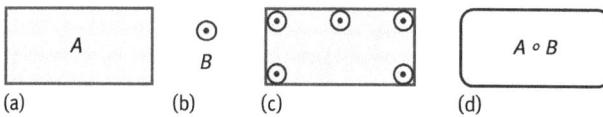

(a)   (b)   (c)   (d)

Figure 8.7: The filling property of opening.

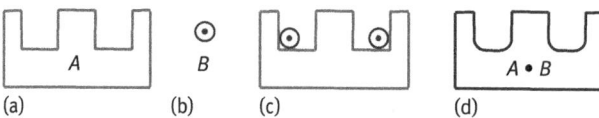

(a)   (b)   (c)   (d)

Figure 8.8: Geometric interpretation for closing.

Table 8.1: The relationship between the set and opening/closings.

Operation	Union		Intersection	
Opening	$\left(\bigcup_{i=1}^{n} A_i\right) \circ B \supseteq \bigcup_{i=1}^{n} (A_i \circ B)$		$\left(\bigcap_{i=1}^{n} A_i\right) \circ B \subseteq \bigcap_{i=1}^{n} (A_i \circ B)$	
Closing	$\left(\bigcup_{i=1}^{n} A_i\right) \bullet B \supseteq \bigcup_{i=1}^{n} (A_i \bullet B)$		$\left(\bigcap_{i=1}^{n} A_i\right) \bullet B \subseteq \bigcap_{i=1}^{n} (A_i \bullet B)$	

8.1.2.4 Relationship Between the Set and Binary Opening and Closing

The relationship between the set as well as binary opening and closings can be represented by four swap properties listed in Tables 8.1.

When the objects of operation are multiple images, the opening and closing can be performed with the help of the property of sets:

1. Opening and union: the opening of union includes the union of opening;
2. Opening and intersection: the opening of intersection is included in the intersection of opening;
3. Closing and union: the closing of union includes the union of closing;
4. Closing and intersection: the closing of intersection is included in the intersection of closing.

Example 8.8 Comparative examples of four basic operations.

A group of results obtained by using the four basic operations for a same image are presented in Figure 8.9 for comparison. The image A is shown in Figure 8.9(a). The structure element consists of the origin point and its 4-neighbors. From Figure 8.9(b, e), the results obtained by using dilation, erosion, opening, and closing are given, respectively. Wherein, the dark pixels in Figure 8.9(b) are expanded pixels, the dark pixels in Figure 8.9(d) are pixels of first eroded and then expanded, and the dark pixels in Figure 8.9(e) are those pixels first expanded out but not eroded after then. It is clear there exists $A \ominus B \subset A \subset A \oplus B$. In addition, the opening operation achieves the goal of producing a more compact and smooth contour of the object by eliminating the spike (or narrow band) on the object; while the closing operation realizes the filling of depressions (or holes) on the object. Both operations reduced the irregular nature of the contour.

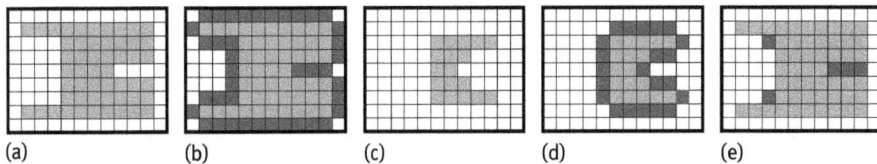

(a) (b) (c) (d) (e)

Figure 8.9: Result comparison of four basic operations.

8.2 Combined Operations of Binary Morphology

Four basic operations of binary mathematical morphology (dilation, erosion, opening, and closing) have been introduced in the previous section. Someone regarded the hit-or-miss transform as also a basic operation of binary mathematical morphology. If the hit-or-miss transform is combined with the above four basic operations, some other composition operations and fundamental algorithms for morphological analysis can be obtained.

8.2.1 Hit-or-Miss Transform

The **hit-or-miss transform** (or hit-or-miss operator) in mathematical morphology is a basic tool for shape detection and is also the basis for many composition operations. Hit-or-miss transform actually corresponds to two operations, so it uses two structure elements. Let A be the original image, E and F are a pair of non-overlapping set (which define a pair of structure elements). Hit-or-miss transform is represented by ⇑ and is defined as:

$$A \Uparrow (E, F) = (A \ominus E) \cap (A^c \ominus F) = (A \ominus E) \cap (A \oplus F)^c \qquad (8.18)$$

In the result of hit-or-miss transform, any pixel z meets $E + z$ is a subset of A, and $F + z$ is a subset of A^c. Conversely, the pixel z met the two conditions must be in hit-or-miss transform results. E and F are called hit structure element and miss structure element, respectively. Figure 8.10(a) is the hit structure element. Figure 8.10(b) is the miss structure element. Figure 8.10(c) gives four examples of the original images. Figure 8.10(d) provides the results obtained by using hit-or-miss transform, respectively. Note that two structure elements need to meet $E \cap F = \emptyset$; otherwise, the hit-or-miss transform will produce empty set.

Example 8.9 Hit-or-miss transform examples.
In Figure 8.11, ● and ○ represent the object and the background pixels, respectively. Let structure element B be as shown in Figure 8.11(a), the arrow points to the pixel corresponding to center (origin) of B. If the object A is as shown in Figure 8.11(b),

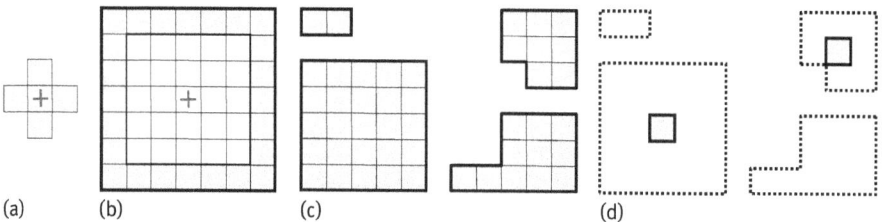

(a)   (b)   (c)   (d)

Figure 8.10: Examples of hit-or-miss transform.

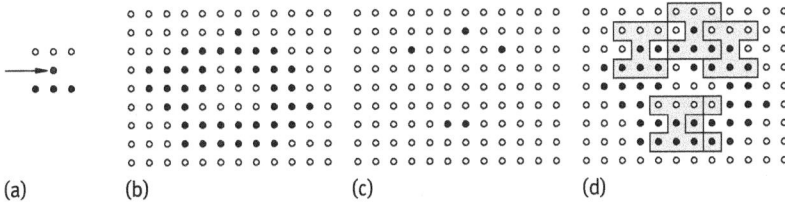

Figure 8.11: Hit-or-miss transform examples.

then the result of $A \Uparrow B$ is shown in Figure 8.11(c). Figure 8.11(d) provides further explanation, the object pixels remained in the result of $A \Uparrow B$ correspond to the pixels whose neighborhood in A matches to structure element B. ▢

8.2.2 Binary Composition Operations

Composition operations combine the basic operations to complete some meaningful actions, or to achieve some specific image processing functions.

8.2.2.1 Convex Hull
The **convex hull** of a region is one of presentation means for this region (Section 3.2.3). Given a set A, it is possible to use a simple morphology algorithm to get its convex hull $H(A)$. Let B_i, $i = 1, 2, 3, 4$, be four structure elements, first constructing

$$X_i^k = \left(X_i^{k-1} \Uparrow B_i \right) \cup A \qquad i = 1, 2, 3, 4 \quad \text{and} \quad k = 1, 2, \cdots \qquad (8.19)$$

In the above equation, $X_i^0 = A$. Let now $D_i = X_i^{\text{conv}}$, where the superscript "conv" indicates the convergence at the sense of $X_i^k = X_i^{k-1}$. According to these definitions, the convex hull of A can be expressed as

$$H(A) = \bigcup_{i=1}^{4} D_i \qquad (8.20)$$

In other words, the construction process for convex hull is as follows. First, performing iteratively hit-or-miss transform to the region A with B_1, when no further changes will be evaluated, the union of the transform result and A is computed, this union result is noted as D_1. Then, B_2 is used for transform, repeating iteration and seeking union, the union result is recorded as D_2. This procedure will be conducted also for B_3 and B_4 to obtain the D_3 and D_4; the union of four results D_1, D_2, D_3, and D_4 provides the convex hull of A.

Example 8.10 Construction of convex hull.
One example of constructing convex hull is presented in Figure 8.12. Figure 8.12(a) gives the four structure elements used, the origin of each structure element is in the

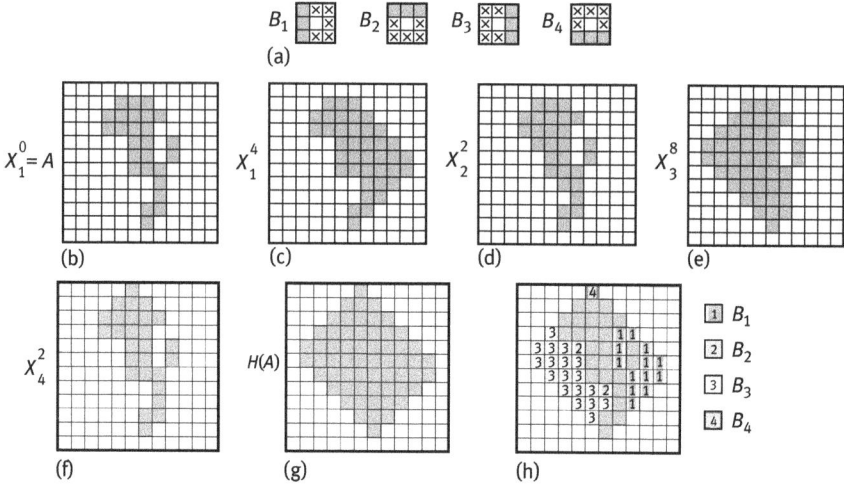

Figure 8.12: Construction of convex hull.

center, "×" means it can be any value. Figure 8.12(b) shows the set A whose convex hull needs to construct. Figure 8.12(c) is the result obtained by starting from $X_1^0 = A$, and after four iterations according to eq. (8.19). Figure 8.12(d–f) are the results obtained by starting from $X_2^0 = A$, $X_3^0 = A$, $X_4^0 = A$, and after two iterations, eight iterations, two iterations according to eq. (8.19), respectively. The finally obtained convex hull is the union of the above four results according to eq. (8.20), as shown in Figure 8.12(g). In Figure 8.12(h), the numbers indicate the contribution of different structure elements in constructing the convex hull.

8.2.2.2 Thinning
In some applications (such as seeking skeleton), it is hoped to erode the object region but do not to split it into multiple subregions. Here, the pixels located around the boundary of the object region are to be checked, to see whether they can be removed without splitting the object region into multiple sub-regions. This task can be achieved by thinning. **Thinning** set A by structure element B is noted as $A \otimes B$, which can be defined (with the help of hit-or-miss transform) as follows:

$$A \otimes B = A - (A \Uparrow B) = A \cap (A \Uparrow B)^c \tag{8.21}$$

In the above equation, hit-or-miss transform is used to determine the pixels that should be thinning out, then they can be removed from the original set A. In practice, a series of masks with small size are used. If a series of structure element is defined as $\{B\} = \{B_1, B_2, \ldots, B_n\}$, in which B_{i+1} represents the result of rotating B_i, then thinning can be defined as:

$$A \otimes \{B\} = A - ((\cdots((A \otimes B_1) \otimes B_2)\cdots) \otimes B_n) \qquad (8.22)$$

In other words, this process is first using B_1 for thinning A, then using B_2 for thinning the previous result, and so on until B_n is used for thinning. The whole process can be repeated until no change produced so far.

The following set of four structure elements (hit-or-miss masks) can be used for thinning ("×" denotes the value 0 or 1):

$$B_1 = \begin{bmatrix} 0 & 0 & 0 \\ \times & 1 & \times \\ 1 & 1 & 1 \end{bmatrix} \quad B_2 = \begin{bmatrix} 0 & \times & 1 \\ 0 & 1 & 1 \\ 0 & \times & 1 \end{bmatrix} \quad B_3 = \begin{bmatrix} 1 & 1 & 1 \\ \times & 1 & \times \\ 0 & 0 & 0 \end{bmatrix} \quad B_4 = \begin{bmatrix} 1 & \times & 0 \\ 1 & 1 & 0 \\ 1 & \times & 0 \end{bmatrix} \quad (8.23)$$

Example 8.11 A thinning process and results.
A set of structure elements and a thinning process using these structure elements are presented in Figure 8.13. Figure 8.13(a) gives a set of structure elements commonly used, in which the origin of a structure element is in its center, white and gray pixels have values 0 and 1, respectively. If the points detected by structure element B_1 are subtracted from the object, the object will be thinned from the upper side, if the points detected by structure element B_2 are subtracted from the object, the object will be thinned from the upper right corner, and so on. With the above set of structure elements, symmetrical thinning results will be obtained. Besides, the four structure elements with odd numbers have stronger thinning capability, while the four structure elements with even numbers have weaker thinning capability.

Figure 8.13(b) shows the original region to be thinned, its origin is located in the upper left corner. Figure 8.13(c, k) provides successively the thinning results with the various structure elements (circles mark the pixels thinned out in the current step). The converged result obtained by using B_6 a second time is shown in Figure 8.13(l). Final result obtained by converting the thinning result to a mixed connectivity to eliminate multi-connection problem is shown in Figure 8.13(m). In many applications (such as seeking skeleton), it is required to erode the object but does not to break the object into several parts. To do this, a number of points on the contour of the object should be first detected. Removing these points will not lead the division of object into two parts. Using the above hit-or-miss masks can satisfy this condition.

◻

8.2.2.3 Thickening
Thickening set A by structure element B is denoted by $A \otimes B$. From the point of morphological view, thickening corresponds to thinning, and can be defined by

$$A \otimes B = A \bigcup (A \Uparrow B) \qquad (8.24)$$

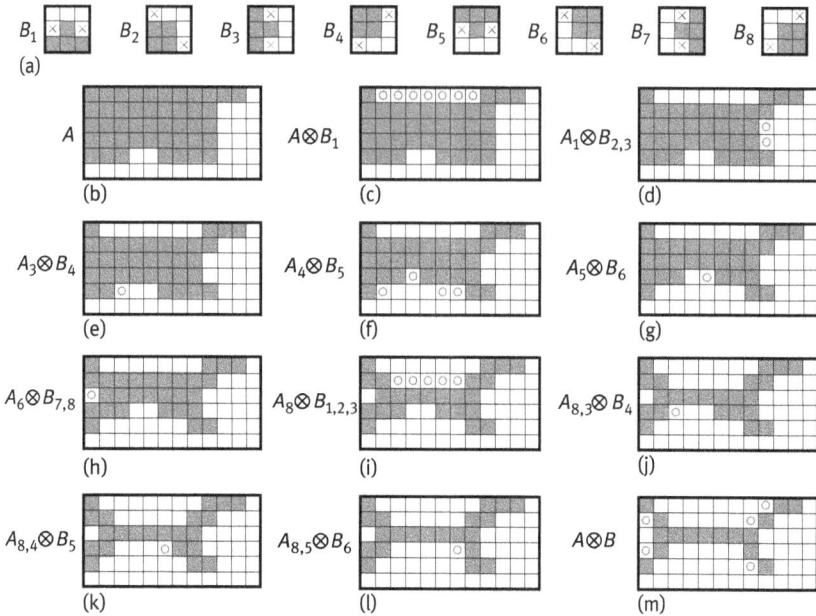

Figure 8.13: A thinning process and results.

Similar to thinning, thickening can also be defined by a series of operations:

$$A\otimes\{B\} = ((\cdots((A\otimes B_1)\otimes B_2)\cdots)\otimes B_n) \tag{8.25}$$

The structure elements used in thickening are similar as in thinning, only 1 and 0 should be exchanged over. In practice, the thickening results are often obtained by first thinning background and then finding the complements of thinning results. In other words, if a set A needs to be thickened, it can be achieved by first constructing $C = A^c$, then thinning C, and finally seeking C^c.

Example 8.12 Using thinning for thickening.
One example of using thinning for thickening is given in Figure 8.14. Figure 8.14(a) is the set A. Figure 8.14(b) is $C = A^c$. Figure 8.14(c) shows the result of thinning. Figure 8.14(d) shows the C^c obtained from the complement Figure 8.14(c), and the thickening result after simple post-treatment for removing disconnected points is shown in Figure 8.14(e).

8.2.2.4 Pruning
Pruning is an important complement to thinning and skeleton extraction operation, or it is often used as a post processing means for thinning and skeleton extraction. Because the thinning and skeleton extraction often leave some excessive parasitic

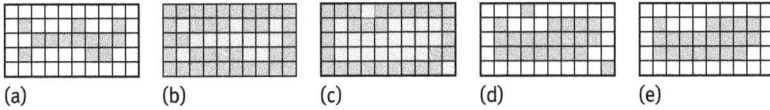

Figure 8.14: Using thinning for thickening.

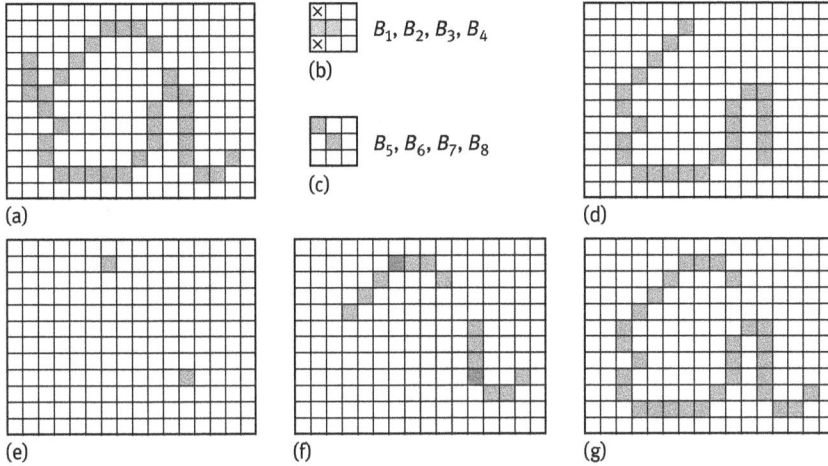

Figure 8.15: Pruning process and result.

components, so it is necessary to use a method such as pruning as postprocessing to eliminate these components. Pruning can make use of a combination of several methods above to achieve. To explain the pruning process, let us considering an example of handwritten character recognition using automatic methods, as shown in Figure 8.15. A skeleton for a handwritten character "a" is given in Figure 8.15(a). The segment at the leftmost of the character is a typical example of a parasitic component. To solve this problem, it is possible to continuously eliminate the endpoint. Of cause, this process will eliminate or shorten other segments in the character. It is assumed that the length of the parasitic segment is not more than three pixels, so the process is designed to eliminate only segments with less than three pixels. For a set A, using a series of structure elements capable of detecting endpoint to thin A can get the desired results. Let

$$X_1 = A \otimes \{B\} \tag{8.26}$$

where $\{B\}$ represent the structure elements series, as shown in Figure 8.15(b, c). This series contains two kinds of structures, rotating each series can obtain four structure elements, so there are a total of eight structure elements.

Thinning set A three times, according to eq. (8.26), gives the result X_1 as shown in Figure 8.15(d). The next step is to restore the character to get the original shape without the parasitic segment. To do this, a set X_2 with all endpoints of X_1 is constructed, as shown in Figure 8.15(e).

$$X_2 = \bigcup_{k=1}^{8}(X_1 \Uparrow B_k)$$

(8.27)

wherein B_k is an endpoint detector. In the following, dilating endpoints three times by taking A as constrain to obtain X_3:

$$X_3 = (X_2 \oplus H)\bigcap A$$

(8.28)

wherein H is a 3×3 structure element with all pixel values 1. Such a conditional dilation can prevent to produce value 1 elements outside of the region of interest, as shown in Figure 8.15(f). Finally, the union of X_1 and X_3 provides the result shown in Figure 8.15(g).

$$X_4 = X_1\bigcup X_3$$

(8.29)

The above pruning process cyclically uses a set of structure elements for eliminating noise pixels. Generally, an algorithm only uses one or two cycles of this set of structure elements; otherwise, it is possible that the object region in image may be undesirably changed.

8.3 Practical Algorithms of Binary Morphology

With the help of the previously described basic operations and composition operations of binary mathematical morphology, a series of practical algorithm of binary morphology may be constituted. The following describes several specific algorithms.

8.3.1 Noise Removal

The binary image obtained by segmentation often consists of some small holes or islands. These holes or islands are generally caused by the system noise, threshold value selected or preprocessing. Pepper and salt noise is a typical noise, which leads to small holes or islands in binary images. Combining the opening and closing operations could constitute a noise filter to eliminate this type of noise. For example, using a structure element comprising a central pixel and its 4-neighborhood pixels for the opening of the image can eliminate the pepper noise, while using this structure element for the closing of the image can eliminate the salt noise (Ritter 2001).

One example of **noise removal** using a combination of opening and closing is given in Figure 8.16. Figure 8.16(a) comprises a rectangular object A, due to the

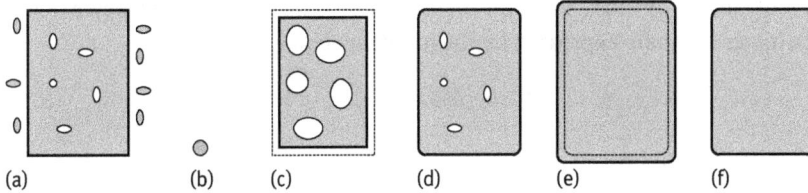

Figure 8.16: Pepper and salt noise removal.

influence of noise, there are some holes inside of the object and some blocks around the object. Now, the structure elements B, as shown in Figure 8.16(b), will be used by morphological operations to eliminate noise. Here, the structure element should be larger than all of the holes and blocks. The first step is to erode A by B, which gives Figure 8.16(c). The second step is to dilate Figure 8.16(c) by B, which gives Figure 8.16(d). The serial combination of erosion and dilation operations is an opening operation, it will eliminate the noise around the object. Now, Figure 8.16(d) is dilated by B, which gives Figure 8.16(e). Finally, Figure 8.16(e) is eroded by B, which gives Figure 8.16(f). The serial combination of these last two operations is a closing operation, the noise inside the object will be eliminated. The whole process is first opening and then closing, it can be written as

$$\{[(A \ominus B) \oplus B] \oplus B\} \ominus B = (A \circ B) \bullet B \tag{8.30}$$

Comparing Figure 8.16(a, f), it is seen that the inside and outside noises are both eliminated, while the object itself has not been changed much, except that the original four right angles turn to be rounded.

8.3.2 Corner Detection

Corner pixel is located at the place where the slope has a sudden change and the absolutely curvature is very big. Corner pixel can be extracted by means of morphological operations. In the first step, a circular structure element with a suitable size should be selected, using the structure element for opening and subtract the result from the original image. Then two new structure elements are selected, one is bigger than the remaining area, another is smaller than the remaining area. Using these two structure elements to erode the remaining area, and comparing the two results obtained. This is equivalent to a band-pass filter.

Corner detection can also be achieved by means of asymmetric closing (Shih 2010). Asymmetrical closing comprises two steps. First a structure element is used to dilate the image, then another structure element is used to erode the image. The idea is to make dilation and erosion complementary. For example, two structure elements with forms of cross "+" and diamond "◊" can be used. The asymmetric closing for image A can be expressed by the following equation:

$$A^c_{+\Diamond} = (A \oplus +) \ominus \Diamond \tag{8.31}$$

Here, the corner strength is

$$C_+(A) = |A - A^c_{+\Diamond}| \tag{8.32}$$

For different corner points, the corner strength after rotating $45°$ can also be calculated (the structure elements are square $\square$ and cross $\times$, respectively):

$$C_\times(A) = |A - A^c_{\times\square}| \tag{8.33}$$

The above four structure elements can be combined (in sequence as shown in Figure 8.17), then the detection of the corners can be written as

$$C_{+\times}(A) = |A^c_{+\Diamond} - A^c_{\times\square}| \tag{8.34}$$

8.3.3 Boundary Extraction

Suppose there is a set A, its boundary is denoted by $\beta(A)$. By first eroding A by a structure element B and then taking the difference between the result of erosion and A, $\beta(A)$ can be obtained:

$$\beta(A) = A - (A \ominus B) \tag{8.35}$$

Let us look at Figure 8.18, wherein Figure 8.18(a) is a binary object A, Figure 8.18(b) is a structure element B, Figure 8.18(c) shows the result of eroding A by B, $A \ominus B$, Figure 8.18(d) is the final result obtained by subtracting Figure 8.18(c) from Figure 8.18(a). Note that when the origin of B is at the edge of A, some parts of B will be outside A, where is generally set to 0. Also note that the structure element is 8-connected, while the resulting border is 4-connected.

8.3.4 Object Detection and Localization

How to use hit-or-miss transform to determine the location of a given size square (object) is explained with the help of Figure 8.19 (Ritter, 2001). Figure 8.19(a) is the

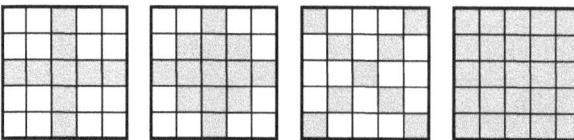

Figure 8.17: Four structure elements, in order of +, ◊, ×, □.

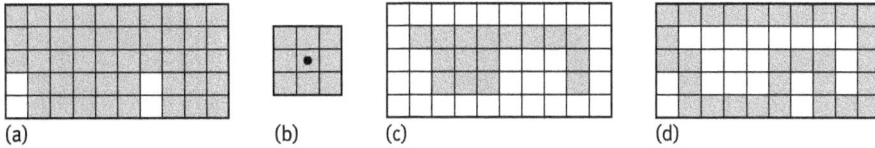

Figure 8.18: An example of border extraction.

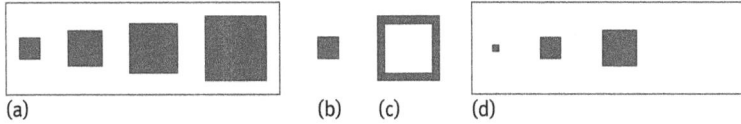

Figure 8.19: Square region detection with a hit-or-miss transform.

original image, including four solid squares of size 3 × 3, 5 × 5, 7 × 7, and 9 × 9, respectively. Taking the 3 × 3 solid square in Figure 8.19(b) as a structure element E, and the 9 × 9 box frame (edge width of one pixel) in Figure 8.19(c) as another structure element F. Their combination constitute a pair of structure elements $B = (E, F)$. In this example, the hit-or-miss transform E is designed to hit region covered by E and to miss the region covered by F. The finally obtained results are shown in Figure 8.19(d).

8.3.5 Region Filling

A region and its boundary can be mutually demanded. When a region is known, its boundary can be obtained according to eq. (8.35). On the other side, when a boundary is known, the region can be obtained by **region filling**. An example is given in Figure 8.20, wherein Figure 8.20(a) provides boundary points for a region A, its complement is given in Figure 8.20(b). The region A can be filled by using the structure element in Figure 8.20(c) for computing dilation, complement, and intersection. First, assigning 1 to a point within the boundary (as shown by the dark pixel in Figure 8.20(d)), then filling the inside of the boundary according to the following iterative equation (Figure 8.20(e, f) provides the situations in two intermediate steps, respectively)

$$X_k = (X_{k-1} \oplus B) \bigcap A^c, \quad k = 1, 2, 3, \ldots \tag{8.36}$$

The iteration will be stopped when $X_k = X_{k-1}$ (in this example $k = 7$, as shown in Figure 8.20(g)). At this moment, the intersection of X_k and A includes the inside and boundary of filled region, as shown in Figure 8.20(h). The dilation process in eq. (8.36) must be controlled to avoid exceeding the boundary, but the computation of intersection with A^c at every step will limit it in the region of interest. This dilation process can be called a conditional dilation process. Note that the structure element here is 4-connected, while the original filled boundary is 8-connected.

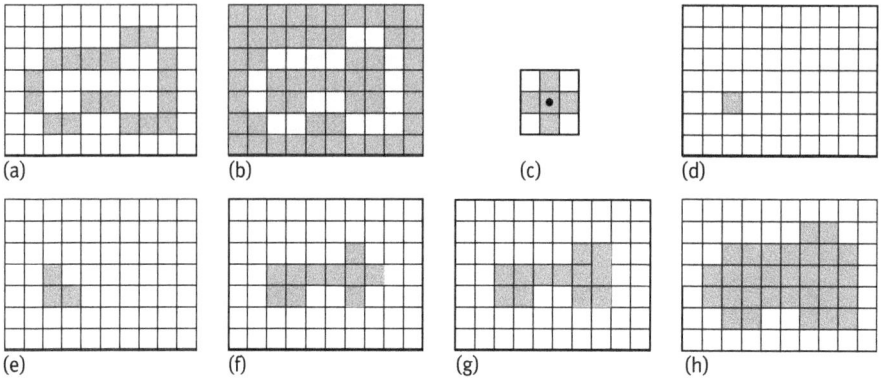

Figure 8.20: Region filling.

8.3.6 Extraction of Connected Components

Let Y represent a connected component in a set A, and one point in Y is known, then all the elements in Y can be obtained by using the following iterative formula:

$$X_k = (X_{k-1} \oplus B) \bigcap A, \quad k = 1, 2, 3, \ldots \tag{8.37}$$

When $X_k = X_{k-1}$, stop iteration, then $Y = X_k$.

Comparing eqs. (8.37)–(8.36), except replacing A^c by A, they are identical. Because here the elements needed to extract have been marked as 1. Computing intersection with A in each iteration can remove the dilation taking the element marked 0 as the center. One example of extraction of connected components is shown in Figure 8.21, where the structure element used is same as in Figure 8.20. In Figure 8.21(a), the values of lightly shaded pixels (i. e., the connected components) are 1, but at this time, these pixels have not yet been discovered by algorithm. The values of darkly shaded pixels in Figure 8.21(a) are also 1 and that they are known to belong Y. These pixels can be taken as the start points of algorithm. The results of first and second iterations are given in Figure 8.21(b, c), respectively. The final result is given in Figure 8.21(d).

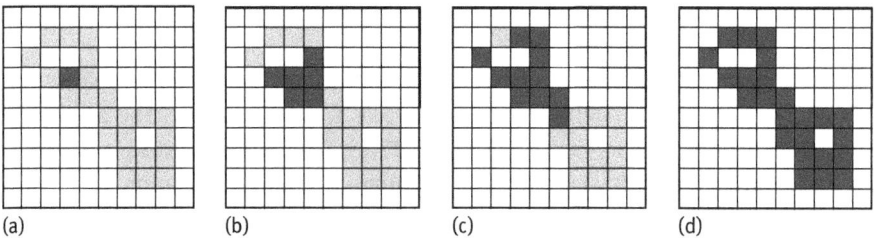

Figure 8.21: Extraction of connected components.

8.3.7 Calculation of Region Skeleton

In Section 3.3.5, the concept of the skeleton and a computational method are introduced. Here, a technique for calculating **skeleton** with mathematical morphological operations is presented. Let $S(A)$ represent the skeleton of A, which can be expressed as

$$S(A) = \bigcup_{k=0}^{K} S_k(A) \tag{8.38}$$

The $S_k(A)$ in the above formula is generally referred to as a subset of the skeleton, which can be written as

$$S_k(A) = (A \ominus kB) - [(A \ominus kB) \circ B] \tag{8.39}$$

where B is a structure element; $(A \ominus kB)$ means eroding A by B for k consecutive times, which can be represented by T_k:

$$T_k = (A \ominus kB) = ((\dots (A \ominus B) \ominus B) \ominus \dots) \ominus B \tag{8.40}$$

K in eq. (8.38) represents the last iteration in eroding A to the empty set, namely

$$K = \max\{k | (A \ominus kB) \neq \emptyset\} \tag{8.41}$$

Example 8.13 Morphological skeleton.
One example of computing the skeleton with morphological operations is shown in Figure 8.22. In the original image (top left), there is a rectangular object that has a little addenda above it. The first row then shows the result sets T_k obtained by sequentially erosions, i. e., T_0, T_1, T_2, T_3, T_4. Because $T_4 = \emptyset$, so $K = 3$. The second line gives a set of sequentially obtained skeleton sets S_k, that is, S_0, S_1, S_2, S_3, and the final skeleton S (including two separately connected portions). ◻

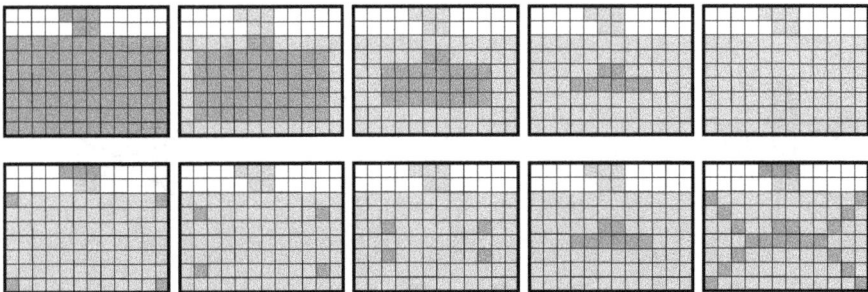

Figure 8.22: The process of skeleton computation.

Equation (8.38) shows that the skeleton A may be obtained by the union of a subset $S_k(A)$. Conversely, A can also be used for the reconstruction of $S_k(A)$:

$$A = \bigcup_{k=0}^{K} (S_k(A) \oplus k\,B) \tag{8.42}$$

wherein B is a structure element; $(S_k(A) \oplus kB)$ represents the dilation of $S_k(A)$ by B for k consecutive times, namely

$$(S_k(A) \oplus kB) = ((\cdots (S_k(A) \oplus B) \oplus B) \oplus \cdots) \oplus B \tag{8.43}$$

Example 8.14 A real example of morphological skeleton.
One example of computing the skeleton of a real image is shown in Figure 8.23. Figure 8.23(a) is a binary image; Figure 8.23(b) shows the skeleton obtained using the structure element of 3×3 in Figure 8.18; Figure 8.23(c) shows the skeleton obtained using a similar structure element of 5×5; and Figure 8.23(d) shows the skeleton obtained using a similar structure element of 7×7. Note that in Figure 8.23(c, d), since the masks are relatively big so the petiole is not preserved. ▣

8.4 Basic Operations of Grayscale Morphology

The four basic operations of binary mathematical morphology, namely dilation, erosion, opening, and closing, can be easily extended to the gray-level space. Here, the extension is made by analogy. Different from binary mathematical morphology, the operands of operations here are no longer seen as sets but as the image functions. In the following, $f(x, y)$ represents the input image, $b(x, y)$ represents a structure element, which itself is a small image.

8.4.1 Grayscale Dilation and Erosion

Grayscale dilation and erosion are the most basic operations of grayscale mathematical morphology. Different from the binary dilation and erosion operations,

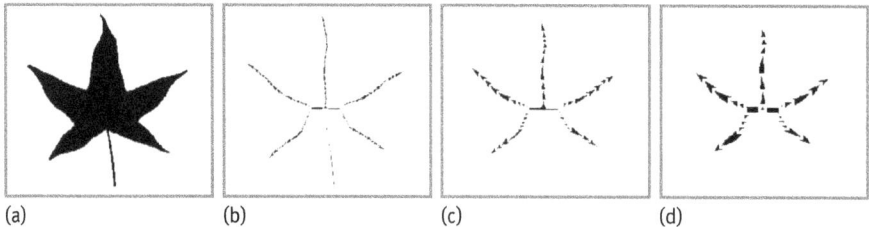

(a)   (b)   (c)   (d)

Figure 8.23: A real example of morphological skeleton.

in which the result is mainly reflected in the image plane, the result of grayscale dilation and erosion operations is mainly reflected in the amplitude axis.

8.4.1.1 Grayscale Dilation

The **grayscale dilation** of input image f by structure element b is denoted by $f \oplus b$, which is defined as

$$(f \oplus b)(s, t) = \max\{f(x, y) + b(s - x, t - y) \mid (x, y) \in D_f \text{ and } [(s - x), (t - y)] \in D_b\} \quad (8.44)$$

where D_f and D_b are the domains of definition of f and b, respectively. Here, limiting $[(s - x), (t - y)]$ inside the domain of definition of b is similar to the binary dilation in which at least two elements from each of two operands have a (nonzero) element in intersection. Equation (8.44) is very similar to the formula of 2-D convolution, the differences here are that the max (maximum) has replaced the summation (or integration) in convolution and the addition has replaced the multiplication in convolution. In the dilated grayscale images, the region that is brighter than the background will be expanded, while the region that is darker than the background will be condensed.

In the following, the meaning and operation mechanism of operation in eq. (8.44) are first briefly explained with 1-D function. When considering the 1-D function, eq. (8.44) is simplified to:

$$(f \oplus b)(s) = \max\{f(x) + b(s - x) \mid x \in D_f \text{ and } (s - x) \in D_b \quad (8.45)$$

Similar as in convolution, $b(-x)$ is the inverted mapping of $b(x)$ respective to the origin of x-axis. For positive s, $b(s - x)$ moves to the right; for negative s, $b(s - x)$ moves to the left. The requirements of x in the domain of definition of f and of $(s - x)$ values in the domain of definition of b is to make f and b coincide.

Example 8.15 Illustration of grayscale dilation.
An illustration of grayscale dilation is shown in Figure 8.24. Figure 8.24(a, b) gives f and b, respectively. In Figure 8.24(c), several locations of the structure element (inverted) during the calculation process are provided. In Figure 8.24(d), the bold line gives the final dilation results. Due to the interchangeable property of dilation, if let f have reverse translation for dilation, it will produce exactly the same result.   ◙

The computation of dilation is to select the maximum value of $f + b$ in the neighborhood determined by the structure element. Therefore, there will be two effects with the dilation of gray-scale images:
1. If the values of the structure element are all positive, then the output image will be brighter than the input image.
2. If the size of dark details in input image is smaller than the structure element, then the visual effect will be weakened. The degree of diminished extent depends on the gray-level values surrounding these dark details and the shape and magnitude values of structure elements.

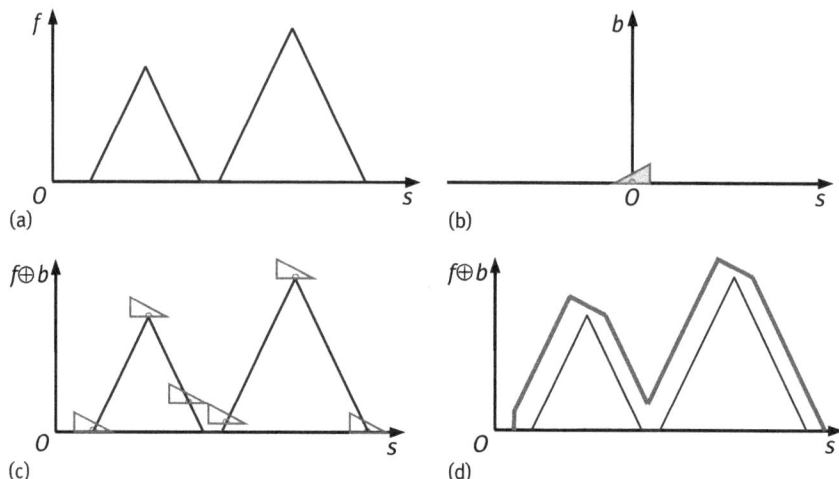

Figure 8.24: Illustration of grayscale dilation.

Example 8.16 The numerical computation of grayscale dilation.
One example for the computation of grayscale dilation is shown in Figure 8.25. Figure 8.25(a) is a 5×5 image A. Figure 8.25(b) is a 3×3 structure element B with its origin at the center. Figure 8.25(c) is its inversion. Since the size of B is 3×3, in order to avoid the coefficient of B falling out of A upon dilation, the edge pixels of A will not be considered, that is, for a 5×5 image A, considering only the 3×3 center portion. The origin of A, $(0, 0)$, is at the upper left corner, so the coordinates of 3×3 center portion are shown as in Figure 8.25(d). If overlapping the origin of B with $(1, 1)$ in image A, and calculating the sum of each value of B with its corresponding pixel value in A, the result is shown in Figure 8.25(e). Taking the highest value as the result of dilation, the A will be updated as shown in Figure 8.25(f). Another example, overlapping the origin of B with $(2, 2)$ and also calculating the sum of each value of B with its corresponding pixel value in A, the results is shown in Figure 8.25(g). Then, taking the highest value as the result of dilation, the A will be updated again as shown in Figure 8.25(h). Dilating all center 3×3 pixels of A as above, the final result is shown in Figure 8.25(i). ◻

8.4.1.2 Grayscale Erosion
The **grayscale erosion** of input image f by structure element b is denoted by $f \ominus b$, which is defined as

$$(f \ominus b)(s, t) = \min\{f(x, y) - b(s + x, t + y) \,|\, (x, y) \in D_f \text{ and } [(s + x), (t + y)] \in D_b\} \quad (8.46)$$

where D_f and D_b are the domains of definition of f and b, respectively. Here, limiting $[(s - x), (t - y)]$ in the domain of definition of b is similar to in the definition of

(a)

10	10	10	10	10
10	11	12	13	10
10	13	15	17	10
10	17	18	19	10
10	10	10	10	10

(b)

	1	
3	5	7
	9	

(c)

	9	
7	5	3
	1	

(d)

(1,1)	(2,1)	(3,1)
(1,2)	(2,2)	(3,2)
(1,3)	(2,3)	(3,3)

(e)

	19	
17	16	15
	14	

(f)

10	10	10	10	10
10	19	12	13	10
10	13	15	17	10
10	17	18	19	10
10	10	10	10	10

(g)

	21	
20	20	20
	19	

(h)

10	10	10	10	10
10	19	12	13	10
10	13	21	17	10
10	17	18	19	10
10	10	10	10	10

(i)

10	10	10	10	10
10	19	19	19	10
10	20	21	22	10
10	22	24	26	10
10	10	10	10	10

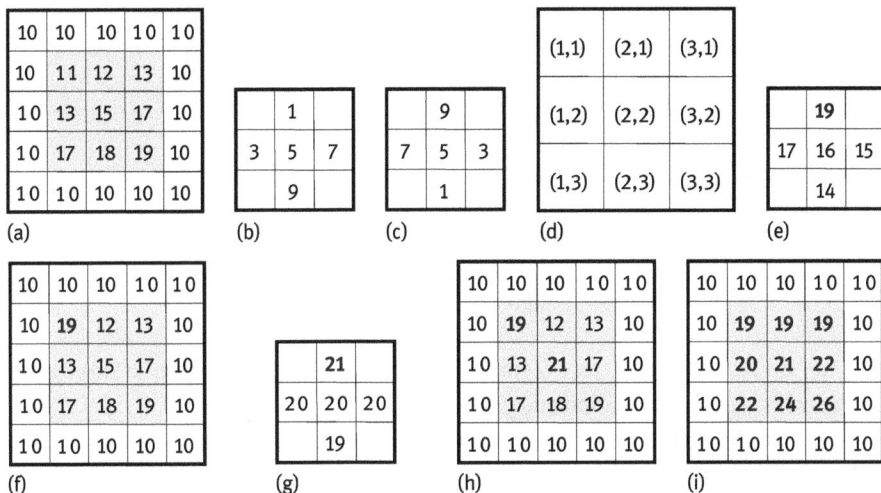

Figure 8.25: The computation of grayscale dilation.

binary erosion in which the structure elements are required to be included completely in eroded set. Equation (8.46) is very similar to 2-D correlation, except that here the summation (or integration) in correlation is replaced by the min (minimum) operation, and the multiplication in correlation is replaced by subtraction. Therefore, in the results of grayscale image erosion, the region that is darker than the background will be expanded and the region that is brighter than the background will be condensed.

For simplicity, similar as in the discussion of the dilation, the meaning and operation mechanism of operation in eq. (8.46) are first briefly explained with 1-D function. When considering the 1-D function, eq. (8.46) can be simplified to

$$(f \ominus b)(s) = \min\{f(x) - b(s + x) \,|\, x \in D_f \text{ and } (s + x) \in D_b\} \qquad (8.47)$$

Similar as in convolution, for positive s, $b(s + x)$ moves to the right; for negative s, $b(s + x)$ moves to the left. The requirements of x in the domain of definition of f and of $(s + x)$ in the domain of definition of b are to make b totally be contained in the definition range of f.

Example 8.17 Illustration of grayscale erosion.
An illustration of grayscale erosion is shown in Figure 8.26. The input image f and the structure element b are the same as in Figure 8.24. Figure 8.26(a, b) provides several locations of the structure element (b shifted) and the final result of erosion, respectively. ◻

The computation of erosion is to select the minimum value of $f - b$ in the neighborhood determined by the structure element. Therefore, there will be two effects with the erosion of grayscale images:

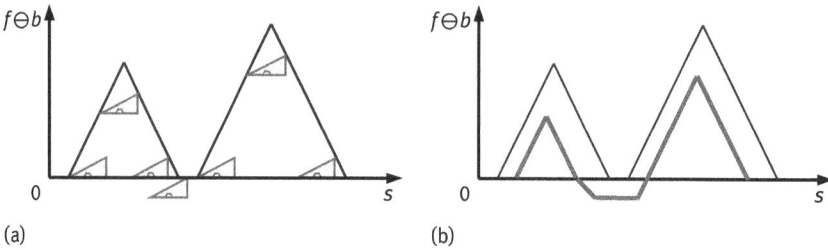

Figure 8.26: Illustration of grayscale erosion.

1. If the values of the structure element are all positive, then the output image will be darker than the input image.
2. If the size of light details in input image is smaller than the structure element, then the visual effect will be weakened. The degree of diminished extent depends on the gray-level values surrounding these light details and the shape and magnitude values of structure elements.

Example 8.18 The numerical computation of grayscale erosion.
One example for the computation of grayscale erosion is shown in Figure 8.27. The images A and structure element B used are still the same as in Figure 8.25. Similar to Example 8.16, if overlapping the origin of B with (1, 1), and calculating the sum of each value of B with its corresponding pixel value in A, the result is shown in Figure 8.27(a). Taking the smallest value as the result of erosion, the image A will be updated as shown in Figure 8.27(b). If overlapping the origin of B with (2, 2) and also calculating the sum of each value of B with its corresponding pixel value in A, the results is shown in Figure 8.27(c). Then, taking the lowest value as the result of erosion, the A will be updated again as shown in Figure 8.27(d). Eroding all center 3×3 pixels of A as above, the final result is shown in Figure 8.27(e). ▢

Example 8.19 Grayscale dilation and erosion.
One example of grayscale dilation and erosion is shown in Figure 8.28. Figure 8.28(a) is the original image. Figure 8.28(b, c) are the results obtained by one dilation and

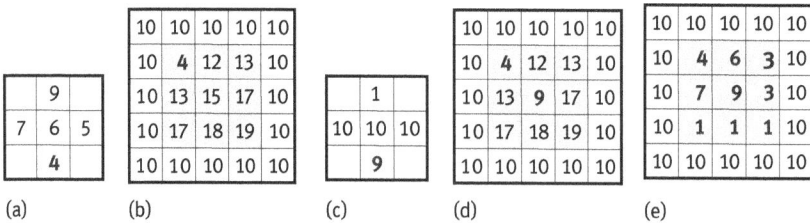

Figure 8.27: The effects of grayscale erosion.

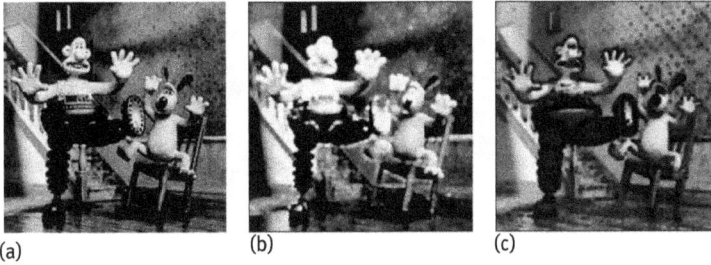

(a)   (b)   (c)

Figure 8.28: Real examples of grayscale dilation and erosion.

one erosion, respectively. The value at the center of structure elements is 2, the four
neighborhood pixels have values 1.   ▣

8.4.1.3 Duality of Grayscale Dilation and Erosion

Grayscale dilation and erosion are dual with respect to the complement of function
(complementary function) and the reflection of function. The **duality of grayscale
dilation and erosion** can be expressed as

$$(f \oplus b)^c = f^c \ominus \hat{b} \tag{8.48}$$

$$(f \ominus b)^c = f^c \oplus \hat{b} \tag{8.49}$$

Complement of function is defined herein by $f^c(x, y) = -f(x, y)$, and the reflection of
function is defined by $\hat{b}(x, y) = b(-x, -y)$.

8.4.2 Grayscale Opening and Closing

The representations of opening and closing in grayscale mathematical morphology
are consistent with their corresponding operations in binary mathematical morphology. **Opening** f by b is denoted as $f \circ b$, which is defined as

$$f \circ b = (f \ominus b) \oplus b \tag{8.50}$$

Closing f by b is denoted as $f \bullet b$, which is defined as

$$f \bullet b = (f \oplus b) \ominus b \tag{8.51}$$

Opening and closing are dual with the complement of function and reflector of
function, their duality can be expressed as

$$(f \circ b)^c = f^c \bullet \hat{b} \tag{8.52}$$

$$(f \bullet b)^c = f^c \circ \hat{b} \tag{8.53}$$

Figure 8.29: Grayscale opening and closing.

Since $f^c(x, y) = -f(x, y)$, eqs. (8.52) and (8.53) can also be written as

$$- (f \circ b) = -f \bullet \hat{b} \tag{8.54}$$

$$- (f \bullet b) = -f \circ \hat{b} \tag{8.55}$$

Example 8.20 Grayscale opening and closing.
Some results obtained by using grayscale mathematical morphology operations are shown in Figure 8.29. Figure 8.29(a, b) is results of grayscale opening and grayscale closing operations, respectively. The grayscale structure elements used are the same as in Example 8.19. Note that in Figure 8.29(a) the operating handle held by photographer is now less obvious, which indicates the operation of grayscale opening eliminated smaller bright details. On the other hand, the blur of photographer's mouth shows that grayscale closing can eliminate smaller dark details. Figure 8.29(c, d) shows the results of grayscale opening and grayscale closing of Figure 8.28, respectively. The grayscale opening make the image to become darker, while the grayscale closing make the image to become brighter. ▢

8.5 Combined Operations of Grayscale Morphology

Based on the basic operations of mathematical morphology introduced in the previous section, a series of combined operations for grayscale mathematical morphology can be obtained.

8.5.1 Morphological Gradient

Dilation and erosion are often combined for computing morphological gradient. The **morphological gradient** of an image is denoted g:

$$g = (f \oplus b) - (f \ominus b) \tag{8.56}$$

Morphological gradient can enhance relatively sharp transition region in the gray-level image. Different from a variety of spatial gradient operators (Section 2.2.1), the morphological gradient obtained by using symmetrical structure elements is less affected by the edge direction, but the amount of computation required for computing morphological gradient is generally large.

Example 8.21 Result of morphological gradient calculation.
Some gradient calculation results are given in Figure 8.30. Figure 8.30(a, b) is the results of morphological gradient calculated for the two images in Figure 1.1. For comparison, Figure 8.30(c) shows the (amplitude) result of using Sobel gradient operator for the second image.   ▣

8.5.2 Morphological Smoothing

One kind of image smoothing methods can be achieved by first opening and then closing. Suppose the result of **morphological smoothing** is g:

$$g = (f \circ b) \bullet b \tag{8.57}$$

The combined effect of the two operations is to remove or reduce all kinds of noise in the relative lighter and darker regions, in which the opening can remove or reduce the noise in lighter region that is smaller in size than the structure elements, while the closing can remove or reduce the noise in darker region that is smaller in size than the structure elements.

Example 8.22 Morphological smoothing.
The result of morphological smoothing for image in Figure 1.1(b) is shown in Figure 8.31, here a grayscale structure element using four neighborhood is taken. After smoothing, the ripple on the tripod has been removed, and the image appears more smoothing.   ▣

(a)   (b)   (c)

Figure 8.30: Result of morphological gradient calculation.

Figure 8.31: Morphological smoothing.

8.5.3 Top Hat Transform and Bottom Hat Transform

Top hat transform gets the name from the structure element it used, which is a cylinder or parallelepiped with the flat upper portion (like a hat). The results of top hat transforming an image f with a structure element b is noted as T_h:

$$T_h = f - (f \circ b) \tag{8.58}$$

The process is to subtract the result of opening an original image from the original image itself. This transform is suitable for images with bright target on a dark background, which can enhance image detail in the bright region.

The corresponding transform of top hat transform is **bottom hat transform**. From its name, it is clear that the structure element used should be a cylinder or parallelepiped with the flat bottom portion (like a hat upside down). In practice, the cylinder or parallelepiped with the flat upper portion can still be used as structure elements, but the operation instead will be first using structural element for closing the original image, and then subtracting the original image from the result. The results of bottom hat transforming an image f with a structure element b is noted as T_b:

$$T_b = (f \bullet b) - f \tag{8.59}$$

This transformation applies to image with dark target on a light background, the image details in the dark region can be enhanced.

Example 8.23 Top hat transform examples.
Using the grayscale structure elements with 8-neighborhood as top hats, and applying them to the two images in Figure 1.1 can provide the top hat transform results as shown in Figure 8.32. ▢

8.5.4 Morphological Filters

Morphological filter is a group of nonlinear signal filters using morphological transformations to locally modify the geometric characteristics of the signal

Figure 8.32: Top hat transform examples.

(Mahdavieh, 1992). If each signal in Euclidean space is seen as a set, then the morphological filtering is the signal set operations to change the shape of signal. Given filtering operation and the filter output, a quantitative description of the geometry of the input signal can be obtained.

Dilation and erosion are not the inverse of each other, so their applying order cannot be interchanged. For example, the discarded information during erosion process cannot be recovered by dilating the erosion result. Among the basic morphological operations, dilation and erosion are rarely used alone. By combining dilation and erosion, the opening and closing can be formed, their various combinations are the most commonly used in the image analysis. One implementation of morphological filters is to combine opening and closing together. Opening and closing can be used for the quantitative study of the geometric characteristics, because they have little effect on the retention or removal of grayscale features.

From the point of view to eliminate the structures that is brighter than the background and that is smaller than the structure elements, the opening operation is roughly like nonlinear low-pass filter. However, the opening is different from low-pass filtering that prevents a variety of high-frequency components in the frequency domain. In case when both big and small structures have high frequency, the opening allows to pass only large structures but to remove small structures. Opening an image will eliminate bright dots in the image, such as island or spike. Closing has the corresponding functions on darker features as the opening has on lighter features, it can remove those structure that is darker than the background and whose size is smaller than the structure elements.

The combination of opening and closing can be used to eliminate noise. If a small structure element is used to first open and then close an image, it is possible to remove from image the noise-like structures whose size are smaller than the structure elements. A commonly used structure element is a small hemisphere. The diagram of a hybrid filter is shown in Figure 8.33. It should be noted, in filtering noise, the median, Sigma and Kalman filter are often better than mixed morphological filters.

Sieve filter is a morphological filter that allows only structures within a narrow size range passing through. For example, to extract the light defects with size of $n \times n$

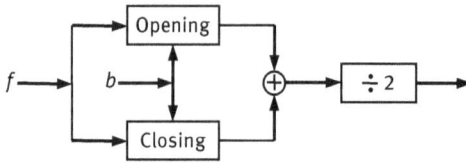

Figure 8.33: A hybrid filter comprising opening and closing.

pixels (n is an odd number), the following filter S (superscript denotes the size of structure elements) can be used:

$$S = (f \circ b^{n \times n}) - [f \circ b^{(n-2) \times (n-2)}] \tag{8.60}$$

Sieve filter is similar to the band-pass filter in frequency domain. In the above formula, the first term can remove all light structures with size smaller than $n \times n$, the second term can remove all light structures with size smaller than $(n-2) \times (n-2)$. The subtraction of two terms leaves the structures between $n \times n$ and the $(n-2) \times (n-2)$.

8.5.5 Soft Morphological Filters

Soft morphological filters and basic morphological filter are very similar, the main difference is that soft morphological filter is not very sensitive to additive noise and to the shape changes of object to be filtered.

Soft morphological filters can be defined on the basis of weighted-order statistics. Two basic morphological filtering operations are soft erosion and soft dilation. It is required to replace the standard structure elements by the structuring systems. A structural system including three parameters is denoted as $[B, C, r]$: B and C are finite plane sets, $C \subset B$, r is a natural number satisfying $1 \leq r \leq |B|$. Set B is called a structure set, C is its hard center, $B - C$ gives its soft contours, and r is the order of the center. Soft morphological filter can convert a gray-level image $f(x, y)$ into another image.

Using the structuring system $[B, C, r]$ to soft dilate $f(x, y)$ is denoted as $f \oplus [B, C, r](x, y)$, which is defined as:

$$f \oplus [B, C, r](x, y) = r\text{th Max of } \{r \Diamond f(c) : c \in C_{(x,y)}\} \cup \{f(b) : b \in (B - C)_{(x,y)}\} \tag{8.61}$$

where $\Diamond$ represents a repeating operation, the part in braces is a multiset that is a set of objects on which the operation can be repeated. For example, $\{1, 1, 1, 2, 3, 3\} = \{3 \Diamond 1, 2, 2 \Diamond 3\}\}$ is a multiset.

Using the structuring system $[B, C, r]$ to soft erode $f(x, y)$ is denoted as $f \ominus [B, C, r](x, y)$, which is defined as:

$$f \ominus [B, C, r](x, y) = r\text{th Min of } \{r \Diamond f(c) : c \in C_{(x,y)}\} \cup \{f(b) : b \in (B - C)_{(x,y)}\} \tag{8.62}$$

Using the structuring system [B, C, r] to soft dilate or soft erode f(x, y) at position (x, y) needs to move B and C to (x, y) and to form multiset with values of f(x, y) inside the moved sets. Wherein the values of f(x, y) at the hard center need to repeat first r times, then for soft dilation the rth maximum value from the multiset is taken while for soft erosion the rth minimum value from the multiset is taken.

One example of soft dilation is given in Figure 8.34. Figure 8.34(a) shows the structure element B and its center C, in which light points represent the origin of coordinate system and also C (an element of B), and dark points represent elements belong to B. Figure 8.34(b) is the original image. Figure 8.34(c) shows f ⊕ [B, C, 1]. Figure 8.34(d) shows f⊕[B, C, 2]. Figure 8.34(e) shows f⊕[B, C, 3]. Figure 8.34(f) shows f⊕[B, C, 4]. In the last four figures, the light points in every figure represent dilated points that were not belong to original image.

One example of soft erosion is given in Figure 8.35. Figure 8.35(a) shows the structure element B and its center C, in which light points represent the origin of coordinate system and also C (an element of B), and dark points represent elements belong to B. Figure 8.35(b) is the original image. Figure 8.35(c) shows f ⊖ [B, C, 1]. Figure 8.35(d) shows f⊖[B, C, 2]. Figure 8.35(e) shows f⊖[B, C, 3]. Figure 8.35(f) shows f⊖[B, C, 4]. In the last four figures, the light points in every figure represent eroded points that were belong to original image.

In extreme cases, a soft morphological operation is simplified into the standard morphological operation. For example, r = 1 or C = B, a soft morphological operation

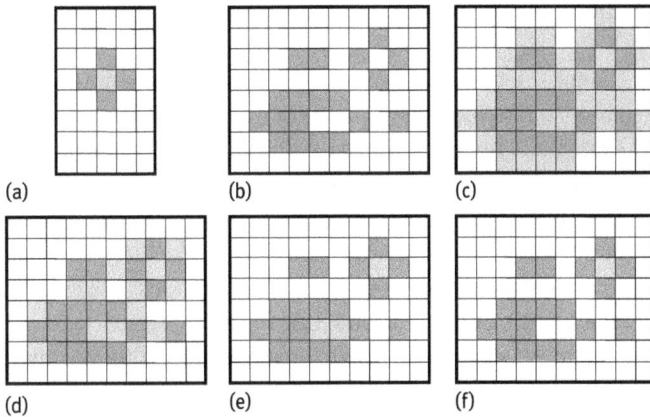

(a) (b) (c) (d) (e) (f)

Figure 8.34: Soft dilation for a binary image.

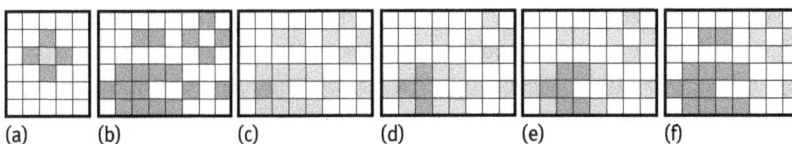

(a) (b) (c) (d) (e) (f)

Figure 8.35: Soft erosion for a binary image.

becomes the standard operation with structure element B. If $r > |B - C|$, a soft morphological operation becomes the standard operation by using the center of structure set C.

For a grayscale image, let the structuring set $B = \{(-1, 0), (0, 1), (0, 0), (0, -1), (1, 0)\}$, and its center $C = \{(0, 0)\}$, then the soft erosion with structuring system $[B, C, 4]$ is defined as:

$$f \ominus [B, C, 4](x, y) = 4\text{th Min of } \{f(x - 1, y), f(x, y + 1), f(x, y), f(x, y),$$
$$f(x, y), f(x, y), f(x, y - 1), f(x + 1, y)\} \tag{8.63}$$

The output at point (x, y) will be $f(x, y)$, unless all values of the set $\{f(b) : b(B - C)_{(x,y)}\}$ are smaller than the value of $f(x, y)$. In the latter case, the output is the maximum value of the set $\{f(b) : b(B - C)_{(x,y)}\}$.

8.6 Practical Algorithms of Grayscale Morphology

With the help of the previously described basic operations and composition operations of grayscale mathematical morphology, a series of practical algorithms may be constituted. Here, the operating target and the processing result are both gray-scale images. The general effect is often more obvious in the brighter or darker regions.

8.6.1 Background Estimation and Elimination

Morphological filter can change the gray value of image, but the change of the gray value depends on the geometry of the filter and can be controlled by means of structure elements. A typical example is the **background estimation and elimination**. Morphological filtering can detect weak (less obvious) objects well, especially for images with low-contrast transition regions. Opening operation can remove regions brighter than the background and smaller than the size of the structure elements. Therefore, by selecting a suitable structure element for opening, it is possible to leave only the background estimates in the image. If subtracting the estimated background from the original image the required object will be extracted (see top hat transforms).

$$\text{Background Estimation} = f \circ b \tag{8.64}$$
$$\text{Background Elimination} = f - (f \circ b) \tag{8.65}$$

Closing operation can remove regions darker than the background and smaller than the size of the structure elements. Therefore, by selecting a suitable structure element for closing, it is also possible to leave only the background estimates in the image. If subtracting the original image from the estimated background, the required object will also be extracted (see bottom hat transforms).

$$\text{Background Estimation} = f \bullet b \qquad\qquad (8.66)$$

$$\text{Background Elimination} = (f \bullet b) - f \qquad\qquad (8.67)$$

In the estimation of background, the effect of cylindrical structure elements would be good. But in the filtering process, the hemispherical structure elements would be better than that of the cylindrical structure elements. Due to relatively sharp edges, cylinder will remove a lot of useful grayscale information, while hemisphere grazed the original image surface with more moderate edges.

Example 8.24 Comparison of hemispherical and cylindrical structure elements. One illustration of using the cylinder and hemispherical structure elements of same radius for the opening of 1-D cross-section is shown in Figure 8.36. Figure 8.36(a, b) is the schematic diagram and operating result of using a hemispherical structure element, respectively. Figure 8.36(c, d) is the schematic diagram and operating result of using a cylindrical structure element, respectively.

It is seen that the result obtained by using the cylindrical structure element is better than that of the hemispherical one. The cylindrical structure element allows minimal residual structure. When the opening is performed with the hemispherical structure element, its upper portion may be fitted to the spike structure and could make the estimated background surface not really reflect the gray-level of background. Thus, when the background surface is subtracted from the original gray level, the gray level near the peak structure will be reduced. If the cylindrical structure element is used, this problem will be substantially eliminated. Besides, it is seen from the figure: to remove the light regions from the image, a cylindrical structure element with diameter larger than these regions should be used for opening; while to remove the dark regions from the image, a cylindrical structure element with diameter larger than these regions should be used for closing. Of course, if using a hemispherical structure element with large enough radius, similar results can also be obtained. However, the computing time will be greatly increased. ◻

8.6.2 Morphological Edge Detection

Many commonly used edge detectors work by calculating the partial differential in local region. In general, this type of edge detector is sensitive to noise and thus will

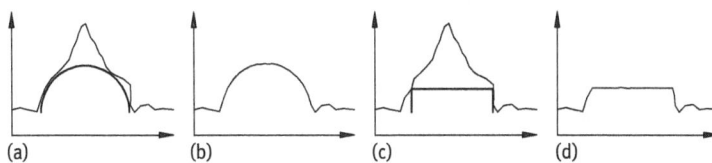

Figure 8.36: Comparison of hemispherical and cylindrical structure elements.

strengthen the noise. **Morphological edge detector** is also more sensitive to noise but does not enhance or amplify noise. Here, the concept of morphological gradient is used. The basic morphological gradient can be defined as follows (see eq. (8.56)):

$$\text{grad}_1 = (f \oplus b) - (f \ominus b) \tag{8.68}$$

Example 8.25 Morphological gradient grad_1.
One example of using eq. (8.68) to detect edges in a binary image is shown in Figure 8.37 (an 8-neighborhood structure element is used). Figure 8.37(a) is the image f. Figure 8.37(b) shows $f \oplus b$. Figure 8.37(c) shows $f \ominus b$. Figure 8.37(d) is grad_1 image. It is seen that $f \oplus b$ expands the bright region in f with one pixel in all directions, while $f \ominus b$ shrinks the dark region in f with one pixel in all directions, so the edge contour provided by grad_1 is two pixels wide. ◙

Relatively sharp (thin) contours can be obtained by using the following two equivalent (strictly speaking, not equivalent in discrete domain) forms of morphological gradients:

$$\text{grad}_2 = (f \oplus b) - f \tag{8.69}$$
$$\text{grad}_2 = f - (f \ominus b) \tag{8.70}$$

Example 8.26 Morphological gradient grad_2.
One example of using eq. (8.69) to detect edges in a binary image is shown in Figure 8.38 (an 8-neighborhood structure element is used). Figure 8.38(a) is the image f. Figure 8.38(b) shows $f \oplus b$. Figure 8.38(c) is grad_2 image $(f \oplus b - f)$. The edge contour thus obtained is one pixel wide. As shown in Figure 8.38, the single pixel edge

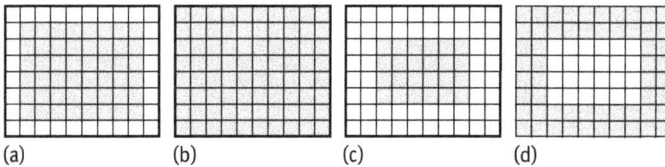

(a)　　　　(b)　　　　(c)　　　　(d)

Figure 8.37: Morphological gradient grad_1.

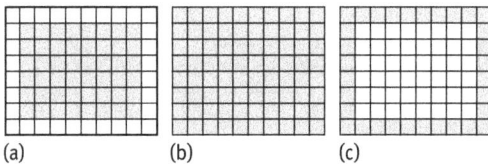

(a)　　　　(b)　　　　(c)

Figure 8.38: Morphological gradient grad_2.

contour thus obtained is actually inside the background. If the eq. (8.70) is used, the resulting single pixel edge contour will belong to the object. ◙

Note that the above $grad_1$ and $grad_2$ will not magnify the noise in image, but still retain the original noise. Here is another form of morphological gradient:

$$grad_3 = \min\{[(f \oplus b) - f], [f - (f \ominus b)]\} \tag{8.71}$$

This morphological gradient is not sensitive to isolated point noise. If it is used over the ideal ramp edge, it will detect the edge with good results. Its disadvantage is that the ideal step edge cannot be detected. However, in this case, the image may be blurred first to convert the step edge to ramp edge, and then, $grad_3$ can be used. It should be noted here that the scope of blurring masks and the scope of structure elements for the morphological operation should be consistent. When using 4-neighborhood cylinder structure elements, for a given image f, its corresponding blurred image is h:

$$h(i, j) = [f(i, j) + f(i + 1, j) + f(i, j + 1) + f(i - 1, j) + f(i, j - 1)]/5 \tag{8.72}$$

The morphological gradient thus obtained is:

$$grad_4 = \min\{[(h \oplus b) - h], [h - (h \ominus b)]\} \tag{8.73}$$

As a result of blurring, the edge strength thus obtained will be weakened, so if the noise in the image is not too strong, it is better to directly use $grad_3$ while not to blur image. When choosing one of $grad_3$ and $grad_4$, it must take into account both requirements of the large signal to noise ratio and the sharp edges.

8.6.3 Cluster Fast Segmentation

The combination of conditional dilation and final erosion (also known as the ultimate erosion) can be used to achieve image segmentation.

The standard dilation may have two extensions: one is the conditional dilation; other is repeated conditional dilation. The dilation of f by b under the condition X (X can be regarded as a limiting set) is denoted by $f \oplus b; X$, which is defined as

$$f \oplus b; X = (f \oplus b) \bigcap X \tag{8.74}$$

Repeated conditional dilation is an extension of the above operation. It is denoted by $f \oplus \{b\}; X$ (here $\{b\}$ means to iteratively dilate f by b until no further change).

$$f \oplus \{b\}; X = \left[\left[(f \oplus b) \bigcap X\right] \oplus b\right] \bigcap X\right] \oplus b \cdots \tag{8.75}$$

Final erosion refers to repeatedly erode an object until it disappears, then to keep the result before the last step (this result is also known as the seeds of object). Let $f_k = f \ominus kb$, where b is the unit circle, kb is the circle with radius k. The set of final erosion g_k is the element of f_k, if $l > k$, g_k will be disappear in f_l. The first step of the final erosion is the conditional dilation:

$$U_k = (f_{k+1} \oplus \{b\}); f_k \tag{8.76}$$

The second step of the final erosion is to subtract the above dilation result from the erosion of f, namely

$$g_k = f_k - U_k \tag{8.77}$$

If the image has multiple objects, the final eroded object set g can be obtained by taking the union of each respective g_k. In other words, the final erosion image is:

$$g = \bigcup_{k=1,m} g_k \tag{8.78}$$

where m is the number of erosion.

The cluster fast segmentation (CFS) for image containing the object with convex contour includes three steps:
1. Iterative erosion of f: Using unit circular structure element b to iteratively erode the original image f:

$$f_k = f \ominus kb, \quad k = 1, \dots, m \quad \text{where} \quad \{m : f_m \neq \emptyset\} \tag{8.79}$$

Here, $f_1 = f \ominus b$, $f_2 = f \ominus 2b$, subsequently until next $f_m = f \ominus mb$ and $f_{m+1} = \emptyset$; where m is the maximum number of non-empty graph.

Example 8.27 Repeated conditional dilation.
One example of repeated conditional dilation is shown in Figure 8.39. Figure 8.39(a) is a binary image, where pixels of different gray levels are used to facilitate the following explanation. In this example, a 4-neighborhood structure element is used. First erosion will remove the dark gray region away and make the other two regions shrink and separated, the result is shown in Figure 8.39(b). Figure 8.39(c) provides the result of conditional erosion of Figure 8.39(b) by taking Figure 8.39(a) as condition. Second erosion is made on Figure 8.39(b); it eroded the medium gray region away and made the light gray region shrink, and the result is shown in Figure 8.39(d). Figure 8.39(e) provides the result of conditional dilation of Figure 8.39(d) by taking Figure 8.39(b) as condition. If a third erosion was made, it will give the empty set. The first seed is obtained by subtracting

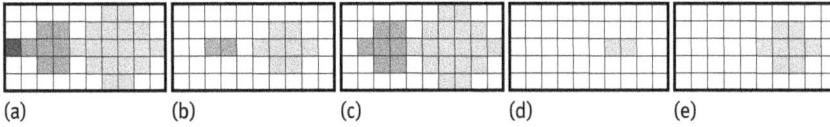

(a)   (b)   (c)   (d)   (e)

Figure 8.39: Steps of cluster fast segmentation.

Figure 8.39(c) from Figure 8.39(a); the second seed is obtained by subtracting Figure 8.39(e) from Figure 8.39(b); Figure 8.39(d) provides the third seed.   ▣

2. Determining the final erosion set g_k: For each f_k, perform the final erosion and subtract the erosion result from f_k:

$$g_k = f_k - (f_{k+1} \oplus \{b\}; f_k) \tag{8.80}$$

Here, g_k is the ultimate erosion set, or the seed of each g_k.

The current step can also be explained with the help of Figure 8.39. If Figure 8.39(b) is iteratively dilated with Figure 8.39(a) as constrain, Figure 8.39(c) can be obtained. Comparing Figure 8.39(c) with Figure 8.39(a), the only difference is the dark gray point. To obtain the dark gray point, the Figure 8.39(a) is subtracted from Figure 8.39(c), and the result obtained is the final erosion of Figure 8.39(a). This is the first seed. If repeating the dilation of Figure 8.39(b) with b and using Figure 8.39(b) as constrain, the region associated with the first seed can be restored. Further, subtracting this result from Figure 8.39(b) will give the seed in medium gray region. Similarly, the seed in light gray region can also be obtained.

3. Determining the object contour: Starting from each seed, the full size of each corresponding original region can be restored back with the following equation:

$$U = \bigcup g_k \oplus (k-1)b \quad \text{for} \quad k = 1 \text{ to } m \tag{8.81}$$

8.7 Problems and Questions

8-1 (1) Prove eqs. (8.1), (8.4), and (8.6) are equivalent.   [?]

     (2) Prove eqs. (8.3), (8.5), and (8.7) are equivalent.

8-2 (1) Draw a schematic representation of a circle of radius r dilating by a circular structure element of radius $r/4$;

     (2) Draw a schematic representation of a square $r \times r$ dilating by the above structural elements;

     (3) Draw a schematic representation of an isosceles triangle with one side length r dilating by the above structural elements;

     (4) Change the dilation operation in (1), (2), and (3) to erosion, and draw the corresponding schematic representations, respectively.

8-3 Prove the holding of eqs. (8.16) and (8.17).

8-4* With the original images as shown in Figure Problem 8-4(a), produce the results of the hit-or-miss transform, with the structure elements (three center pixels corresponding to hit, the pixels around corresponding to miss) given in Figure Problem 8-4(b) and Figure Problem 8-4(c), respectively.

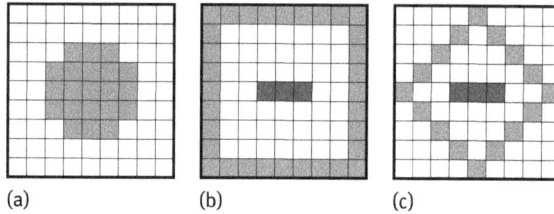

(a)   (b)   (c)

Figure Problem 8-4

8-5 Given the image and the structural element shown in Figure Problem 8-5, provide the steps in obtaining the result of thinning.

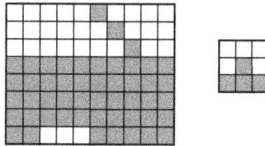

Figure Problem 8-5

8-6 Provide the results of each step in using thickening, instead of thinning, to Figure 8.13(b).

8-7 Given the structure elements shown in Figure Problem 8-7(a) and the image shown in Figure Problem 8-7(b):

(1) Calculate the result of grayscale dilation;

(2) Calculate the results of grayscale erosion.

0	0	0	0	0
0	2	2	4	0
0	1	3	5	0
0	3	4	2	0
0	0	0	0	0

0	1	0
1	1	1
0	1	0

(a)   (b)

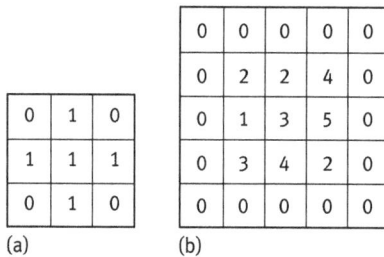

Figure Problem 8-7

8-8* The image $f(x, y)$ and the structural element $b(x, y)$ in eq. (8.44) are all rectangular, D_f is $\{[F_{x1}, F_{x2}], [F_{y1}, F_{y2}]\}$, D_b is $\{[B_{x1}, B_{x2}], [B_{y1}, B_{y2}]\}$:

(1) Supposing $(x, y) \in D_b$, derive the interval satisfied by eq. (8.44) with the translational variables s and t. These intervals on s and t axes define the rectangular region of $(f \oplus b)(s, t)$ on the ST plane;

(2) For the erosion operation, determine the corresponding intervals according to eq. (8.46).

8-9 Prove eqs. (8.48) and (8.49).

8-10 Prove eqs. (8.52) and (8.53).

8-11 (1) A binary morphological gradient operator for edge enhancement is $G = (A \oplus B) - (A \ominus B)$. Using a 1-D edge model, show that the above operator will give a wide edge in the binary image.

(2) If the grayscale dilation ($\oplus$) is achieved by calculating the local maximum of the luminance function in a 3×3 window, grayscale erosion ($\ominus$) is achieved by calculating the local minimum of the luminance function in a 3×3 window, provide the results of using the operator G. Prove that the effect is similar to the Sobel operator when the orientation is neglected (amplitude only):

$$g = (g_x^2 + g_y^2)^{1/2}$$

8-12 The grayscale image $f(x, y)$ is disturbed by nonoverlapping noise which can be modeled with a small cylindrical shape with radius $R_{min} \le r \le R_{max}$ and height $H_{min} \le h \le H_{max}$,

(1) Design a morphological filter to eliminate these noises;

(2) Supposing now the noise is overlapped, repeat (1).

8.8 Further Reading

1. **Basic Operations of Binary Morphology**
   - The introduction to basic operations is also in Serra (1982) and Russ (2016).
   - More discussions on the case where the origin does not belong to the structural elements can be found in Zhang (2005b).
2. **Combined Operations of Binary Morphology**
   - More discussions on the hit-or-miss transform can see Mahdavieh (1992).
   - Details about 3-D topology refinement are visible in Nikolaidis (2001).
3. **Practical Algorithms of Binary Morphology**
   - One example application of detecting news headline based on morphological operation can be found in Jiang (2003).

4. **Basic Operations of Grayscale Morphology**
   - In the mathematical morphology of grayscale images, there are many kinds of structural elements. For example, the stack filters are based on the morphological operators of flat structure elements (Zhang, 2005b).
5. **Combined Operations of Grayscale Morphology**
   - The morphological filter is a nonlinear signal filter that locally modifies the geometrical characteristics of the signal, while the soft morphological filter is a weighted morphological filter. For example, see Mahdavieh (1992).
6. **Practical Algorithms of Grayscale Morphology**
   - The practical grayscale morphological algorithms discussed can make use of the grayscale morphological operations to distinguish the difference between various grayscale pattern areas, so they can be used for clustering segmentation, watershed segmentation, and texture segmentation (Gonzalez, 2008).

Answers to Selected Problems and Questions

Chapter 1 Introduction to Image Analysis

1-3 {hint} Possible sources: Engineering Village, Scopus, Web of Science, etc.

1-7 The two equidistant discs based on the chamfer distance, $\Delta_{3,4}(15)$ and $\Delta_{3,4}(28)$, are shown in Figure Solution 1-7, respectively.

Figure Solution 1-7

Chapter 2 Image Acquisition

2-4 The cost value is: $2 + 2 + 2 + 1 + 3 = 10$.

2-10 The inter-region contrast is calculated according to Eq. (2.38). For the algorithm A, the average gray level of the target area $f_1 = (9 + 0.5 \times 4)/13 = 0.85$ and the average gray level of the background area $f_2 = 0$, so $GC_A = |0.85 - 0|/(0.85 + 0) = 1$. For algorithm B, the average gray level of the target area $f_1 = 1$, and the average gray level of the background area $f_2 = (0 + 0.5 \times 4)/13 = 0.15$, so $GC_B = |1 - 0.15|/(1 + 0.15) = 0.74$. From the perspective of inter-regional contrast, good segmentation results should have a large inter-regional contrast, so the algorithm A is better than algorithm B here.

Then, the intra-region uniformity is calculated according to Eq. (2.37). Here, only the object area is considered, and take $C = 1$. For algorithm A, $UM_A = 1 - [4 \times (0.5 - 0.85)^2 + 9 \times (1 - 0.85)^2] = 0.35$. For algorithm B, $UM_B = 1$. From the perspective of intra-region uniformity, good segmentation results should have large intra-region uniformity, so the algorithm B is better than algorithm A here.

Chapter 3 Object Representation and Description

3-2 Refer to Figure Solution 3-2(a), take X-axis as a reference line, trace the boundary in a counter-clockwise direction, then the signature obtained is shown in Figure Solution 3-2(b).

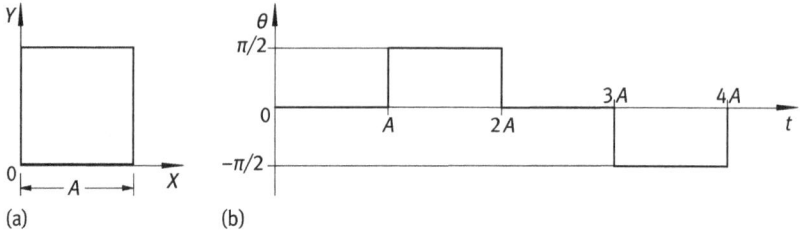

Figure Solution 3-2

3-9 The shape number is 0 0 1 0 1 1 3 3 1 1 0 1. The order of shape number is 12.

Chapter 4 Feature Measurement and Error Analysis

4-4 The mean of the first method is 5.0 and the variance is 5.1. The mean of the second method is 5.7 and the variance is 0.13. Since the mean of the first method is equal to the true value 5, and the mean of the second method is different from the true value, the accuracy of the first method is higher than that of the second method. However, the variance of the second method is much smaller than that of the first method, so the precision of the second method is higher than that of the first method.

4-10 The computed results and errors are given below (the perimeter of the circle should be 12.5664):

$$
\begin{aligned}
L_1 &= 12 & E_1 &= 4.51\%; \\
L_2 &= 13.3284 & E_2 &= 6.06\%; \\
L_3 &= 13.6560 & E_3 &= 8.67\%; \\
L_4 &= 12.9560 & E_4 &= 3.10\%; \\
L_5 &= 12.7360 & E_5 &= 1.35\%.
\end{aligned}
$$

Chapter 5 Texture Analysis

5-3 The co-occurrence matrices of workpiece intact graph and workpiece damage graph are calculated, respectively. Since the damaged region is a block form, the position operator can be defined with a 4-neighborhood relationship. The co-occurrence matrix of workpiece intact graph should have peaks at (100, 100), while the co-occurrence matrix of workpiece damage graph should have also peaks at (100, 50), (50, 100), (100, 150), and (150, 100). Figure Solution 5-3 shows the two co-occurrence matrices in the form of images, where white and light gray represent peaks with different amplitudes. By comparing these two co-occurrence matrices, it is possible to determine the quality of the workpiece.

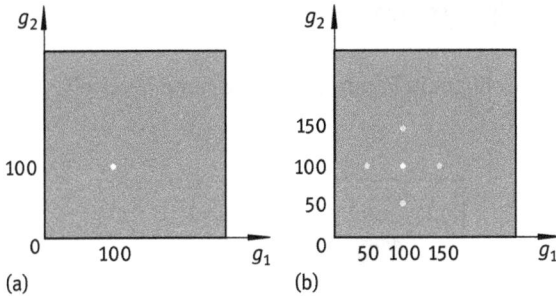

Figure Solution 5-3

5-5 Since the checkerboard image is composed black and white units, alternatively, either the operators of positioning rightward i pixels ($i < m$) or positioning downward j pixels ($j < n$) can give diagonal co-occurrence matrices. More generally, the operators of positioning rightward i pixels ($i < m$) and positioning downward j pixels ($j < n$) at the same time can also give diagonal co-occurrence matrices. Due to the symmetry, the above-mentioned rightward and downward may also be changed to the leftward and upward.

Chapter 6 Shape Analysis

6-3 $F = 1.0787$, $S = 0.8575$, $C = 16.3625$, $E = 1$.
6-12 After N step of dividing and removing, a total of 8^N unit squares will be obtained, which is the result of equally dividing the square side length by 3^N parts, so the fractal dimension of the original square is: $d = \log 8^N / \log 3^N = 1.89$.

Chapter 7 Motion Analysis

7-9 The six-parameter motion model is as follows:

$$\begin{cases} u = k_0 x + k_1 y + k_2 \\ v = k_3 x + k_4 y + k_5 \end{cases}$$

Since the motion of each point is the same, only the values of origin point and the points of (1, 0) and (0, 1) are taken into the model. The results are: $k_2 = 2$, $k_5 = 4$, and $k_0 = k_1 = k_3 = k_4 = 0$.
7-11 {hint} The average operation is used here to reduce the error, so the total variance is not the sum of the parts.

Chapter 8 Color Image Processing

8-4   The results are shown in Figure Solution 8-4, respectively.

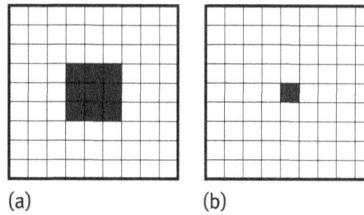

(a)   (b)

Figure Solution 8-4

8-8   (1) According to the definition of dilation, the rectangular area of $(f \oplus b)(s, t)$ on the st plane is $\{[S_{x1}, S_{x2}], [T_{y1}, T_{y2}]\}$, where: $S_{x1} = F_{x1} + B_{x1}$, $S_{x2} = F_{x2} + B_{x2}$, $T_{y1} = F_{y1} + B_{y1}$, $T_{y2} = F_{y2} + B_{y2}$.

   (2) According to the definition of erosion, the rectangular area of $(f \ominus b)(s, t)$ on the st plane is $\{[S_{x1}, S_{x2}], [T_{y1}, T_{y2}]\}$, where: $S_{x1} = F_{x1} - B_{x1}$, $S_{x2} = F_{x2} - B_{x2}$, $T_{y1} = F_{y1} - B_{y1}$, $T_{y2} = F_{y2} - B_{y2}$.

References

[Abdou 1979]   Abdou I E, Pratt W. 1979. Quantitative design and evaluation of enhancement/thresholding edge detectors. Proceedings of IEEE, 67: 753–763.

[ASM 2000]   ASM International. 2000. Practical Guide to Image Analysis. ASM International, Materials Park, OH.

[Ballard 1982]   Ballard D H, Brown C M. 1982. Computer Vision. Prentice-Hall, New Jersey.

[Beckers 1992]   Beckers A L D, Smeulders A W M. 1992. Optimization of length measurements for isotropic distance transformations in three dimensions. CVGIP-IU, 55(3):296–306.

[Beddow 1997]   Beddow J K. 1997. Image Analysis Sourcebook. American Universities Science and Technology Press, Iowa City.

[Besl 1988]   Besl P C, Jain R C. 1988. Segmentation using variable surface fitting. IEEE-PAMI, 10:167–192.

[Borgefors 1984]   Borgefors G. 1984. Distance transformations in arbitrary dimensions. CVGIP, 27:321–345.

[Borgefors 1986]   Borgefors G. 1986. Distance transformations in digital images. CVGIP, 34: 344–371.

[Bowie 1977]   Bowie J E, Young I T. 1977. An analysis technique for biological shape – II. Acta Cytological 21(3): 455–464.

[Brodatz 1966]   Brodatz P. 1966. Textures: A Photographic Album for Artists and Designer. Dover Publications, New York.

[Canny 1986]   Canny J. 1986. A computational approach to edge detection. IEEE-PAMI, 8: 679–698.

[Castleman 1996]   Castleman K R. 1996. Digital Image Processing. Prentice-Hall, New Jersey.

[Chaudhuri 1995]   Chaudhuri B B, Sarkar N. 1995. Texture segmentation using fractal dimension. IEEE-PAMI, 17(1): 72–77.

[Chen 1995]   Chen J L. Kunda A. 1995. Unsupervised texture segmentation using multichannel decomposition and hidden Markov models. IEEE-IP, 4(5): 603–619.

[Chen 2006]   Chen W, Zhang Y-J. 2006. Tracking ball and players with applications to highlight ranking of broadcasting table tennis video. Proc. CESA, 2: 1896–1903.

[Chen 2010]   Chen Z H, Zhang Y-J. 2010. Global motion estimation based on the reliability analysis of motion vectors. Journal of Tsinghua University, 50(4): 623–627.

[Comaniciu 2000]   Comaniciu D, Ramesh V, Meer P. 2000. Real-time tracking of non-rigid objects using mean shift. Proc. CVPR, 2: 142–149.

[Correia 2003]   Correia P L, Pereira F M B. 2003. Objective evaluation of video segmentation quality. IEEE-IP, 12(2): 186–200.

[Costa, 1999]   Costa L F. 1999. Multidimensional scale-space shape analysis. Proc. International Workshop on Synthetic-Natural Hybrid Coding and Three Dimensional Imaging, 241–247.

[Costa 2001]   Costa L F, Cesar R M. 2001. Shape Analysis and Classification: Theory and Practice. CRC Press, Boca Raton.

[Dai 2003]   Dai S Y, Zhang Y-J. 2003. Color image segmentation with watershed on color histogram and Markov random fields. Proc. 4PCM, 1: 527–531.

[Dai 2006]   Dai S Y, Zhang Y-J. 2006. Color image segmentation in both feature and image space. Advances in Image and Video Segmentation, Zhang Y J (ed.), Chapter 10 (209–227).

[Dalai 2005]   Dalai N, Triggs B. 2005. Histograms of oriented gradients for human detection. Proc. CVPR, 1: 886–893.

[Davies 2005]   Davies E R. 2005. Machine Vision: Theory, Algorithms, Practicalities (3rd Ed.). Elsevier, Amsterdam.

[Davies 2005]   Davies E R. 2012. Computer and Machine Vision: Theory, Algorithms, Practicalities (4th Ed.). Elsevier, Amsterdam.

[Dorst 1987]   Dorst L, Smeulders A W M. 1987. Length estimator for digitized contours. CVGIP, 40: 311–333.

[Ekin 2000]   Ekin A, Mehrotra R, Tekalp A M. 2000. Parametric description of object motion using EMUs. Proc. ICIP, 2: 570–573.

[Forsyth 2003]   Forsyth D, Ponce J. 2003. Computer Vision: A Modern Approach. Prentice Hall, New Jersey.

[Forsyth 2012]   Forsyth D, Ponce J. 2012. Computer Vision: A Modern Approach (2nd Ed.). Prentice Hall, New Jersey.

[Fu 1981]   Fu K S, Mui J K. 1981. A survey on image segmentation. PR, 13:3–16.

[Gauch 1992]   Gauch J. M. 1992. Multiresolution Image Shape Description. Springer-Verlag, Heidelberg.

[Gonzalez 1987]   Gonzalez R C, Wintz P. 1987. Digital Image Processing (2nd Ed.). Addison-Wesley, New Jersey.

[Gonzalez 1992]   Gonzalez R C, Woods R E. 1992. Digital Image Processing (3rd Ed.). Addison-Wesley, New Jersey.

[Gonzalez 2002]   Gonzalez R C, Woods R E. 2002. Digital Image Processing, (2nd Ed.). Prentice Hall, New Jersey.

[Gonzalez 2008]   Gonzalez R C, Woods R E. 2002. Digital Image Processing, (3rd Ed.). Prentice Hall, New Jersey.

[Haralick 1985]   Haralick R M, Shapiro L G. 1985. Image segmentation techniques. CVGIP, 29: 100–132.

[Haralick 1992]   Haralick R M, Shapiro L G. 1992. Computer and Robot Vision, Vol.1. Addison-Wesley, New Jersey.

[Haralick 1993]   Haralick R M, Shapiro L G. 1993. Computer and Robot Vision, Vol.2. Addison-Wesley, New Jersey.

[Harms 1984]   Harms H, Aus H M. 1984. Estimation of sampling errors in a high-resolution TV microscope image-processing system. Cytometry, 5: 228–235.

[He 1993]   He Y, Zhang Y-J. 1993. An analytical study of the effect of noise on the performance of differential edge operators. Proc. 2ICSP, 2: 1314–1317.

[Howard 1998]   Howard C V, Reed M G. 1998. Unbiased Stereology – Three-Dimensional Measurement in Microscopy. Springer, Heidelberg.

[Hu 1992]   Hu G Z, Cao X P. 1992. A subpixel edge detector using expectation of first-order derivatives. SPIE 1657: 415–425.

[Huang 2003]   Huang X Y, Zhang Y-J, Hu D. 2003. Image retrieval based on weighted texture features using DCT coefficients of JPEG images. Proc. 4PCM, 3: 1571–1575.

[Jain 1991]   Jain A K, Farrokhinia F. 1991. Unsupervised texture segmentation using Gabor filters. PR, 24(12): 1167–1186.

[Jain 1996]   Jain A K, Vailaya A. 1996. Image retrieval using color and shape. PR, 29(8): 1233–1244.

[Jäin 1997]   Jähne B. 1997. Digital Image Processing – Concepts, Algorithms and Scientific Applications. Springer, Heidelberg.

[Jähne 1999]   Jähne B, Haußecker H, Geißler P. 1999. Handbook of Computer Vision and Applications: Volume 1: Sensors and Imaging. Academic Press, Washington, D.C.

[Jeannin 2000]   Jeannin, S, Jasinschi R, She A, et al. 2000. Motion descriptors for content-based video representation. Signal Processing: Image Communication, 16(1–2): 59–85.

[Jensen 1995]   Jensen K, Anastassiou D. 1995. Subpixel edge localization and the interpolation of still images. IEEE-IP, 4: 285–295.

[Jia 2008]   Jia H X, Zhang Y-J. 2008. Human detection in static images. Pattern Recognition Technologies and Applications: Recent Advances. Verma B, Blumenstein M, (eds.). IGI Global, Hershey PA, 227–243.

[Jia 2009]   Jia H X, Zhang Y-J. 2009. Multiple kernels based object tracking using histograms of oriented gradients. ACTA Automatica Sinica, 35(10): 1283–1289.

[Jiang 2003]   Jiang F, Zhang Y-J. 2003. A caption detection algorithm based on morphological operation. Journal of Electronics & Information Technology, 25(12): 1647–1652.

[Joyce 1985]   Joyce Loebl. 1985. Image Analysis: Principles and Practice. Joyce Loebl, Ltd, Gateshead.

[Kass 1988]   Kass M, Witkin A, Terzopoulos D. 1988. Snakes: Active contour models. IJCV, 1(4): 321–331.

[Keller 1989]   Keller J M, Chen S. 1989. Texture description and segmentation through fractal geometry. CVGIP, 45: 150–166.

[Khotanzad 1989]   Khotanzad A, Chen J Y. 1989. Unsupervised segmentation of textured images by edge detection in multidimensional features. IEEE-PAMI, 11(4): 414–421.

[Kim 2004]   Kim K, Chalidabhongse T H, Harwood D, et al. 2004. Background modeling and subtraction by codebook construction. Proc. ICIP, 5: 3061–3064.

[Kiryati 1993]   Kiryati N, Szekely G. 1993. Estimating shortest paths and minimal distances on digitized three-dimensional surfaces. PR, 26(11): 1623–1637.

[Kitchen 1981]   Kitchen L, Rosenfeld A. 1981. Edge evaluation using local edge coherence. IEEE-SMC, 11(9): 597–605.

[Kitchen 1989]   Kitchen L J, Malin J A. 1989. The effect of spatial discretization on the magnitude and direction response of simple differential edge operators on a step edge. CVGIP, 47: 243–258.

[Kropatsch 2001]   Kropatsch W G, Bischof H, eds. 2001. Digital Image Analysis – Selected Techniques and Applications. Springer, Heidelberg.

[Levine 1985]   Levine M D. 1985. Vision in Man and Machine. McGraw-Hill, New York.

[Li 2006]   Li X P, Yan Y, Zhang Y-J. 2006. Analysis and comparison of background modeling methods. Proc. 13NCIG, 469–473.

[Li 2010]   Li S, Zhang Y-J. 2010. A novel system for video retrieval of frontal-view indoor moving pedestrians. Proc. 5ICIG, 266–271.

[Li 2011a]   Li L, Zhang Y-J. 2011a. FastNMF: Efficient NMF algorithm for reducing feature dimension. Advances in Face Image Analysis: Techniques and Technologies, Zhang Y J (ed.). Medical Information Science Reference, Hershey – New York. Chapter 8 (137–162).

[Li 2011b]   Li X, Zhang Y-J, Liu B D. 2011b. Robust moving cast shadow detection and removal with reliability checking of the color property. Proc. 6ICIG, 440–445.

[Lin 2003]   Lin K H, Lam K M, Sui W C. 2003. Spatially eigen-weighted Hausdorff distances for human face recognition. PR, 36: 1827–1834.

[Liu 2007]   Liu W J, Zhang Y-J. 2007. Real time object tracking using fused color and edge cues. Proc. 9th ISSPIA, (1–4).

[Liu 2008b]   Liu W J, Zhang Y-J. 2008b. Edge-color-histogram and Kalman filter-based real-time object tracking. Journal of Tsinghua University, 48(7): 1104–1107.

[Lohmann 1998]   Lohmann G. 1998. Volumetric Image Analysis. John Wiley & Sons and Teubner Publishers, New Jersey.

[Luo 1999]   Luo H T, Zhang Y-J. 1999. Evaluation based segmentation algorithm optimization: Idea and system. Journal of Electronics, 16(2): 109–116.

[Mahdavieh 1992]   Mahdavieh Y, Gonzalez R C. 1992. Advances in Image Analysis. SPIE Optical Engineering Press, Bellingham.

[Mandelbrot 1982]   Mandelbrot B. 1982. The Fractal Geometry of Nature. W.H. Freeman and Company, New York.

[Marcello 2004]   Marcello J, Marques F, Eugenio F. 2004. Evaluation of thresholding techniques applied to oceanographic remote sensing imagery. SPIE, 5573: 96–103.

[Marchand 2000]   Marchand-Maillet S, Sharaiha Y M. 2000. Binary Digital Image Processing – A Discrete Approach. Academic Press, Washington, D.C.

[Medioni 1987]   Medioni G, Yasumoto Y. 1987. Corner detection and curve representation using cubic B-Splines. CVGIP, 39: 267–278.

[Medioni 2000]   Medioni G, Lee M S, Tang C K. 2000. A Computational Framework for Segmentation and Grouping. Elsevier, Amsterdam.

[Murray 1995]   Murray J D. 1995. Use and abuse of fractal theory in neuroscience, The Journal of Comparative Neurology, 361:369–371.

[Nazif 1984]   Nazif A, Levine M. 1984. Low-level image segmentation: an expert system. IEEE-PAMI, 6(5): 555–577.

[Nikolaidis 2001]   Nikolaidis N, Pitas I. 2001. 3-D Image Processing Algorithms. John Wiley & Sons, Inc, New Jersey.

[Ohm 2000]   Ohm J R, Bunjamin F, Liebsch W, et al. 2000. A set of visual feature descriptors and their combination in a low-level description scheme. SP: IC, 16(1–2): 157–179.

[Otterloo 1991]   Otterloo P J. 1991. A Contour-Oriented Approach to Shape Analysis. Prentice Hall, New Jersey.

[Pal 1993]   Pal N R, Pal S K. 1993. A review on image segmentation techniques. PR, 26:1277–1294.

[Panda 1997]   Panda R, Chatterji B N. 1997. Unsupervised texture segmentation using tuned filters in Gaborian space. PRL, 18: 445–453.

[Paulus 2009]   Paulus C, Zhang Y-J. 2009. Spatially adaptive subsampling for motion detection. Tsinghua Science and Technology, 14(4): 423–433.

[Pavlidis 1988]   Pavlidis T. 1988. Image analysis. Annual Review of Computer Science, 3:121–146.

[Pratt 2007]   Pratt W K. 2007. Digital Image Processing: PIKS Scientific inside (4th Ed.). Wiley Interscience, New Jersey.

[Qin 2003]   Qin X, Zhang Y-J. 2003. A tracking method based on curve fitting prediction of IR object. Infrared Technology, 25(4): 23–25.

[Ragnemalm 1993]   Ragnemalm I. 1993. The Euclidean distance transform in arbitrary dimensions. PRL, 14: 883–888.

[Rao 1990]   Rao A R. 1990. A Taxonomy for Texture Description and Identification. Springer-Verlag, Heidelberg.

[Rao 1990]   Ritter G X, Wilson J N. 2001. Handbook of Computer Vision Algorithms in Image Algebra. CRC Press, Boca Raton.

[Roberts 1965]   Roberts L G. 1965. Machine perception of three-dimensional solids. Optical and Electro-Optical Information Processing, 159–197.

[Rosenfeld 1974]   Rosenfeld A. 1974. Compact figures in digital pictures. IEEE-SMC, 4: 221–223.

[Rosenfeld 1976]   Rosenfeld A, Kak A C. 1976. Digital Picture Processing. Academic Press, Washington, D.C.

[Rosenfeld 1984]   Rosenfeld A. 1984. Image analysis: Problems, progress and prospects. PR, 17: 3–12.

[Russ 2002]   Russ J. C. 2002. The Image Processing Handbook (4th Ed.). CRC Press, Boca Raton.

[Russ 2016]   Russ J. C. 2016. The Image Processing Handbook (7th Ed.). CRC Press, Boca Raton.

[Sahoo 1988]   Sahoo P K, et al. 1988. A survey of thresholding techniques. CVGIP, 41: 233–260.

[Samet 2001]   Samet H. 2001. Object representation. Foundations of Image Understanding. Kluwer Academic Publishers, Dordrecht.

[Serra 1982]   Serra J. 1982. Image Analysis and Mathematical Morphology. Academic Press, Washington, D.C.

[Shapiro 2001]   Shapiro L, Stockman G. 2001. Computer Vision. Prentice Hall, New Jersey.

[Shen 2009a]   Shen B, Zhang Y-J. 2009a. Subsequence-wise approach for online tracking. Auto-Identification and Ubiquitous Computing Applications: RFID and Smart Technologies for Information Convergence. Information Science Reference, Symonds J, Ayoade J, Parry D, (eds.). Information Science Reference, Hershey – New York. Chapter IX (157–166).

[Shen 2009b]   Shen B, Zhang Y-J. 2009b. Foreground segmentation via sparse representation. Proc. International Conference on Systems, Man, and Cybernetics, 3363–3367.

[Shih 2010]   Shih F Y. 2010. Image Processing and Pattern Recognition – Fundamentals and Techniques. IEEE Press, Piscataway, N.J.

[Sonka 1999]   Sonka M, Hlavac V, Boyle R. 1999. Image Processing, Analysis, and Machine Vision (2nd Ed.). Brooks/Cole Publishing, Pacific Grove, California.

[Sonka 2008]   Sonka M, Hlavac V, Boyle R. 2008 Image Processing, Analysis, and Machine Vision (3rd Ed.). Thomson Learning, Toronto.

[Strasters 1991]   Strasters K, Gerbrands J J. 1991. Three-dimensional image segmentation using a split, merge and group approach. PRL, 12: 307–325.

[Tabatabai 1984]   Tabatabai A J, Mitchell O R. 1984. Edge location to subpixel values in digital imagery. IEEE-PAMI, 6: 188–201.

[Tan 2006]   Tan H C, Zhang Y-J. 2006. A novel weighted Hausdorff distance for face localization. Image and Vision Computing, 24(7): 656–662.

[Tan 2011]   Tan H C, Zhang Y-J, Wang W H, et al. 2011. Edge eigenface weighted Hausdorff distance for face recognition. International Journal of Computational Intelligence Systems, 4(6): 1422–1429.

[Tang 2011]   Tang D, Zhang Y-J. 2011. Combining mean-shift and particle filter for object tracking. Proc. 6ICIG, 771–776.

[Taylor 1994]   Taylor R I, Lewis P H. 1994. 2D shape signature based on fractal measurements. IEE Proc. Vis. Image Signal Process. 141: 422–430.

[Theodoridis 2003]   Theodoridis S, Koutroumbas K. 2003. Pattern Recognition (2nd Ed.). Elsevier Science, Amsterdam.

[Tomita 1990]   Tomita F, Tsuji S. 1990. Computer Analysis of Visual Textures. Kluwer Academic Publishers, Dordrecht.

[Toyama 1999]   Toyama K, Krumm J, Brumitt B, et al. 1999. Wallflower: Principles and practice of background maintenance. Proc. ICCV, 1: 255–261.

[Umbaugh 2005]   Umbaugh S E. 2005. Computer Imaging – Digital Image Analysis and Processing. CRC Press, Boca Raton.

[Viola 2001]   Viola P, Jones M. 2001. Rapid object detection using a boosted cascade of simple features. Proc. CVPR, 511–518.

[Verwer 1991]   Verwer B J H. 1991. Local distances for distance transformations in two and three dimensions. PRL, 12: 671–682.

[Wang 2010]   Wang Y X, Zhang Y-J, Wang X H. 2010. Mean-shift object tracking through 4-D scale space. Journal of Electronics & Information Technology, 32(7): 1626–1632.

[Weszka 1978]   Weszka J S, Rosenfeld A. 1978. Threshold evaluation technique. IEEE-SMC, 8: 622–629.

[Wu 1999]   Wu G H, Zhang Y-J, Lin X G. 1999. Wavelet transform-based texture classification using feature weighting. Proc. ICIP, 4: 435–439.

[Xu 2006]   Xu F, Zhang Y-J. 2006. Comparison and evaluation of texture descriptors proposed in MPEG-7. International Journal of Visual Communication and Image Representation, 17: 701–716.

[Xue 2011]   Xue F, Zhang Y-J. 2011. Image class segmentation via conditional random field over weighted histogram classifier. Proc. 6ICIG, 477–481.

[Xue 2012]   Xue J H, Zhang Y-J. 2012. Ridler and Calvard's, Kittler and Illingworth's and Otsu's methods for image thresholding. PRL, 33(6): 793–797.

[Yan 2009]   Yan Y, Zhang Y-J. 2009. State-of-the-art on video-based face recognition. Encyclopedia of Artificial Intelligence, Mehdi Khosrow-Pour (ed.). Information Science Reference, Hershey – New York. 1455–1461.

[Yan 2010]   Yan Y, Zhang Y-J. 2010. Effective discriminant feature extraction framework for face recognition. Proc. 4th International Symposium on Communications, Control and Signal Processing, 1–5.

[Yao 1999]   Yao Y R, Zhang Y-J. 1999. Shape-based image retrieval using wavelets and moments. Proc. International Workshop on Very Low Bitrate Video Coding, 71–74.

[Yasnoff 1977]   Yasnoff W A, et al. 1977. Error measures for scene segmentation. PR, 9: 217–231.

[Young 1988]   Young I T, Renswoude J. 1988. Three-dimensional image analysis. SPIN Program, 1–25.

[Young 1993]   Young I T. 1993. Three-dimensional image analysis. Proc. VIP'93, 35–38.

[Young 1995]   Young I T, Gerbrands J, Vliet L J. 1995. Fundamental of Image Processing. Delft University of Technology, The Netherlands.

[Yu 2001a]   Yu T L, Zhang Y-J. 2001a. Motion feature extraction for content-based video sequence retrieval. SPIE, 4311: 378–388.

[Yu 2001b]   Yu T L, Zhang Y-J. 2001b. Retrieval of video clips using global motion information. IEE Electronics Letters, 37(14): 893–895.

[Yu 2002]   Yu T L, Zhang Y-J. 2002. Summarizing motion contents of the video clip using moving edge overlaid frame (MEOF). SPIE 4676: 407–417.

[Zhang 1990]   Zhang Y-J, Gerbrands J J, Back E. 1990. Thresholding three-dimensional image. SPIE, 1360: 1258–1269.

[Zhang 1991a]   Zhang Y-J. 1991a. 3-D image analysis system and megakaryocyte quantitation. Cytometry, 12: 308–315.

[Zhang 1991b]   Zhang Y-J, Gerbrands J J. 1991b. Transition region determination based thresholding. PRL, 12: 13–23.

[Zhang 1992a]   Zhang Y-J, Gerbrands J J. 1992a. On the design of test images for segmentation evaluation. Signal Processing VI, Theories and Applications, 1: 551–554.

[Zhang 1992b]   Zhang Y-J, Gerbrands J J. 1992b. Segmentation evaluation using ultimate measurement accuracy, SPIE, 1657: 449–460.

[Zhang 1993a]   Zhang Y-J. 1993a. Comparison of segmentation evaluation criteria. Proc. 2ICSP, 870–873.

[Zhang 1993b]   Zhang Y-J. 1993b. Image synthesis and segmentation comparison. Proc. 3rd International Conference for Young Computer Scientists, 8.24–8.27.

[Zhang 1993c]   Zhang Y-J. 1993c. Quantitative study of 3-D gradient operators. IVC, 11: 611–622.

[Zhang 1993d]   Zhang Y-J. 1993d. Segmentation evaluation and comparison: A study of several algorithms. SPIE, 2094: 801–812.

[Zhang 1994]   Zhang Y-J. 1994. Objective and quantitative segmentation evaluation and comparison. SP, 39: 43–54.

[Zhang 1995]   Zhang Y-J. 1995. Influence of segmentation over feature measurement. PRL, 16: 201–206.

[Zhang 1996a]   Zhang Y-J. 1996a. Image engineering in China: 1995. Journal of Image and Graphics. 1(1): 78–83.

[Zhang 1996b]   Zhang Y-J. 1996b. Image engineering in China: 1995 (Supplement). Journal of Image and Graphics. 1(2): 170–174.

[Zhang 1996c]   Zhang Y-J. 1996c. A survey on evaluation methods for image segmentation. PR, 29(8): 1335–1346.

[Zhang 1996d]   Zhang Y-J. 1996d. Image engineering and bibliography in China. Technical Digest of International Symposium on Information Science and Technology, 158–160.

[Zhang 1997a]   Zhang Y-J. 1997a. Image engineering in China: 1996. Journal of Image and Graphics. 2(5): 336–344.

[Zhang 1997b]   Zhang Y-J. 1997b. Evaluation and comparison of different segmentation algorithms. PRL, 18(10): 963–974.

[Zhang 1997c]   Zhang Y-J. 1997c. Quantitative image quality measures and their applications in segmentation evaluation. Journal of Electronics, 14: 97–103.

[Zhang 1998a]   Zhang Y-J. 1998a. Image engineering in China: 1997. Journal of Image and Graphics. 3(5): 404–414.

[Zhang 1998b]   Zhang Y-J. 1998b. Framework and experiments for image segmentation characterization. Chinese Journal of Electronics, 7(4): 387–391.

[Zhang 1999a]   Zhang Y-J. 1999a. Image engineering in China: 1998. Journal of Image and Graphics. 4(5): 427–438.

[Zhang 1999b]   Zhang Y-J. 1999b. Image Engineering (1): Image Processing and Analysis. Tsinghua University Press, Beijing.

[Zhang 1999c]   Zhang Y-J. 1999c. Two new concepts for N_{16} space. IEEE Signal Processing Letters, 6(9): 221–223.

[Zhang 2000a]   Zhang Y-J. 2000a. Image engineering in China: 1999. Journal of Image and Graphics.
5(5): 359–373.

[Zhang 2000b]   Zhang Y-J. 2000b. Image Engineering (2): Image Understanding and Computer
Vision. Tsinghua University Press, Beijing.

[Zhang 2000c]   Zhang Y-J, Luo H T. 2000c. Optimal selection of segmentation algorithms based on
performance evaluation. Optical Engineering, 39(6): 1450–1456.

[Zhang 2001a]   Zhang Y-J. 2001a. Image engineering in China: 2000. Journal of Image and Graphics.
6(5): 409–424.

[Zhang 2001b]   Zhang Y-J. 2001b. A review of recent evaluation methods for image segmentation.
Proc. 6th International Symposium on Signal Processing and Its Applications,
148–151.

[Zhang 2001c]   Zhang Y-J. 2001c. Image Segmentation. Science Press, Beijing.

[Zhang 2002a]   Zhang Y-J. 2002a. Image engineering in China: 2001. Journal of Image and Graphics.
7(5): 417–433.

[Zhang 2002b]   Zhang Y-J. 2002b. Image Engineering (3): Teaching References and Problem
Solutions. Tsinghua University Press, Beijing.

[Zhang 2002c]   Zhang Y-J. 2002c. Image engineering and related publications. International Journal
of Image and Graphics, 2(3): 441–452.

[Zhang 2003a]   Zhang Y-J. 2003a. Image engineering in China: 2002. Journal of Image and Graphics.
8(5): 481–498.

[Zhang 2003b]   Zhang Y-J. 2003b. Content-Based Visual Information Retrieval. Science Press,
Beijing.

[Zhang 2004]   Zhang Y-J. 2004. Image engineering in China: 2003. Journal of Image and Graphics.
9(5): 513–531.

[Zhang 2005a]   Zhang Y-J. 2005a. Image engineering in China: 2004. Journal of Image and Graphics.
10(5): 537–560.

[Zhang 2005b]   Zhang Y-J. 2005b. Image Engineering (2): Image Analysis (2nd Ed.). Tsinghua
University Press, Beijing.

[Zhang 2005c]   Zhang Y-J. 2005c. New advancements in image segmentation for CBIR. Encyclopedia
of Information Science and Technology, Mehdi Khosrow-Pour (ed.). Idea Group Reference,
Hershey – New York. Chapter 371 (2105–2109).

[Zhang 2006a]   Zhang Y-J. 2006a. Image engineering in China: 2005. Journal of Image and Graphics.
11(5): 601–623.

[Zhang 2006b]   Zhang Y-J. (Ed.). 2006b. Advances in Image and Video Segmentation. IRM Press,
Hershey – New York.

[Zhang 2006c]   Zhang Y-J. 2006c. A summary of recent progresses for segmentation evaluation.
Advances in Image and Video Segmentation, Zhang Y J (ed.)., IRM Press, Hershey – New York.
Chapter 20 (423–440).

[Zhang 2006d]   Zhang Y-J. 2006d. An overview of image and video segmentation in the last 40 years.
Advances in Image and Video Segmentation, IRM Press, Hershey – New York. Chapter 1 (1–15).

[Zhang 2007a]   Zhang Y-J. 2007a. Image engineering in China: 2006. Journal of Image and Graphics.
12(5): 753–775.

[Zhang 2007b]   Zhang Y-J. 2007b. Image Engineering (2nd Ed.). Tsinghua University Press, Beijing.

[Zhang 2008a]   Zhang Y-J. 2008a. Image engineering in China: 2007. Journal of Image and Graphics.
13(5): 825–852.

[Zhang 2008b]   Zhang Y-J. 2008b. A study of image engineering. Encyclopedia of Information
Science and Technology (2nd Ed.), Mehdi Khosrow-Pour (ed.). Information Science Reference,
Hershey – New York. Vol. VII, 3608–3615.

[Zhang 2008c]   Zhang Y-J. 2008c. Image segmentation evaluation in this century. Encyclopedia of
Information Science and Technology (2nd Ed.), Mehdi Khosrow-Pour (ed.). Information Science
Reference, Hershey – New York. Chapter 285 (1812–1817).

[Zhang 2008d]   Zhang Y-J. 2008d. Image segmentation in the last 40 years. Encyclopedia of Information Science and Technology (2nd Ed.), Mehdi Khosrow-Pour (ed.). Information Science Reference, Hershey – New York. Chapter 286 (1818–1823).

[Zhang 2008e]   Zhang Y-J. 2008e. Recent progress in image and video segmentation for CBVIR. Encyclopedia of Information Science and Technology (2nd Ed.), Mehdi Khosrow-Pour (ed.). Information Science Reference, Hershey – New York. Chapter 515 (3224–3228).

[Zhang 2009]   Zhang Y-J. 2009. Image engineering in China: 2008. Journal of Image and Graphics. 14(5): 809–837.

[Zhang 2010]   Zhang Y-J. 2010. Image engineering in China: 2009. Journal of Image and Graphics. 15(5): 689–722.

[Zhang 2011a]   Zhang Y-J. 2011a. Image engineering in China: 2010. Journal of Image and Graphics. 16(5): 693–702.

[Zhang 2011b]   Zhang Y-J (ed.). 2011b. Advances in Face Image Analysis: Techniques and Technologies. Medical Information Science Reference, IGI Global, Hershey – New York.

[Zhang 2011c]   Zhang Y-J. 2011c. Face, image, and analysis. In: Advances in Face Image Analysis: Techniques and Technologies, Zhang Y J (ed.). Information Science Reference, Hershey – New York. Chapter 1 (1–15).

[Zhang 2012a]   Zhang Y-J. 2012a. Image engineering in China: 2011. Journal of Image and Graphics. 17(5): 603–612.

[Zhang 2012b]   Zhang Y-J. 2012b. Image Engineering (1): Image Processing (3rd Ed.). Tsinghua University Press, Beijing.

[Zhang 2012c]   Zhang Y-J. 2012c. Image Engineering (2): Image Analysis (3rd Ed.). Tsinghua University Press, Beijing.

[Zhang 2012d]   Zhang Y-J. 2012d. Image Engineering (3): Image Understanding (3rd Ed.). Tsinghua University Press, Beijing.

[Zhang 2013a]   Zhang Y-J. 2013a. Image engineering in China: 2012. Journal of Image and Graphics. 18(5): 483–492.

[Zhang 2013b]   Zhang Y-J. 2013b. Image Engineering (3rd Ed.). Tsinghua University Press, Beijing.

[Zhang 2014]   Zhang Y-J. 2014. Image engineering in China: 2013. Journal of Image and Graphics. 19(5): 649–658.

[Zhang 2015a]   Zhang Y-J. 2015a. Image engineering in China: 2014. Journal of Image and Graphics. 20(5): 585–598.

[Zhang 2015b]   Zhang Y-J. 2015b. A comprehensive survey on face image analysis. Encyclopedia of Information Science and Technology (3rd Ed.), Mehdi Khosrow-Pour (ed.). Information Science Reference, Hershey – New York. Chapter 47 (491–500).

[Zhang 2015c]   Zhang Y-J. 2015c. A review of image segmentation evaluation in the 21st century. Encyclopedia of Information Science and Technology (3rd Ed.), Mehdi Khosrow-Pour (ed.). Information Science Reference, Hershey – New York. Chapter 579 (5857–5867).

[Zhang 2015d]   Zhang Y-J. 2015d. Half century for image segmentation. Encyclopedia of Information Science and Technology (3rd Ed.), Mehdi Khosrow-Pour (ed.). Information Science Reference, Hershey – New York. Chapter 584 (5906–5915).

[Zhang 2015e]   Zhang Y-J. 2015e. Statistics on image engineering literatures. Encyclopedia of Information Science and Technology (3rd Ed.), Mehdi Khosrow-Pour (ed.). Information Science Reference, Hershey – New York. Chapter 595 (6030–6040).

[Zhang 2016]   Zhang Y-J. 2016. Image engineering in China: 2015. Journal of Image and Graphics. 21(5): 533–543.

[Zhang 2017]   Zhang Y-J. 2017. Image engineering in China: 2016. Journal of Image and Graphics. 22(5): 563–573.

[Zhu 2011]   Zhu Y F, Torre F, Cohn J F, et al. 2011. Dynamic cascades with bidirectional bootstrapping for action unit detection in spontaneous facial behavior. IEEE-AC, 2(2): 79–91.

Index